WASTEWATER TREATMENT with MICROBIAL FILMS

HOW TO ORDER THIS BOOK

BY PHONE: 800-233-9936 or 717-291-5609, 8AM–5PM Eastern Time

BY FAX: 717-295-4538

BY MAIL: Order Department
Technomic Publishing Company, Inc.
851 New Holland Avenue, Box 3535
Lancaster, PA 17604, U.S.A.

BY CREDIT CARD: American Express, VISA, MasterCard

PERMISSION TO PHOTOCOPY–POLICY STATEMENT

CHAPTER 1

Basic Problems Concerning the Microbial Film Process

1.1 DEFINITION AND CLASSIFICATION OF MICROBIAL FILM PROCESSES

The microbial film process is one of the biological wastewater treatment processes which uses immobilized microbes and purifies wastewater by utilizing microbes attached to the solid surface kept in contact with wastewater continuously or intermittently. The immobilization methods of microbes are classified broadly into three categories as carrier binding method, entrapment and cross-linking method, with the microbial film process considered as a kind of carrier binding method. However, a biological treatment process utilizing biomass immobilized with the other two methods might be regarded as a microbial film process, because of having common purification mechanism and treating characteristics, etc.

In this text, the microbial film process is discussed mainly in its narrow sense.

Based upon its configuration, microbial film processes are divided roughly into three kinds, i.e., submerged biological filter, rotating biological contactor, and the trickling filter process. The submerged biological filter is also divided into fixed beds, expanded beds and fluidized beds, depending upon the state of submerged solids (called filter media, contact media, packing media or carrier, etc.) on which microbial film grows, as shown in Figure 1.1. Rotating biological contactors utilize, as shown in Figure 1.2, many numbers of rotating disks submerged partly or wholly in water, and purify wastewater biologically by the action of microbial films grown on both surfaces.

Trickling filters (see Figure 1.3) consist of circular or rectangular beds of well-graded media (stones, plastics tips, corrugated plastics plates and the

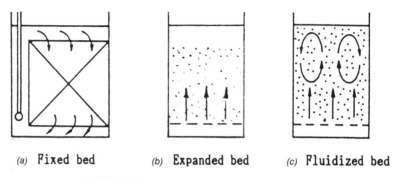

(a) **Fixed bed**   (b) **Expanded bed**   (c) **Fluidized bed**

FIGURE 1.1. *Fixed, expanded and fluidized bed.*

like), with wastewater sprinkled continuously or intermittently from a suitable distributor set above the bed. As feeding of wastewater to the bed is continued, microbial films grow on the media, and wastewater is purified biologically while it trickles down the surface of the media. Unlike the two former microbial film processes, trickling filters are applied solely as aero-ᵇic treatment processes.

Microbial film processes are also classified widely into aerobic (oxic) anaerobic (anoxic) processes. When submerged biological filters are ᵈd as aerobic treatment processes, some oxygen supplying device is ᵇsable as an air diffuser or mechanical aerator. The aerobic sub-ᵇlogical filter process with fixed bed is sometimes called a con-ᵗ process, contact aeration process, contact filter process, or ᵗrocess, etc. When utilizing submerged biological filter as ᵇ, the only operation needed is often just passing the w   filter bed. Rotating biological contactors are used as aerᵗ   ᵈiscs are partly exposed to the atmosphere, and as an anᵗ   discs entirely submerged.

Thus,   ᵗrocess is largely classified into aerobic and an-aerobic oneᵇ.   in microbial films in aerobic treatment facil-ities, not only aerᵤ   ᵗobic microorganisms exist together. Such a fact will often take plaᵤ   ᵃctivated sludge flocs where the outer side is aerobic and the inner   ᵗrobic, especially when the atmospheric dissolved oxygen level is ᵢ   ᵗ. However, coexistence of aerobic and anaerobic microbes occurs   ᵗquently in microbial film processes, because the depth of microbiᵤ   ᵗuch larger than the diameter of bio-logical floc, so generally dissoᵤ   ᵗn penetrates by diffusion only near the surface of the film. It will   ᵤnderstood from the above men-tioned fact that extremely differᵤ   ᵗtances coexist and microbes of many different constituents are oᵤ   a single microbial film treat-ment facility.

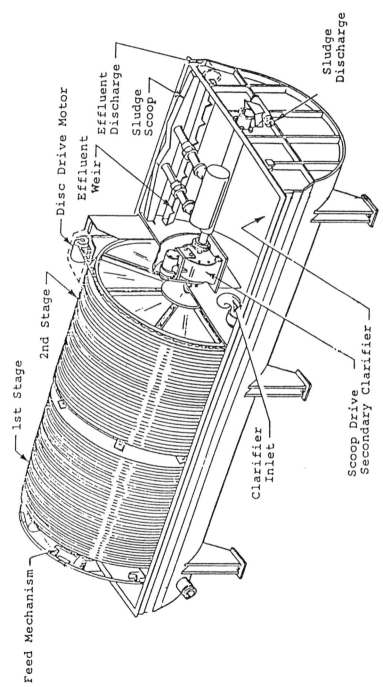

FIGURE 1.2. *The apparatus of rotating biological contactor.*

3

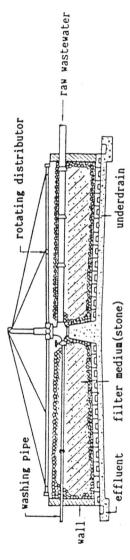

FIGURE 1.3. Cross section of a trickling filter.

1.2 PRINCIPLES OF THE MICROBIAL FILM PROCESS

1.2.1 Ingestion of Substrate and Purification of Wastewater

As shown in Figure 1.4, microbial film grown on media surface ingests substances as organic matter, oxygen, trace elements, etc., which is required for biological activity, from the liquid phase with which it is in contact. These substances, reached on the microbial film surface, then travel into the film by molecular diffusion and are ingested and then metabolized by film microorganisms. If organic substances in the wastewater are colloidal or suspended matter, they are not able to diffuse into microbial film directly, and must be hydrolyzed to low molecules at the film surface before they take a course similar to lower molecular weight organic substances. End products of metabolism are transferred into the liquid phase, moving in the reverse direction of the substrates. Therefore, the reaction in microbial films is indicated by the following formulas:

Aerobic film

$$\text{Organic matter } + \text{ Oxygen } + \text{ Trace nutrients } \rightarrow$$

$$\text{Cellular material of microbes } + \text{ End products} \qquad (1.1)$$

Anaerobic film

$$\text{Organic matter } + \text{ Trace nutrients } \rightarrow$$

$$\text{Cellular material of microbes } + \text{ End products} \qquad (1.2)$$

These Equations (1.1) and (1.2) are applied not only to microbial films, but are exactly general formulas of aerobic and anaerobic reactions. Comparing Equations (1.1) and (1.2), anaerobic film systems are simpler in their reaction kinetics than aerobic systems because of the lower number of chemical species in microbial reactions.

When any one of components essential to microbes is not supplied, biological reactions will not proceed steadily. Therefore, if any one of these components is exhausted at a certain depth of microbial film, then biological reactions will not occur in the deeper portion. Thus, the substance exhausted first determines the effective depth of a microbial film, so such substance is called the limiting factor. Trace nutrients such as nitrogen, phosphorus and trace metals generally will not become limiting factors only if they are contained in wastewater as much as required stoichiomet-

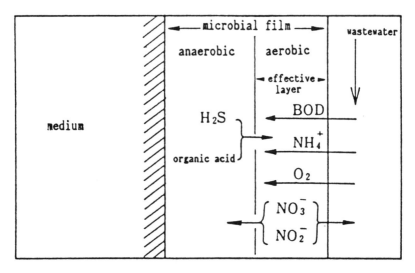

FIGURE 1.4. *Schematic diagrams of a microbial film.*

rically in biological reaction. Consequently, it is possible for either organic matter or oxygen to be the limiting factor in aerobic films; on the other hand, organic matter alone can be the limiting factor in anaerobic film. Usually, as dissolved oxygen concentration in the bulk liquid being contact with microbial film is much lower than organic matter concentration, dissolved oxygen becomes the limiting factor in most cases. That is to say, even in an aerobic film, the portion deeper than a certain depth is kept anoxic or anaerobic, so the film consists of an aerobic layer on the surface side and an anoxic (anaerobic) layer on the opposite side. Anaerobic layers make no direct contribution to the purification of wastewater, so the depth of the effective layer coincides with that of the aerobic layer. Nevertheless, even in an anoxic layer, liquefaction and/or acid fermentation of particulate organic solid, oxidation of organic matter and sulfide formation by sulfate reduction, or reduction of nitrite and/or nitrate (i.e., denitrification) produced in aerobic layer will take place in many cases. Thus, the fact that both aerobic and anaerobic action coexist in a single film is one of the very important features of the microbial film process, and denitrification is a function of great importance for anaerobic layers.

On the other hand, in an anaerobic film, organic matter is the sole potential limiting factor, so the depth at which organic matter is able to reach is really the effective layer.

1.2.2 Growth and Degradation of Microbial Film

The general rule concerning the process of microbial film growth by ingesting substrate from wastewater is as follows: The process of microbial film growth on the surface of fresh solid is divided into three stages. The first stage shows logarithmic growth where the film is thin and frequently does not cover all of the solid surface. In such conditions, all microorganisms grow under the same conditions, the growth behavior being similar to that of suspended culture. Next, when the film thickness becomes larger than the effective depth, the second stage begins. In the second stage, the growth rate remains constant, because the depth of the effective layer is constant regardless of total film depth, and so the total amount of growing microbes is also constant throughout this stage. Even in this stage, microbial film does not grow under lower substrate concentration than a certain value. The reason is that the substrate ingested by microbes is consumed only to maintain the lives of microbes present (maintenance metabolism), making no contribution to bacterial growth when the supply of substrate is very scarce. Further, if the supply of substrate is less than the required quantity for maintenance metabolism, microbial film will begin to thin out. By decreasing the film thickness, the balance will be kept between supply of substrate and consumption rate by maintenance metabolism.

In the third stage, film thickness reaches the plateau, where film growth rate is balancing the rate of decrease by endogenous respiration, destruction by food chain, or washing by shear force. However, frequently no clear plateau is observed and film thickness continues to increase to result in the clogging of the filter. Characklis et al. [3] summarizes methods available for measuring biofilm accumulation in Table 1.1. In the course of biofilm growth, microorganisms show changes not only in their quantity but also in their composition. As shown in Figure 1.5, early in the growth process, most of the biomass is constituted of bacteria, later protozoas and then still later metazoas begin to grow, and an ecosystem builds up. The higher the position a microorganism occupies in an ecosystem and the smaller its specific growth rate is, the later its growth is. Therefore, it is often experienced that nitrification bacteria or bacteria assimilating slowly degradable substrates grow at much smaller rates than bacteria which utilize easily degradable substrates. It is effective to decrease excess sludge production that protozoas and metazoas eat biofilm to a desired degree. Under certain environmental conditions, involving, e.g., water temperature or water quality, metazoas will grow enormously, so they prey on biofilm excessively to damage purification performance and cause a large amount of peeling off of biofilm. It is very difficult to mount an effective counter-

TABLE 1.1. *Measurement of Biofilm Accumulation.*

Classification	Analytical Method	References
(A) Direct measurement of biofilm quantity	Biofilm thickness	[3-1-3-6]
	Biofilm mass	[3-5-3-8]
(B) Indirect measurement of biofilm quantity: specific biofilm constituent	Polysaccharide	[3-9,3-10]
	Total organic carbon	[3-2,3-11]
	Chemical oxygen demand	[3-9,3-12]
	Protein	[3-13]
(C) Indirect measurement of biofilm quantity: microbial activity within the biofilm	Viable cell count	[3-14-3-16]
	Epifluorenence microscopy	[3-11,3-17]
	ATP	[3-2,3-18,3-19]
	Lipopolysaccharide	[3-20,3-21]
	Substrate removal rate	[3-5,3-6,3-15]
(D) Indirect measurement of biofilm quantity: effects of biofilm on transport properties	Frictional resistance	[3-3,3-5-3-8]
	Heat transfer resistance	[3-22-3-25]

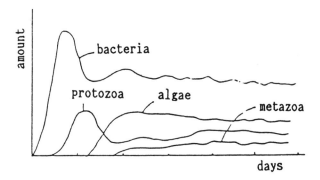

FIGURE 1.5. *Succession of microbes constituting microbial film.*

measure against this problem. Inamori et al. [4] indicated that two kinds of predators live in a biofilm. One of them eats suspended bacteria and excretes aggregated excrement, consequentially accelerating purification of water, the other preys on flocculated bacteria in microbial film and so prompts dispersion of biomass. They also pointed out that highly transparent effluent is gained together with a high extent of mineralization of organic matters, when two kinds of predators act in keeping a desirable balance.

1.2.3 Attachment Characteristics of Microbial Film [5]

The attachability of microbial film on a solid surface is one of the important factors of microbial film processes because it severely affects the growth rate and the relative difficulty of scouring off of microbial film. There are two physicochemical factors affecting attachability of a microbial film: (1) The first is electrostatic action. The electric charge on the surface of microorganisms is caused by electrical dissociation of radicals as amino, carboxyl, phosphate and the like, so the state of electric charge is affected by the pH of solution. The surface of a microorganism has a positive electric charge if the pH of the solution is lower than the value at which the surface has no electric charge (isoelectric point), and vice versa [6]. Since the isoelectric points of the surface of microorganisms range in the acidic region (pH 4–5), a microorganism cell in almost neutral water is considered to be a colloidal particle with some negative electric charge. Therefore, an electrostatic drag force works between a microorganism cell and a particle or surface with positive electric charge, and so they can adhere easily. On the other hand, adhesion of a microorganism on a particle or a surface with a negative electric charge will be difficult to observe. The detail of those electrostatic phenomena are referred to in texts of physical chemistry, but only important points for microbial attachment will be discussed here. In the neighborhood of a charged surface, an electric double layer is formed, and so when two particles with the same kind of charge come close to each other, the repulsive force caused by the overlapping of two double layers and van der Waal's force will work together. The theory of this problem is called DLVO (Deraguin, Landau, Verwey, Overbeek) theory on which the total energy, V_T of interaction of two particles is equal to the sum of repulsion energy, V_R of the overlapping of electric double layers and energy by van der Waal's drag force, V_A. That is,

$$V_T = V_R + V_A \qquad (1.3)$$

Figure 1.6 shows V_R and V_A as functions of the distance between particles, and the V_T curve is determined as the sum of these two energies. Actually, if distance between the particles is too small, the surfaces will repulse each other (Born's repulsion force), so that V_T will increase rapidly with a decrease in distance. The maximum of V_T (V_{max}) is the energy barrier which a particle must get beyond, and whose height decreases drastically when the thickness of the electric double layer is reduced by adjustment of pH or electrolyte concentration of the solution.

When the interaction energy between a microbial cell and a solid surface containing Born's repulsion force's energy is illustrated as in Figure 1.6(b'), there exists two points at which total energy is minimum, and the one nearer to the surface is called the first minimum point, with the other being called the second minimum point. At the first minimum energy point, the total energy being extremely low, microbial cells are adsorbed in a highly stable manner, but to reach this point, microbial cells have to go beyond the energy barrier V_{max}. As the height of this energy barrier is much higher than the energies of a Brownian motion or a flagellum motion of a bacterium, it is impossible to get beyond. The energy level of the second minimum point is not so low as to cause a stable adsorption, but microbial cells are adsorbed temporarily at this point and then the adsorption is strengthened with cilia or extra-cellular polymer. According to this hypothesis, the fact that whether a species of microorganism has the ability to excrete extra-cellular polymer or not is the controlling factor for the adsorption of this microorganism [7,8] (2). The second physicochemical factor of the adsorption of a microbial cell on a surface is the degree of hydrophilicity of them. As a general rule, the binding of two hydrophobic substances to each other or to two hydrophilic substances, also is stable from the viewpoint of free energy. The binding by such a mechanism is called a "hydrophobic interaction," and it is needless to say that this rule might be applied to the microbial adsorption on a solid surface. For example, blue-green algae are classified into two types, planctonic type suspended in water and benthic type ad-

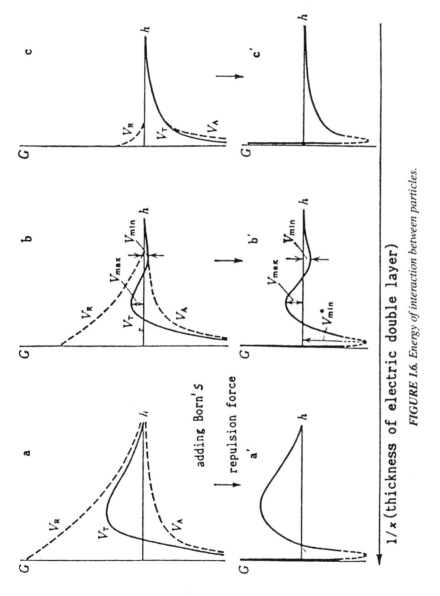

FIGURE 1.6. Energy of interaction between particles.

11

sorbed on a solid surface. Fattom et al. [4] have reported that among blue-green algae samples gained from various circumstances, all the benthic types were hydrophobic, and all the planctonic types hydrophilic. The spread of hydrophilicity of microbial cell surface is random, some being strongly hydrophobic and others being hydrophilic. Therefore, considering the free energy change by adsorption, it is concluded that the surfaces of highly hydrophobic materials such as polystyrene, polyethylene, and polyamide easily adsorb hydrophobic microorganisms, and conversely, the hydrophilic surface of silicon dioxide etc. adsorbs hydrophilic microorganism.

Among physical characteristics of solid surfaces that would affect attachability of microbial film, roughness is one which can be recognized. Observations have been reported that surface roughness has an important effect in the primary film-forming period and the attached amount on a rough surface is larger than that on a smooth surface [11], but that surface roughness is not a significant factor in the total amount of microbial film formed [12,13]. The authors, however, compared microbial film attachment characteristics of nontreated sheets of polyvinylchloride (hard type) and polyethylene (A), the same materials polished with fine sandpaper (B), and coarse sandpaper (C) by sticking them on the same disk in a rotating biological contactor unit, with the results obtained indicating that both the attaching rates and the total attached amount were in the order; $A < B < C$. As surface roughness should affect at least the strength of joining of a microbial film to the surface, it is given that the surface roughness has some effect on the upper limit of attached film amounts. It is reported that the attachment characteristics changed depending upon the species or the physiological conditions of microorganism [1]. This is easily realized based on the above mentioned attachment mechanism.

Concerning the effect of flow velocities past the film surface, Heukelekian [11] observed that higher velocities retard primary film formation, but once established, faster flow results in greater growth. Sanders [15] and Characklis [16] obtained maximum growth at the upper limit of the velocity range of 0.1–1.0 ft/sec. This effect has been attributed to the greater flux of substrate from the bulk liquid to the film surface. Slime film could withstand shear forces exceeding 10–15 dyn/cm² [13,16] and films grown at higher velocities adhere to the surface more firmly [16]. It should be noted that species distribution of microorganisms in a film varies with the level of turbulence [17].

1.3 CHARACTERISTICS OF THE MICROBIAL FILM PROCESS

The reason why the general name "microbial film process" is widely used despite its wide variety of configurations, results from the commonness in purification mechanism and characteristics. All the characteristics of this process are, needless to say, attributed to the fact that the microbes responsible for purification are maintained as biofilms (i.e., biomass layer with a certain thickness) and act in that state. Thus, to know the advantages and disadvantages of the microbial film process, and to practice correct application, design, operation, and maintenance of this process, it is essentially important to recognize the differences brought about by the state of microbes in a bioreactor.

1.3.1 Biological Characteristics

1.3.1.1 THE BIOMASS QUANTITY HELD IN REACTOR AND DIVERSITY OF BIOTA

Generally speaking, the amount of biomass held in unit volume of a microbial film unit is fairly large. The biomass concentrations if shown as MLSS are as high as 20–40 kg/m³ in rotating biological contactors, 10–20 kg/m³ in ordinary fixed bed submerged filters, and 5–7 kg/m³ even in conventional trickling filters. On the other hand, it is universally known that under appropriate operations, microbial film processes produce much smaller quantities of excess sludge than the activated sludge process because of their longer food chains in microbial films [18]. The quotient of total biological solids held in a reactor (S) divided by the average daily excess sludge production (ΔS) is equal to sludge retention time (or mean sludge residence time or sludge age; A_s). That is,

$$A_s = S/\Delta S \qquad (1.4)$$

where A_s shows how many days sludge is retained in a treatment system. In a steady state, the excess sludge production in a system is equal to the amount of sludge withdrawn from it. In such a system, the change in the number (n) of a certain species of microorganism in biological sludge is given by the following equation

$$\frac{dn}{dt} = \mu n - \frac{n}{A_s} = \left(\mu - \frac{1}{A_s} \right) n \qquad (1.5)$$

where

μ = the specific growth rate of the microorganism
t = time

From Equation (1.5), if $\mu < 1/A_s$, n will decrease with time until it disappears. In other words, for microorganisms with small specific growth rates to be able to grow, the value of SRT, A_s must be large enough. The values of specific growth rates of various microorganisms are listed in Table 1.2. As mentioned before, in microbial film processes, the amount of microbial solid held in the reactor is large, and that of excess sludge withdrawn is small moreover, so the value of A_s will be lengthened extremely. Therefore, the species of microbes in films becomes rich, and microbes occupying higher ranks in food chains abound in addition to that. Hawkes [19] comparing the biotas of trickling filter slime and activated sludge, have shown in Figure 1.7. As metazoas (Rotatoria, Nematoda, Insecta, Shellfish, Oligochaeta, etc.) are all large-sized, whose length ranges from several millimeters to a few centimeters, and who prey on film microbes diligently, resulting in a remarkable decrease of excess sludge productions. Moreover, an ecosystem with highly diverse biota is a stable system which enables a stable treatment effect inevitably. Those bacteria, who utilize substrates slowly assimilated or substrates with low value of growth yield, always have relatively small specific growth rates. Therefore, microbial film processes have excellent performance in removing such substrates.

1.3.1.2 COEXISTENCE OF AEROBIC AND ANAEROBIC MICROORGANISMS

Microbial films are often anaerobic partly or mostly even when used for aerobic processes. Generally, as shown in Figure 1.8, the outer layer of microbial film is aerobic, the inner layer anaerobic. Evidently the thickness of the aerobic layer is constant under a constant operational condition, because if the thickness of aerobic layer increases owing to microbial growth, the bottom aerobic layer becomes anaerobic in the same length. Anaerobic layers seem to contribute to the liquefaction of film solids resulting in the decrease of excess sludge production. Moreover, coexistence of aerobic and anaerobic layers is convenient to biological nitrogen removal which is caused by the coupling of aerobic (nitrification) and anaerobic (denitrification) reactions. In a microbial film in which both aerobic and anaerobic layers exist, the concentration profiles of ammoniacal and nitrate nitrogen are presumed to be as shown in Figure 1.9. Though the anaerobic layer does not give any direct effect on the concentration profile of ammoniacal nitrogen in the aerobic layer, that of nitrate nitrogen is influenced deeply by the existence of an anaerobic layer. One portion of nitrate nitrogen produced in aerobic layer is transported into bulk liquid, the other to the anaerobic layer where it is changed to nitrogen gas by denitrification. Thus, the concentration of nitrate nitrogen has the maximum value at a certain depth of aerobic layer, and nitrate nitrogen produced in the outer por-

TABLE 1.2. Growth Rates of Microbes.

Microbes		μ (1/day)	t_d (hr)	Temperature (°C)	Dry Weight of a Cell (mg)
Bacteria	Bacillus megatherium	31.8	0.52	30	3.8×10^{-9}
	Escherichia coli	59.1	0.28	37	4.0×10^{-10}
	Rhodopseudomonas spheroides	6.9	2.4	34	—
	Nitrosomonas sp.	1.3	12.7	25	—
	Staphylococcus aureus	37.6	0.44	37	1.5×10^{-10}
Algae	Anabaena cylindrica	0.66	25.0	25	—
	Microcystis aeruginosa	0.64	25.9	25	—
	Navicula minima	0.97	17.1	25	—
	Chlorella ellipsoidea	2.5	6.7	25	—
	Selenastrum capriconnutum	1.9	8.7	25	1.9×10^{-8}
Fungi	Saccharomyces cerevisiae	8.3	2.0	30	7.1×10^{-8}
Protozoa	Vorticella microstoma	3.3	5.0	20	3.9×10^{-6}
	Epistylis plicatilis	1.6	10.2	20	—
	Colpidium campylun	3.6	4.7	20	1.6×10^{-6}
	Paramecium caudatum	1.4	12.0	20	3.0×10^{-4}
	Tetrahymena pyriformis	5.3	3.1	25	1.4×10^{-6}
	Colpoda steinii	5.5	3.0	30	1.2×10^{-6}
	Stentor coeruleus	0.75	22.1	19	5.0×10^{-3}
	Aspidisca costata	1.2	13.6	20	—
Metazoa	Rotaria sp.	0.28	59.1	20	—
	Philodina sp.	0.23	72.0	20	1.8×10^{-4}
	Lecane sp.	0.31	54	20	—
	Aeolosoma hemprichi	0.35	47.3	20	3.8×10^{-4}
	Nais sp.	0.12	138	20	6.6×10^{-3}
	Pristina sp.	0.12	138	20	—
	Dero sp.	0.07	238	20	—

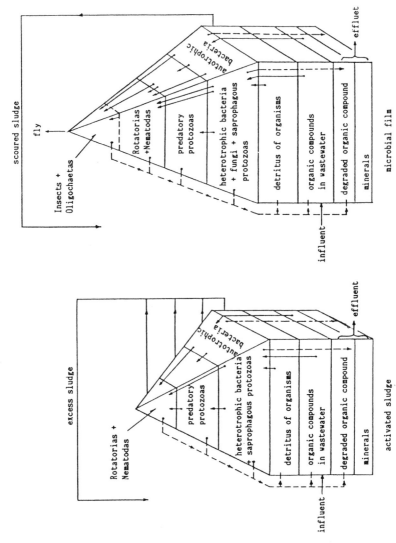

FIGURE 1.7. *Comparison of biotas between activated sludge and microbial film (trickling filter).*

16

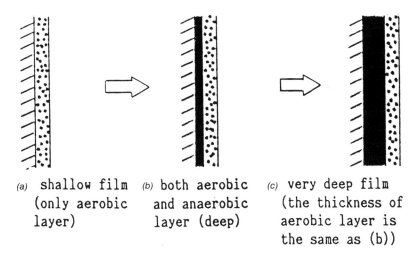

(a) shallow film (only aerobic layer) (b) both aerobic and anaerobic layer (deep) (c) very deep film (the thickness of aerobic layer is the same as (b))

FIGURE 1.8. *Aerobic and anaerobic layers in a biological film.*

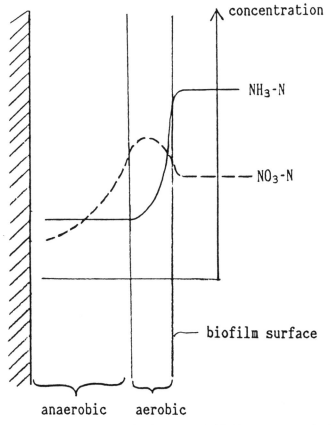

concentration

NH$_3$-N

NO$_3$-N

biofilm surface

anaerobic   aerobic

FIGURE 1.9. *Concentration profile of NH$_3$-N and NO$_3$-N across a microbial film.*

tion of the aerobic layer than the peak concentration level is transferred into bulk liquid, nitrate nitrogen produced in the inner portion being transported into the anaerobic layer and denitrified there. Generally, the higher the dissolved oxygen concentration in the bulk liquid, the thicker the aerobic layer and the higher the nitrification rate become, but the ratio of nitrate nitrogen denitrified decreases because the position of peak concentration of nitrate nitrogen shifts to the inner side because of decrease of the thickness of the anaerobic layer. Contrarily, if the dissolved oxygen concentration in bulk liquid is too low, it is conducive to denitrification, but nitrification will retard. Therefore, there exists the optimum bulk liquid dissolved oxygen concentration that provides maximum nitrogen removal. It is the reason why a proper degree of aeration is required to obtain maximum nitrogen removal.

1.3.2 Characteristics in Suspended Matter Capture Function

Highly transparent (i.e., containing very little suspended matter) effluent is frequently obtained with a microbial film process, especially with a submerged filter process. Usually, suspended matter accounts for a considerable portion of pollutant remaining in ordinary secondary effluent of municipal sewage. For example, it is often reported that 50 percent or higher of remaining BOD in secondary effluent of municipal sewage can be removed with rapid sand filtration or microscreening. The reason why effluent almost free of suspended matter can be obtained is due to the excellent capability to capture particulate substances by microbial film, i.e., a kind of filtering function. In submerged filter units, filter media with a diameter of several tens to one hundred millimeter are most often used, and a filter packed with such large sized media seemingly has a very poor filtering performance. However, the mechanisms involved in removal of suspended solids by a filter are very complex, including not only sieving effects, i.e., the phenomena in which particles with larger size than openings in filter bed is captured, but also various functions by which particles with much smaller size than openings are able to be caught. Ives [20] has shown in Figure 1.10 the actions by which suspended particles are transported onto the media surface. Among these actions, the effects of interception, inertia and gravity will become greater as the particle size becomes larger and the media size become smaller, and so they are effective in removing relatively large particles (larger than 1 μm in diameter). On the other hand, the effects of diffusion and hydrodynamic force is important to catch smaller particles than 1 μm in diameter, and increasing media size, therefore, does not always lower the effect of filtration because of these actions. If there is a proper velocity gradient in the flow through the filter bed, flocculation of particles

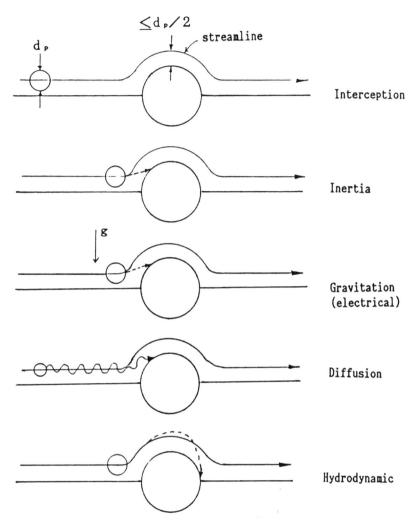

FIGURE 1.10. *Simplified diagram of particle transport mechanism.*

will be hastened. It should be noted, that even though the filter coefficient in a conventional submerged biological filter is remarkably small, highly transparent effluent could be obtained by repeated filtration many times, caused by the circulation flow in the bed. Moreover, it might be considered a kind of biological filtration that large-sized microbes as small metazoas inhabiting microbial films prey on small ones as bacteria. On the contrary, in a microbial film unit as well as in a sand filter, when the accumulated amount of sludge in the unit grow, the quantity of sludge falling off will increase correspondingly, and the net removal of suspended solid by the unit will decrease. Moreover, peeling off of large amounts of sludge occasionally occurs at one time by the activity of large-sized microbes. Much attention should be paid to this phenomena because many kinds of predating microbes feed on aggregated biomass and excrete fine particulate solids, having a deleterious effect on the effluent quality. Many problems remain unsolved about the mechanisms of particle capture or slime peeling off in microbial film processes.

1.3.3 Characteristics in Substrate Removal

Substrate removal characteristics in a microbial film process differ far from that of suspended growth processes as activated sludge process, mainly in the two following points of view:

One is that the microbial reaction is regulated by two factors, i.e., the diffusion and the ingestion of substrate in the film. Then, the process of diffusion will be the rate-limiting step if the film thickness is larger than a certain value [21]. Since diffusion is a physicochemical phenomena, it is less affected by temperature than biological actions such as ingestion or metabolism. In microbial film processes, therefore, the dependencies of substrate removal rates on temperature are frequently smaller than in suspended growth processes, and consequently, more stable treatment is obtained. These problems are discussed in detail in the following paragraph.

The other point of view is the problem concerning the behavior of colloidal or suspended particles removed, and also this problem relates to the transport of substrates by diffusion. In a treatment process utilizing suspended biomass, evidently even colloidal or suspended substrate is easily and homogeneously mixed with biological solid, ingested, and metabolized immediately. In a microbial process, however, suspended solids are unable to move in the film, and colloidal solids, if not unable, can move only at a practically insignificant rate. Because the molecular diffusion coefficients of substances are proportional to the square root of their molecular weight, the molecular diffusion coefficient of a macromolecular compound with molecular weight in the hundreds of thousand would be far smaller than compounds with low molecular weights, say in the hundreds. Colloidal or suspended substrates, therefore, must be attacked only by the film surface microbes before they might take the same path as substrates originally with low molecular weights. When colloidal or suspended substrates are removed by microbial film processes, therefore, the step of hydrolysis of such substrate is apt to be the rate-limiting step.

The substrate removal characteristics in microbial film processes will be discussed in detail in 1.4, so only the outline has been mentioned previously.

1.3.4 Various Practical Characteristics

1.3.4.1 MAINTENANCE OF THE FACILITY

As compared with activated sludge process, the most important advantage of microbial film processes is easiness of maintenance. In the maintenance of activated sludge processes, many operations are frequently required for activated sludge conditioning, i.e., conditioning to maintain concentration, settleability, condensability, biota of activated sludge in proper conditions, and for settling tank adjustment, it is indispensable to control treated flow, to recycle settled sludge to aeration tanks, and to withdraw proper volume of excess sludge, etc. Especially, it is well known that bulking of activated sludge caused by overgrowing of filamentous bacteria as *Sphaelotilus natans, Beggiatoa,* etc. sometimes makes the settleability of activated sludge very poor, and so continuance of the operation of the process itself becomes difficult. On the other hand, concerning microbial film processes, these worrisome operations for conditioning are almost unnecessary. While the final settling tank in the activated process is essential to maintain the function of aeration tank, settling tanks in microbial film processes are installed only for the removal of settleable solids from reactor effluent. The presence itself or the performance of the final settling tank has no effect on the performance of a microbial film type bioreactor. Therefore, the maintenance of the final settling tank in microbial film process does not need undue care as in the activate sludge process. The settleability of biosolids responsible to wastewater purification, moreover, offers no problem because it is adhering on a solid surface, and so no problem ought not to occur concerning bulking of biosolid. The less excess sludge production owing to the effect of food chain previously mentioned, also contributes to ease of maintenance. Because of these advantages, microbial film processes can be regarded to fit especially well in small-to-medium scale treatment plants.

Ease of maintenance, on the other hand, leads to little ability to adjust the condition of treatment unit by maintenance operations. For example, in an activated sludge plant, the shortage of aeration tank volume to a certain extent is able to be covered by means of increasing MLSS through the regulation of excess sludge withdrawal. Also, if the acceleration of growth of nitrifying bacteria is required, SRT (Sludge Retention Time) should be lengthened, decreasing the amount of excess sludge withdrawal, and the contrary operation should be made to repress the growth of nitrifying bacteria. But, in microbial film processes, it is impossible to control accurately the amount of biomass retained in the system, much less the species of microbes, because no effective method has been developed to withdraw a part of microbial film easily and by any quantity. It may be safely said that the controllable operation conditions in microbial film processes are only the feeding rate and the aeration intensity in a submerged filter process.

1.3.2.1 QUICKNESS OF START-UP

In the activated sludge process which is the typical biological treatment process, the period required for the unit to show its full performance by the accumulation of activated sludge, i.e. "the start-up period" is at least a month or more, and usually about two-months. As compared with this, the start-up periods for microbial film processes are only one to two weeks in submerged filters or rotating biological contactors, though they show some seasonal fluctuations and are a little longer in trickling filters. The reason for such quick start-ups is that almost all biomass produced in the reactor is accumulated there with no peeling off early in the start-up period when the film thickness is thin. By the same reason, the restoration of function is also very fast, even after a high degree of decrease of slime from one cause or another. Also, the process may be able to tolerate an extreme seasonal fluctuation of loading.

1.3.2.2 CAPABILITY TO REMOVE SLOWLY DECOMPOSABLE SUBSTRATES

It can be explained from two points of view that microbial film processes excel in removing slowly decomposable substrates. Such substrates include Polyvinyl Alcohol (PVA), Linear Alkylbenzene Sulfonate (LAS), lignin, chlorinated organic compounds, etc. as organic compounds, and ammoniacal nitrogen, amines, cyanides, etc. as inorganic ones. These compounds are biologically decomposable, but the decomposition rates are very low, and the specific growth rates of microbes which utilize these compounds as main substrates are extremely small, correspondingly. For example, as shown in Figure 1.2, specific growth rate of nitrifying bacteria *Nitrosomonas* sp. is only one tenth as small as that of *Escherichia coli*. It was explained previously why microbes having small specific growth rates are able to grow in microbial films. Thus, one reason is that microbial film processes are suited to the cultivation of slowly growing microbes. The other reason is related to the ratio of the depth of effective layer to total microbial film depth. Generally, the slower the ingestion rate of a substrate is relatively compared to its transport by diffusion, the deeper it reaches in a microbial film, and so the deeper the depth of the effective layer becomes, and, of course, vice versa. In other words, even if the ingestion rate of a substrate is small, more biological solids will contribute to the removal of the substrate, compensating the nature of the substrate as a result. On the contrary, if the substrate is easily biodegraded, smaller quantities of biological solids will work, thus the difference of biodegradability will not affect directly the overall ingestion rate by a microbial film. Hence, microbial film processes are theoretically, suitable to be applied to wastewaters containing slowly degradable substrates, except macromolecular ones.

1.3.4.2 TOLERANCE TO VARIATION OF TEMPERATURE AND LOADING RATE

Both the rates of diffusion and biological reaction will decrease with decreasing water temperature. However, the latter is usually much more intensely affected than the former by water temperature. Activation energy is used as an index to indicate quantitatively the dependence of reaction rate on temperature, and the larger this value is, the higher the dependence is. The activation energy of molecular diffusion in a solution is several kcal/mol, and that of biological reaction, 20–30 kcal/mol, so temperature dependence of the latter is rather larger than that of the former. Therefore, even if water temperature is lowered, the substrate ingestion rate by microbial film is not so much intensely affected as intrinsic biological reaction rate owing to the same mechanism as that of slowly degradable substrate, because the decrease of molecular diffusion rate by temperature drop is smaller than that of biological reactions. On the contrary, when water temperature increases, the reverse change as above will occur, and the substrate ingestion rate by microbial film will not increase so much as the intrinsic biological reaction rate does. That is, a stable treatment efficiency can be expected to temperature variation.

On the other hand, a moderating influence on the treatment efficiency can be expected also to the fluctuation of loading. If increment of influent loading arises, the substrate concentration on a biofilm surface will increase correspondingly, thus the depth from film surface to which substrate reaches, i.e., the depth of effective layer will also increase, resulting in the suppression of increment of substrate concentration, and vice versa.

As mentioned here, in microbial film processes, a change in the system is more easily moderated than in activated sludge process. That is to say, an important advantage of microbial film processes is that a negative feedback control system is contained in them to stabilize treatment efficiency.

1.3.4.3 HIGH EFFECT ON LOW CONCENTRATION WASTEWATER [21]

It is practically impossible to treat low strength wastewater with BOD lower than 20 mg/l by the activated sludge process, because of the difficulty of maintaining normal value of MLSS and conventional efficiency. However, with microbial film processes, only if substrate concentration in the wastewater is higher than the level required for maintenance metabolism (very low concentration), wastewater of a very wide range of strengths can be treated effectively, because microbial films with properties suitable to environmental conditions of substrate concentration would be formed. Moreover, low strength wastewater is considered to be more easily treated with microbial film processes for the following reasons:

(1) Low strength wastewater tends to give high hydraulic loading to treatment plants, but the influence of hydraulic loading on microbial film processes is smaller than on activated sludge process since the final settling tank in the former process is less important than in the latter (as mentioned in section 1.3.4.1).

(2) Dissolved oxygen would frequently be the limiting factor−the substance which control overall purification rate−in microbial film processes [22–24]. That is to say, the substrate is usually supplied superfluously compared with oxygen, so the quantitative balance of these two substances will be improved if the influent substrate concentration is low.

(3) In microbial film processes, when influent substrate concentration is high, it is feared that the contact of treated water with microbial film will be disturbed, because of the depression of flow velocity past the microbial film surface, owing to overgrowing of microbial film. In such cases, excessive amounts of microbial film must be peeled off by washing. When the influent substrate concentration is low, however, washing is unnecessary or, even if necessary, the frequency required is very small. Thus, the amount of sludge withdrawn is also very small.

1.3.4.4 GREAT VARIETY OF FACILITIES

In spite of its common name or purification characteristics, the microbial film process has a wide variety of types of facilities. In any of the submerged biological filters, rotating biological contactors, and trickling filters, the shape, size, material and the installation method of microbial film carriers are extremely diverse. However, though there is not so much difference in microbial film area per unit volume of bioreactor (specific

area) among these three processes, there is a much greater specific film area, and thus correspondingly much greater organic loading in fluidized bed or expanded bed processes. Moreover, these processes are able to be applied as both aerobic and anerobic treatment processes, except for trick-
ling filters. Hence, microbial film processes can be applied to a wide variety of wastewaters. Roughly speaking, fluidized or expanded bed reactors are better applied to wastewaters with influent organic concentration of several hundreds to several thousands mg/l because of their tolerability to high loading; submerged biological filters, rotating biological contactors and trickling filters units are suitable to wastewater with organic concentration of several tens to hundreds mg/l or the secondary effluent of such wastewater. Nevertheless, fluidized beds and expanded beds are frequently ap-
plied to the treatment of low strength wastewaters at high hydraulic loading rates. Microbial film processes do not only include a wide variety of configurations, usages, and operation conditions as mentioned above, but they also have a great possibility of advancement in configuration and/or operations, and of wider application in the future. All the characteristics of sec-
tions 1.3.1.1–1.3.4.6 are the advantages of microbial film processes; however, following disadvantages also need to be pointed out.

1.3.4.5 UNCONTROLLABILITY OF BIOMASS AMOUNT

It is usually impossible to easily control the correct amount of biofilm. Moreover, the increase in biofilm thickness over a certain value – effective layer depth – does not only make a contribution to the performance of the treatment unit, but it also decreases the effective area of biofilm and/or effective hydraulic retention time in the unit [25]. In most cases, biofilm amount is determined by the rates of attachment, and growth and mechanical detachment [29], so that washing by air, water or mechanical forces is practiced only when over-accumulation of biofilm has been observed. Because of lack of information about timely washing, methods and operation of washing, proper frequency and operations are not always practiced in washing. Moreover, it is difficult to maintain a steady state in the reactor, and effluent quality tends to vary periodically, since the quality and quantity of biomass will fluctuate owing to repeated growth and detachment of slime.

The uncontrollability of biomass, in turn, results in uncontrollable SRTs, and so the species of microorganisms in a microbial film will also be uncon-
trollable. In the activated sludge process, if washout of nitrifying bacteria is desired to repress nitrification, SRT should be shortened. If acceleration of nitrification or growth of protozoa and metazoa is wanted, lengthen SRT by decreasing the amount of excess sludge withdrawal. Thus the control of

species of microbes is attainable to some extent. The diversity of microbial species in a biofilm is generally high, and this fact leads to the elongation of food chain, thus excess sludge production is decreased. No means have been developed to control the number of a specific species of microbe, however, overgrowing of large-sized microorganisms as *Daphnia* or *Nais,* which occupies the highest ranks in food chains, leads to an extreme decline of the performance of the treatment unit, because they eat large amounts of other microbes and excrete fine, unsettleable particles in the effluent.

Thus, the microbial film process has few controllable factors, which, on the one hand, means ease of maintenance, but on the other, difficulty in maintaining a microbial film process in good condition. The extent of such characteristics differ depending on the sort and the configuration of microbial film processes, but it should be noted that the ease of maintenance is always antinomic to the controllability of performance by adjustment of operating conditions.

1.3.4.6 DIFFUSION LIMITING NATURE OF PURIFICATION RATE

In a microbial film process, the purification rate-controlling steps are the transfer of essential substances as substrates and oxygen into microbial film, and biological reaction by the film microbes. In most cases, the transfer of essential substances by diffusion being the rate-limiting step (diffusion limiting), the concentration of such substances is the controlling factor of purification rate. In the activated sludge process, for example, low levels of dissolved oxygen concentration in the aeration tank are sufficient (0.2– 0.5 mg/l), only if enough oxygen can be supplied to metabolic reactions; but in the microbial film process, the higher the dissolved oxygen concentration in the bulk liquid, the deeper the depth to which dissolved oxygen reaches, and the greater the purification rate becomes respectively. Therefore, to maintain high purification efficiency in an aerobic microbial film unit, dissolved oxygen concentration must be kept high, which decreases electric power efficiency for aeration. Therefore, an optimum condition must be arrived at considering the efficiencies of both biological reaction and electric power. For the same reason, purification efficiency will obviously be reduced in an anaerobic microbial film process, when low-strength wastewater is treated or highly treated effluent is required. Diffusion limiting properties bring about not only advantages mentioned in sections 1.3.4.1 and 1.3.4.2 but also the disadvantage described here.

To compensate for this disadvantage, it is effective to make the microbial film area as large as possible and thus keep the total amount of effectively-functioning biomass as large as possible under a definite effective layer depth. For this purpose, the specific area of microbial film carrier should be large enough by using small sized and/or complex shaped carriers. However, a filter bed packed with such carriers has a serious defect in that it will be easily clogged, so utilization of such a carrier does not necessarily seem to be the best option from an overall point of view. To attenuate the diffusion limiting nature of microbial film process, it is important to keep total film depth small enough by maximizing film area per unit reactor volume. It is of importance to increase the flow velocity past film surface as much as possible to reduce mass transport resistance (see section 1.4.1). It is unreasonable, however, to increase flow velocity to a value higher than that required for ordinary operation of the unit, because increasing flow velocity demands much energy consumption. It is important to design the bioreactor so as to keep a uniform flow velocity everywhere within it.

1.4 REACTION KINETICS IN MICROBIAL FILM PROCESS

Theoretical analysis of microbial film processes of reaction kinetics is much more complicated than that of the activated sludge process. That is, in a suspended microbial system, it may be concluded that metabolic reaction rate alone controls the overall biological purification rate, but the rates of both the transference of substrates into film and biological reaction participate in microbial film processes. Moreover, such an analysis can be made only about the behavior of low molecular weight dissolved substrates which can move into film by molecular diffusion, and high molecular weight compounds with very low diffusion rates or suspended solids incapable of entering into film have to be converted into low molecular weight ones before they can proceed into the same pathway as the original low molecular weight ones. Moreover, it is pointed out, that while the primary substrate is decomposed to intermediates with lower molecular weights, the overall diffusion rate of substrates increases, and the removal efficiency of the primary substrate is raised as a result [30]. The behavior of macromolecules or particulate matters is extremely complicated, and so little has been elucidated theoretically or experimentally about them. It may be safely said that the whole content of this paragraph might be applied only to low molecular weight substrates. The knowledge of reaction kinetics on substrate removal by biofilm hereafter described, however, will directly contribute to the appreciation of the characteristic of microbial film processes.

1.4.1 Rate Processes of Substrate Removal in a Microbial Film Process

A reaction system containing microbial film is able to be considered as a sort of unhomogeneous catalytic reaction and to be dealt with similarly as the reactions of immobilized enzymes in kinetics. In such reactions, at least two phases – liquid phase (treated water) and solid phase (microbial film) – participate in the reaction, and also gas phase (air) in an aerobic system.

Figure 1.11 schematically shows the mass-transfer across the surface of microbial film. From the solution (bulk liquid) flowing over the microbial film with a uniform velocity, substrate of low molecular weight (abbreviated to only substrate hereafter) is transferred onto the microbial film surface, then it moves into the film by diffusion and is ingested and metabo-

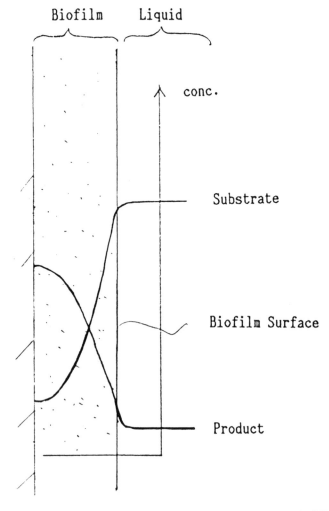

FIGURE 1.11. *Substrate or product transportation across a microbial film.*

lized by film microbes at the same time. End products of biological reaction go in the opposite direction, and come out in the bulk liquid. The end product sometimes undergoes another biological reaction, but such cases are not discussed here.

When consumption of substrate proceeds in the above mentioned manner, the reaction rate observed, i.e., overall reaction rate, generally differs from the intrinsic reaction rate, and the former is affected by the following steps.

(1) The transfer of substrate in the liquid contacted with the microbial film

(2) Diffusive transport of substrate into the microbial film

(3) Consumption of substrate by biological reaction within the microbial film

(4) Transport of end product out of the microbial film

Step (4) is usually able to be excluded from the rate-determining steps, because end product does not generally accumulate in so high a concentration as it inhibits biological reaction. But, Harremoës et al. [31] reported that nitrogen supersaturation and bubble formation inhibited the performance of the biological filter for denitrification, and the inhibition of nitrification has also been reported by the pH decrease by nitrate production in a microbial film for nitrification [32]. Besides, heat being a kind of end product, a temperature gradient is supposed to be made in the microbial film to enable the heat produced in the film to go out of the film. But, the effect of such temperature rises on biological reaction rates would also be negligible in general. In a reaction including a series of rate-affecting steps, if the overall reaction rate is controlled by a single step, the step, i.e., the most slowly proceeding step, is called "the rate-limiting step (or process)." Then, if diffusive transports as (1) and (2) are the rate-limiting steps, the reaction is diffusion-limiting, and if biological reaction (3) is the rate-limiting process, the reaction is said to be reaction-limiting. When two or more substances are participating in the reaction, and one of them controls the overall reaction, then this substance—the substance whose supply is the smallest compared with its consumption—is called the rate-limiting substance. Dissolved oxygen is often the rate-limiting substance in an aerobic treatment system. Hereafter in this chapter, the term substrate means rate-limiting substrate provided no exception is made.

1.4.2 Concentration Profile of Substrate in the Neighborhood of Microbial Film

Consider a water flow along the microbial film surface as in Figure 1.12. In the liquid flowing at a sufficient distance or more from the film surface,

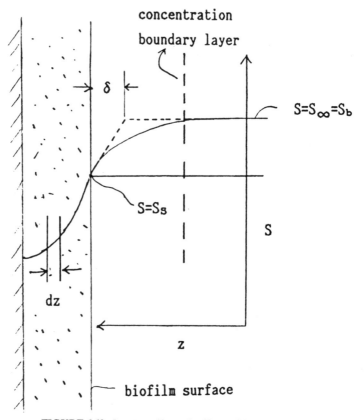

FIGURE 1.12. *Laminar film at biofilm and liquid interface.*

i.e., in bulk liquid or in bulk water, the mixing in the direction perpendicular to the film surface being active, the substrate concentration is uniform in that direction. That is to say, there is no resistance of transportation from bulk liquid to microbial film. On the other hand, in the neighborhood of microbial film, the flow velocity is zero at the surface of the microbial film, and increases with the distance from the film surface, approaching the velocity of bulk liquid (v_∞). Theoretically, flow velocity becomes equal to v_∞ infinitely apart from the film surface, so the distance from it to the point at which flow velocity is equal to 0.99 v_∞ is regarded as the depth of the layer in which the effect of viscosity must be considered. This layer is called boundary layer (see Figure 1.13). In this case, since the boundary layer is decided based on flow velocity profile, it is called momentum boundary layer. Similarly, concentration boundary layer and thermal boundary layer are defined based on concentration and temperature profile, respectively.

Resistance to diffusion occurs in the concentration boundary layer. In the boundary layer, supposing a hypothetic liquid film in which there is no velocity component perpendicular to the film surface or no turbulence, the liquid film is called a laminar film. Since mass transport occurs by molecular diffusion alone in a laminar film, the concentration gradient will be kept constant. Substrate flux by molecular diffusion, or quantity of substrate transported per unit area, unit time N is given by the following equation.

$$N = -D_w \frac{dS}{dz} = -D_w/\delta(S_b - S_s) = -k_F(S_b - S_s) \qquad (1.6)$$

where

N = substrate flux through laminar layer
D_w = diffusion coefficient of substrate through water
S_b = substrate concentration in bulk liquid
S_s = substrate concentration at the surface of microbial film
δ = thickness of laminar layer
k_F = mass transfer coefficient of substrate $(=D_w/\delta)$

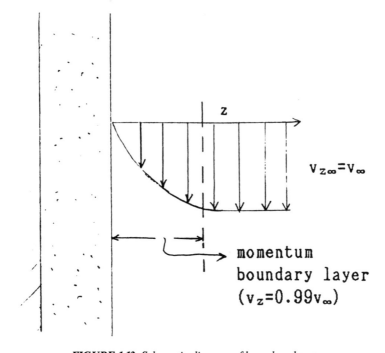

FIGURE 1.13. *Schematic diagram of boundary layer.*

Flux N is positive when the transfer is in the positive direction of z. The thickness of laminar layer δ varies depending upon the position on the film surface unless the area of film is infinitive, but it could be practically considered to be constant anywhere.

Generally, since δ decreases with increasing flow velocity or decreasing viscosity of liquid, S_s and the depth of effective microbial film increase correspondingly. That is, flux of substrate N and overall reaction rate increase.

Williamson and McCarty [24] assumed that liquid laminar layer consists of two parts L_1 and L_2, and the depth of the outer layer may be reduced to zero with adequate mixing. The inner layer L_2 cannot be removed by mixing and is believed to result from the uneven or spongelike nature of the liquid – biofilm interface, and the value of L_2 was calculated as 56 μ. They proposed to use empirical formulas in Table 1.3 summarized by Welty et al. [33] as L_1. In these formulas the value of Y is given as a function of Re, Re' or Re'' in Figure 1.14.

Many empirical relationships have been given for spherical biofilms. Carbbery's formula is the most general formula for k_F in a packed bed of spherical particles, and is applicable under Re of 1–1000, and shown as,

$$\frac{k_F}{u/\epsilon} = \left(\frac{\mu}{\varrho D_w}\right)^{2/3} = 1.15 \left(\frac{d_p u \varrho}{\mu \epsilon}\right)^{-0.5} \tag{1.7}$$

where

$u =$ superficial flow velocity (m³/hr·m²)
$d_p =$ particle diameter (m)
$\epsilon =$ porosity of packed bed ($-$)

When packed particle is fine, and Re based on packed particles is small, then the formula by Wilson and Geankoplis [35] is reliable.

$$\epsilon \left(\frac{k_F}{u}\right)\left(\frac{\mu}{\varrho D_w}\right)^{2/3} = 1.09 \, Re^{-2/3} \tag{1.8}$$

$0.016 < Re < 55$

Kataoka et al. [36] have shown the following equation as the relationship between the modified Reynold's number Re' and k_F,

$$\left(\frac{1-\epsilon}{\epsilon}\right)^{1/3} \frac{k_F Sc}{u/\epsilon} = 1.85 \, Re'^{-2/3} \tag{1.9}$$

TABLE 1.3. Values of Reynolds Number (R_e) and L_1.

Kind of Flow	R_e	L_1
Flow in a pipe	$DV\varrho/\mu$ (R_e)	$\dfrac{D_w^{1/3}}{VY}(\mu/\varrho)^{2/3}$
Flow on a flat plate	$LV'\varrho/\mu$ (R_e')	$\dfrac{D_w^{1/3}}{V'Y}(\mu/\varrho)^{2/3}$
Flow between packing media	$dV''\varrho/\mu\epsilon$ (R_e'')	$\dfrac{D_w^{1/3}}{V''Y}(\mu/\varrho)^{0.58}(10^{-1})$

Y = value determined by Figure 1.14
μ = viscosity (g/cm-sec)
ϱ = density (g/cm³)
D_w = diffusion coefficient in water (cm²/sec)
V = mean velocity (cm/sec)

V' = maximum velocity (cm/sec)
V'' = superficial velocity (cm/sec)
D = diameter of pipe (cm)
L = distance between plates (cm)
d = size of packing medium (cm)

and pointed out that Equation (1.9), both Equations (1.9) and (1.7), and Equation (1.7) best fit the observed values under the conditions of $Re' < 10$, $10 < Re' < 100$, and $Re' > 100$, respectively.

Re' is given by

$$Re' = \frac{d_p u \varrho}{(1 - \epsilon)\mu} \tag{1.10}$$

and Sc is Schmidt number.

$$Sc = \mu/\varrho D_w \tag{1.11}$$

For a single spherical particle put in a liquid flow, Ranz-Marshall Equation can be applied.

$$Sh \equiv \frac{k_F d_p}{D_w} = 2.0 + 0.6\, Re^{1/2} Sc^{1/3} \tag{1.12}$$

where Sh = Sherwood number defined by Equation (1.12).

Dewalle et al. [37] have given the following simplified equations as the relationships between flow rate Q and mass-transfer coefficient in a packed bed.

$$k_F = kQ^{1/3} \tag{1.13}$$

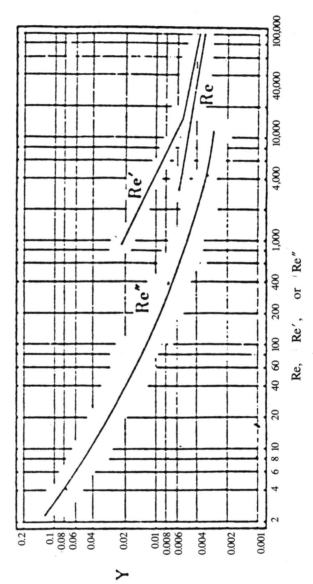

FIGURE 1.14. Y and Reynolds number.

for $Re < 10$

$$k_F = kQ^{1/2} \tag{1.14}$$

for $Re > 10$

Since Q is proportional to u, Equations (1.13) and (1.14) imply the same meanings as Equations (1.8) and (1.7), respectively. Therefore, generally speaking, k_F is proportional to the cube root of flow rate or flow velocity in the small region of Re, and is proportional to the square root of them if $Re > 10$.

1.4.3 Diffusion in Microbial Film – Fundamental Equation of Reaction

Suppose the microbial density is uniform and there is no liquid flow in a certain microbial film. Substrate moves, therefore, in the void between microbial cells only by molecular diffusion. Needless to say, it is only in the void portion that diffusion of substrate actually occurs, so the diffusion rate – or diffusion coefficient in other words – is smaller in microbial film than in water. The diffusion coefficient in such a nonuniform field is called effective diffusion coefficient, if calculated as if the field were uniform.

As in Figure 1.12, z axis is set perpendicularly to the microbial film, and a differential depth dz is considered in the film, then the mass balance of substrate for unit area with depth dz is stated as follows,

Input

$$D_e(\partial S_c / \partial z)$$

Output

$$-D_e(\partial S_c / \partial z) + \partial\left(\frac{\partial(\partial S_c / \partial z)}{\partial z} \cdot dz\right)$$

Consumption

$$(Mdz)Q$$

where

D_e = effective diffusion coefficient of substrate in microbial film (cm²/day)
S_c = substrate concentration in microbial film (mg/cm³)

M = microbial density of film (g/cm^3)
Q = metabolic rate of substrate per unit amount of microbe $(mg/g \cdot day)$

General formulation of mass balance is shown as

$$[\text{Accumulation}] = [\text{Input}] - [\text{Output}] - [\text{Consumption}] \quad (1.15)$$

So the following equation

$$(\partial S_c / \partial t)dz = D_e(\partial^2 S_c / \partial z)dz - (Mdz)Q \quad (1.16)$$

is obtained. Dividing Equation (1.16) by dz, following fundamental equation of mass balance is derived.

$$\frac{\partial S_c}{\partial t} = D_e \frac{\partial^2 S_c}{\partial z^2} - MQ \quad (1.17)$$

Monod's Equation can be used for Q which is most often applied to describe the substrate ingestion rate by microbes.

$$Q = \frac{kS_c}{K_s + S_c} \quad (1.18)$$

where

K_s = saturation constant (or half-saturation constant) (mg/cm^3)
k = constant equivalent to maximum ingestion rate Q_{max} $(mg/g \cdot day)$

Substituting Equation (1.18) into Equation (1.17),

$$\frac{\partial S_c}{\partial t} = D_e \frac{\partial^2 S_c}{\partial z^2} - \frac{MkS_c}{K_s + S_c} \quad (1.19)$$

Equation (1.19) is solved under appropriate boundary conditions, and the concentration profile of substrate in microbial film may be determined, then the overall rate of substrate ingestion might be calculated based on the profile.

The steady state equation is obtained by setting the left-hand side of Equation (1.19) equal to zero and rewriting partial differential to ordinary differential.

$$\frac{d^2 S_c}{dz^2} = \frac{MkS_c}{D_e(K_s + S_c)} \quad (1.20)$$

Equation (1.20) being a second-order, nonlinear differential equation, has no explicit solution, but it can be simplified as follows according to the relative value of S_c and K_s.

for $S_c \gg K_s$

$$\frac{d^2 S_c}{dz^2} = \frac{kM}{De} \tag{1.21}$$

for $S_c \ll K_s$

$$\frac{d^2 S_c}{dz^2} = \frac{kM}{D_e K_s} \tag{1.22}$$

Using Equations (1.21) and (1.22), the flux of substrate N is determined as follows.

The Case of $S_c \gg K_s$

It would be more proper to think that K_s is very small than to think S_c is large enough in this case. Then, microbial reaction would be of zero-order in S_c.

THE CASE IN WHICH SUBSTRATE CAN PENETRATE TO THE BOTTOM OF MICROBIAL FILM

In this case, disregarding Equation (1.21), the flux N is immediately decided by Equation (1.23), because the substrate ingestion rate is uniform all over the biofilm, and N must be equal to the ingestion rate per unit biofilm area.

$$N = kML \tag{1.23}$$

where L = the thickness (or depth) of microbial film

THE CASE IN WHICH SUBSTRATE REACHES ONLY TO THE DEPTH $L_e (L_e < L)$, $(L_e = EFFECTIVE\ DEPTH\ OF\ A\ MICROBIAL\ FILM)$

Equation (1.21) should be solved under the following boundary conditions.

$$
\begin{aligned}
&[\text{B.C. I}] \quad z = 0 \ : \ S_c = S_s \\
&[\text{B.C. II}] \quad z = L_e : S = 0
\end{aligned}
\right\} \tag{1.24}
$$

From Equations (1.21) and (1.24)

$$S_c = \frac{kM}{2D_e}z^2 - \left(\frac{S_s}{L_e} + \frac{kM}{2D_e}L_e\right)z + S_s \qquad (1.25)$$

Besides, since the substrate fluxes on both sides of the biofilm surface are equal, and these are equal to the substrate consumption rate per unit biofilm area N,

$$N = k_F(S_b - S_s) = -\left.\frac{dS_c}{dz}\right|_{z=0} \cdot D_e$$

$$= \left(\frac{S_s}{L_e} + \frac{kM}{2D_e}L_e\right)D_e = kML_e$$

$$(1.26)$$

From Equation (1.26) L_e can be determined.

$$L_e = \sqrt{\left(\frac{D_e}{k_F}\right)^2 + \frac{2D_e}{kM}S_b} - \frac{D_e}{k_F} = \sqrt{\frac{2D_e}{kM}S_s} \qquad (1.27)$$

Combining Equations (1.26) and (1.27), Equation (1.28) is obtained.

$$N = kML_e = kM\left[\sqrt{\left(\frac{D_e}{k_F}\right)^2 + \frac{2D_e}{kM}S_b} - \frac{D_e}{k_F}\right] = \sqrt{2D_e kMS_s} \quad (1.28)$$

That is, the substrate flux N is proportional to the square root of the substrate concentration at the biofilm surface, and if $(D_e/k_F)^2 = (D_e\delta/D_w)^2 \ll 2D_eS_b/kM$, then it can be approximated that $S_s \cong S_b$, so it could be regarded that the overall reaction rate is almost proportionate to square root of bulk liquid substrate concentration.

The Case of $S_c \ll K_s$

This case corresponds to the condition that either substrate concentration is low anywhere in biofilm, or saturation constant K_s for the substrate is large enough, and biological reaction is of first-order in substrate concentration. In any event, the bulk liquid substrate concentration is not so high in this case, and the analyses are of no practical meaning, presupposing that the substrate would reach to the bottom of the biofilm. Equation (1.22) is solved, therefore, under the following boundary conditions.

$$[\text{B.C. I}] \quad z = 0 \ : \ S_c = S_s \left.\vphantom{\begin{matrix}1\\1\end{matrix}}\right\}$$
$$[\text{B.C. II}] \quad z = L_e \ : \ dS_c/dz = 0 \left.\vphantom{\begin{matrix}1\\1\end{matrix}}\right\} \tag{1.29}$$

$$S_c = S_s \frac{\cosh\ [(kM/D_eK_s)^{1/2}\ (L_e - z)]}{\cosh\ [(kM/D_eK_s)^{1/2}L_e]} \tag{1.30}$$

$$N = \int_0^{L_e} \frac{kS_cM}{K_s}\ dz = (D_e kM/K_s)^{1/2}S_s \tag{1.31}$$

From Equations (1.6) and (1.31),

$$k_F(S_b - S_s) = (D_e kM/K_s)^{1/2}S_s \tag{1.32}$$

Rearranging Equation (1.32), S_s is obtained.

$$S_s = \frac{1}{1 + (D_e kM/K_s)^{1/2}/k_F}\ S_b \tag{1.33}$$

Substituting Equation (1.33) into Equation (1.31), N is determined as,

$$N = \frac{(D_e kM/K_s)^{1/2}}{1 + (D_e kM/K_s)^{1/2}/k_F}\ S_b \tag{1.34}$$

That is, overall reaction is of first-order in both bulk liquid and biofilm surface substrate concentrations. Further, overall reaction rate is also proportional to bulk liquid substrate concentration S_b in the case where substrate can penetrate to the bottom of biofilm, and for which case the deviation of formula of N was omitted by reason of lacking practical value. Overall reaction will always be of the first-order, provided intrinsic reaction is so. The above can be summarized as follows:

(1) When intrinsic reaction is of zero-order:
- overall reaction is also of zero-order, if substrate is supplied anywhere in biofilm (1a)
- overall reaction is of half-order, if substrate supply is limited to a portion of biofilm (1b)

(2) When intrinsic reaction is of first-order, overall reaction is also of first-order.

Harremoës [39] has demonstrated similar conclusions by micropore models based on the assumptions that on a biofilm surface, cylindrical pores with uniform radii and depths are distributed uniformly, and that reactants have to diffuse through pores before biochemical reaction takes place on the walls of pores.

Since K_s can be fixed if the substrate is unchanged, it can be considered that the substrate concentration of bulk liquid is in the order of (1a) > (1b) > (2). The relationship between bulk liquid substrate concentration and overall reaction rate, therefore, is schematically shown in Figure 1.15, which gives a similar relationship to Monod's Equation. It has been mentioned that Equation (1.20) has no explicit solution. Such a solution is needed, if it were possible to gain, in the region between Equations (1.21) and (1.22), i.e., in the region that is shown by a broken line. Though the

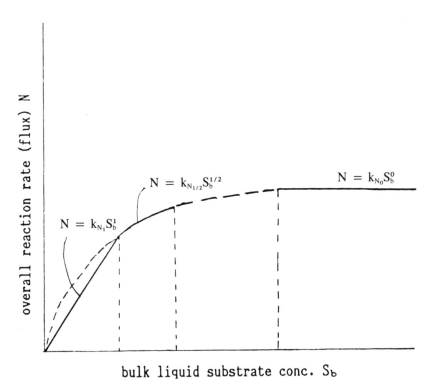

FIGURE 1.15. *Substrate flux dependence on bulk liquid concentration.*

overall reaction rate is not able to be strictly shown as a simple function of S_b^n in this region, the value of n should be between 0 and 0.5 if a function proportional to S_b^n is used as an approximate formula. Hence, the general expression of overall reaction rate containing the approximate formulation is shown as follows,

$$N = k_N S_b^n \qquad n = 0\text{--}1.0 \qquad (1.35)$$

where k_N = constant.

It should be noted that k_N is not a universal constant but a constant which changes depending on n.

Next, the relationship between kM – product of parameter of microbial activity, k and microbial density in biofilm, M – and overall reaction rate is discussed;

(1) When intrinsic reaction is of zero-order:
   • overall reaction rate is proportional to kM, if the substrate can penetrate to the bottom of the biofilm. Therefore, overall reaction rate is the same as one of suspended biomass (1a).
   • overall reaction rate is proportional to $\sqrt{kM}$ by Equation (1.28), if substrate supply is limited to a portion of biofilm (1b).

(2) When intrinsic reaction is of the first-order, overall reaction rate is proportional to $\sqrt{kM}$ from Equation (1.31), or by neglecting the second term of the denominator on the right-hand side of Equation (1.34).

As previously mentioned, Equation (1.20) has no explicit analytical solution, when it cannot be simplified to the form of Equation (1.21) or (1.22); nevertheless, it can easily be estimated that the overall reaction rate would be proportional to $\sqrt{kM}$ even in the intermediate region of S_c, that is, in the region where neither of K_s and S_c can be neglected compared to the other, provided that the substrate cannot penetrate throughout the biofilm, taking into account the continuity of the problem. In brief, overall reaction rate is proportional to $\sqrt{kM}$ whenever substrate cannot penetrate throughout biofilm.

Therefore, except the case of (1a) where the behavior of biofilm is the same as that of suspended biomass, microbial activity and/or density of biofilm do not directly influence the overall reaction rate, but in a moderated manner by the effect of diffusion barrier given by biofilm. That is, in many cases, overall reaction rate is proportional not to kM but to $\sqrt{kM}$, and this is exactly the mathematical explanation of the characteristics of microbial film processes that they are relatively advantageous treatment processes if applied to those wastewaters with low temperature or slowly degradable substrates.

1.4.4 Transformation of Fundamental Equation into Dimensionless Form

To discuss mathematical problems generally, it is effective to transform equations into dimensionless forms. Into the fundamental equation [Equation (1.20)] and the boundary conditions;

$$dS_c/dz \big|_{z=L} = 0 \qquad (1.36)$$

$$-D_e dS_c/dz \big|_{z=0} = k_F(S_b - S_s) \qquad (1.26)$$

The following dimensionless variables and parameters are introduced:

$$\left. \begin{aligned} s &= S_c/S_b, \quad \varrho = z/L, \quad q = \frac{Q(S_c)}{Q(S_b)} \\[2mm] \phi^2 &= \frac{L^2}{D_e S_b} Q(S_b)M, \quad Bi = \frac{k_F L}{D_e} \end{aligned} \right\} \qquad (1.37)$$

where, q is the dimensionless reaction rate which is equal to unity at $s = 1$. ϕ and Bi are called Thiele modulus and Biot number respectively. Then, Equations (1.20), (1.36) and (1.26) are transformed.

$$\left. \begin{aligned} \frac{d^2 s}{d\varrho^2} - \phi^2 q &= 0 \\[2mm] \frac{ds}{d\varrho}\bigg|_{\varrho=1} &= 0 \\[2mm] -\frac{ds}{d\varrho}\bigg|_{\varrho=0} &= Bi(1 - s) \end{aligned} \right\} \qquad (1.38)$$

1.4.5 Effectiveness Factor

It is of no direct benefit to solve Equation (1.38) to determine the substrate concentration profile. The final objective is to obtain the substrate flux N represented by Equation (1.26). That is, to determine N which can be described as follows is most important.

$$N = -D_e \frac{dS_c}{dz}\bigg|_{z=0} = g(D_e, Mk, K_s, L, S_b) \qquad (1.39)$$

The effectiveness factor is the index representing how effective the activity of microbes constituting biofilm is compared to that of suspended microbes. Effectiveness factor η is defined by the following equation:

$$\eta = \frac{N}{N_b} = \frac{\text{(actual rate of substrate removal by microbial film)}}{\text{(substrate removal rate when substrate concentration anywhere in microbial film were equal to } S_b)} \qquad (1.40)$$

Substrate concentration in microbial films being smaller than S_b, η is usually smaller than unity, but sometimes it exceeds unity when substrate is inhibitive or S_b is too high as to inhibit microbial activity. Substrate flux N is formulated by the equation,

$$N = \eta N_b \qquad (1.41)$$

where N_b is represented by Equation (1.42)

$$N_b = MLQ(S_b) = \frac{MLkS_b}{K_s + S_b} \qquad (1.42)$$

Therefore, combining Equations (1.39), (1.41) and (1.42),

$$\eta = (-D_e dS_c/dZ|_{z=0})/N_b = -D_e \frac{S_b}{L} \frac{ds}{d\varrho}\bigg|_{\varrho=0} /MLQ(S_b)$$

$$= -\{D_e S_b/L^2 Q(S_b)M\} \frac{ds}{d\varrho}\bigg|_{\varrho=0} = -\left(\frac{1}{\phi}\right)^2 \frac{ds}{d\varrho}\bigg|_{\varrho=0} \qquad (1.43)$$

is obtained. The practical merit of the effectiveness factor is that the overall reaction rate can very easily be calculated from S_b only if η is known, though it is very complicated to determine concentration profile of substrate in microbial film and calculate overall reaction rate by Equation (1.26).

Equation (1.43) can be rewritten as follows:

$$\eta = \left[\frac{1 + k_3 S_b}{(k_2 L)^2}\right]\left(-\frac{ds}{d\varrho}\bigg|_{\varrho=0}\right) \qquad (1.44)$$

where

$$k_2^2 = \frac{kM}{D_e K_s}, \qquad k_3 = \frac{1}{K_s}$$

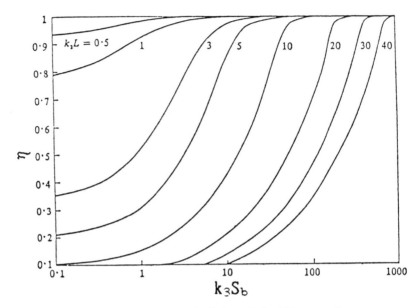

FIGURE 1.16. Numerical solution of η by Atkinson et al.

Effectiveness factor defined by Equation (1.44) has been represented by Atkinson et al. [40,41] numerically solving similar solutions as Equation (1.38), in the following form:

$$\eta = g(k_2 L, K_3 S_b) \qquad (1.45)$$

The results of calculation are graphically represented in Figure 1.16.

Atkinson has given Table 1.4 as the perfect formulation of microbial film. These formulae can be simplified as Table 1.5 under a limited condition.

1.4.6 Biological Rate Equation Coefficients and Characteristics of Microbial Film Reactions

From Equations (1.38), (1.41), and (1.42), N can be represented in the following form:

$$N = g(k_1, k_2, k_3, S_b, L) \qquad (1.46)$$

k_1, k_2 and k_3 are called biological rate equation coefficients. Hence, overall substrate removal rate by microbial films is perfectly prescribed by three parameters k_1–k_3, bulk liquid substrate concentration, and thickness of

TABLE 1.4. Exact Equation for Microbial Film by Atkinson et al.

$$N = \lambda \frac{k_1 L S_b}{1 + k_3 S_b}$$

where

$$\lambda = 1 - \frac{\tanh k_2 L}{k_2 L} \left(\frac{\phi_p}{\tanh \phi_p} - 1 \right), \qquad \phi_p \leq 1$$

$$\lambda = \frac{1}{\phi_p} - \frac{\tanh k_2 L}{k_2 L} \left(\frac{1}{\tanh \phi_p} - 1 \right), \qquad \phi_p \geq 1$$

and

$$\phi_p = \frac{(k_2 L)(k_3 S_b)}{\sqrt{2(1 + k_3 S_b)}} [k_3 S_b - \ln(1 + k_3 S_b)]^{-1/2}$$

TABLE 1.5. Simplified Equations for Microbial Film.

$$N = \frac{\tanh \phi_R}{\phi_R} \frac{k_1 L S_b}{1 + k_3 S_b} \qquad \text{valid for all } k_3 S_b \text{ if } k_2 L < 1$$
$$\text{valid for all } k_2 L \text{ if } k_3 S_b < 0.1$$

$$N = \frac{k_1 L S_b}{1 + k_3 S_b}, \quad \phi_R \leq 1 \qquad \text{valid for all } k_3 S_b$$
$$\text{if } k_2 L > 20$$

$$N = \frac{(1 + 2k_3 S_b)^{1/2}}{(1 + k_3 S_b)} \frac{k_1 S_b}{k_2}, \quad \phi_R \geq 1 \qquad \text{valid for all } k_2 L$$
$$\text{if } k_3 S_b > 100$$

where

$$\phi_R = \frac{k_2 L}{(1 + 2k_3 S_b)^{1/2}}$$

biofilm, i.e., five factors in total. k_2 and k_3 are coefficients presented in Equation (1.44), and k_1 is the coefficient in the following equation which is obtained rewriting Equation (1.42).

$$N_b = \frac{MLkS_b}{K_s + S_b} = \frac{k_1 LS_b}{1 + k_3 S_b} = N_{max}\left(\frac{k_3 S_b}{1 + k_3 S_b}\right) \qquad (1.47)$$

where k_1 has dimension of T^{-1} and is represented as

$$k_1 = kM/K_s \qquad (1.48)$$

N_{max} is equal to $k_1 L/k_3$ and is equivalent to the value of flux N_b ($=N$) at $S_b = \infty$.

Obviously, k_1 is a coefficient relating to microbial density in biofilms and the coefficient in Monod's equation−an equation for biological reaction rate; and k_2, being equal to $\sqrt{k_1/D_e}$, includes effective diffusion coefficients as well as factors included in k_1; and k_3, being $1/K_s$, includes the coefficient of the equation of biological reaction rate alone.

k_2 is the coefficient relating to the diffusion resistance in biofilm, and k_2 is equal to zero when D_e is infinitive and then no diffusion resistance appears; k_2 can be considered to be equal to zero also when L is small enough, because there is no diffusion resistance in such a case. Under these conditions, the rate equation is represented as follows;

$$N = g(k_1, k_3, S_b, L) = \frac{k_1 LS_b}{1 + k_3 S_b}$$

$$= N_{max}\frac{k_3 S_b}{1 + k_3 S_b} \qquad (1.49)$$

When L is large enough so that the diffusion resistance in the biofilm cannot be neglected, then N is represented as follows, if S_b is small.

$$N = k_1/k_2 S_b \qquad (1.50)$$

And when S_b is large,

$$N = \frac{k_1}{k_3} L = N_{max} \qquad (1.51)$$

Equations (1.49)–(1.51) are depicted as Figure 1.17. In Figure 1.17, Equation (1.49), i.e., the same type of equation as one applied to suspended

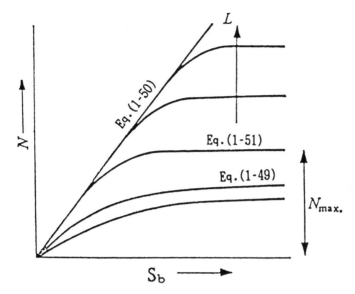

FIGURE 1.17. *Biological reaction rate of a microbial film (Atkinson, B.).*

microbe, can be applied for a small L. If L increases, N is proportional to S_b as Equation (1.50) for a small S_b, but as S_b grows, N gets out of proportion to S_b and asymptotically approaches a straight line parallel to the abscissa, i.e., to N_{max} in Equation (1.51). From this figure, it can be understood that overall reaction rate N can be approximated by a first-order reaction kinetic in a much wider range of S_b than one which might be anticipated from K_s value in Monod's equation. The fact stated here is a very important feature of microbial film processes on substrate removal characteristics, and it should be noted that the rate constant in this case (k_1/k_2) includes not only biochemical rate-limiting factors but also microbial density in biofilm M and effective diffusion coefficients.

Though the above discussion was presented for biofilms attached on plane surfaces, almost the same mathematical analyses are able to be applied to biofilms on spherical or cylindrical surfaces, and the results of analyses were obtained in a similar trend as above [42,43].

1.4.7 Empirical Confirmation of Substrate Concentration Profile in Microbial Film

Substrate concentration profile of parabola or a similar curve represented by Equation (1.25) or (1.30) has already been given many theoretical or indirect confirmation, but the direct confirmation of it by actual survey is very rare.

Chen and Bungay [44] probed microbial slime samples from two trickling filters for treatment of municipal wastes with microelectrodes, and determined DO variations with time or with location under various conditions. They showed a typical DO profile as Figure 1.18. At 200 μm or more above the film surface, DO concentration in the bulk liquid was 8.5 mg/l, and it decreased rapidly as the microprobe entered into the film. DO gradient, as shown by a dotted line, decreased linearly with the distance from the slime surface, meaning that the DO profile was parabolic. Figure 1.19 shows DO profiles under three substrate concentration levels, and DO gradients above and below the slime surface differed implying that the effective diffusion coefficient in the microbial film D_e was significantly smaller than that in water D_w. Typical profiles are shown in Figure 1.20 when photosynthetic slimes were probed under illumination and in the dark. It is evidently shown that photosynthetic microbes were concentrated within 130 μm from the slime surface, and DO rather increased in this region with depth. Moreover, it is observed that the depth of concentration boundary layer is 50–100 μm in any of the Figures 1.18–1.20. From these determinations, it could be considered that the contents theoretically predicted was certified experimentally.

1.4.8 Decision of the Rate Limiting Substance

Among various substances required for a biological reaction, the substance whose relative supply is minimized compared to its consumption (or demand) will possibly limit the reaction rate. When reaction rate is controlled by such a substance, it is called the limiting substance or limiting species. If the limiting substance is the growth-limiting substrate, it is called the limiting substrate, but the term limiting substrate can be regarded as almost the synonym of limiting substance because substrate includes dissolved oxygen, nutrient salts, trace metals, etc. in the widest sense. To find what is the limiting substance of a biochemical reaction is of much technological importance, for an increase in the supply of the substance directly results in an increment of reaction efficiency. In microbial reactions, it is the balance between the supply of hydrogen donor and hydrogen acceptor which is frequently in question, so the following discussions are made taking this problem as an example. But, such discussions are of use for any two substances required in a certain biochemical reaction.

Suppose that the microbial reaction is expressed as follows.

$$\nu_d D + \nu_a A + \text{nutrient material} \rightarrow \text{End products} + \text{cells} \quad (1.52)$$

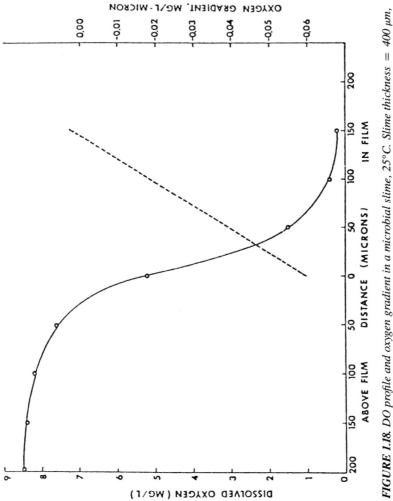

FIGURE 1.18. DO profile and oxygen gradient in a microbial slime, 25°C. Slime thickness = 400 μm, nutrient broth = 500 ppm [44].

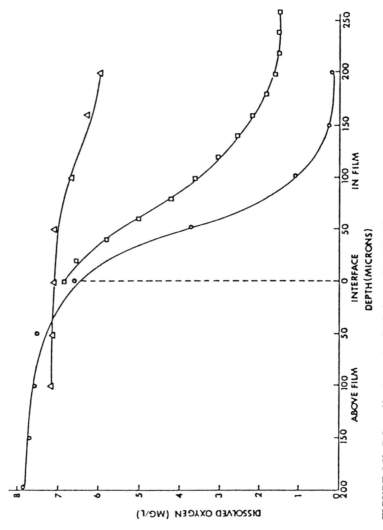

FIGURE 1.19. DO profiles in a microbial slime: (△) 1/15 dilution of sewage, 22°C; (□) 1/5 dilution of sewage, 22°C; (○) nutrient broth, 500 mg/l, 27°C [44].

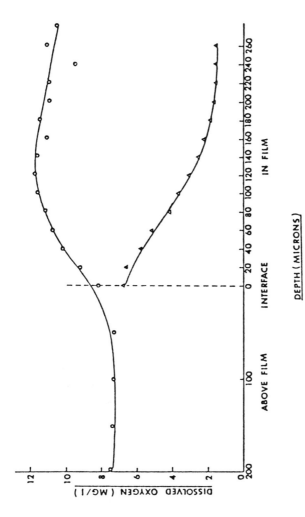

FIGURE 1.20. *DO profiles in a microbial slime:* (○) *light intensity 1000 F.C., 22°C;* (△) *in the dark, 22°C [44].*

51

where ν_d, ν_a = stoichiometric reaction coefficients for the electron donor (D) and the electron acceptor (A), respectively.

Where suspended biomass is concerned, the reaction rate is usually in the state of reaction limiting, so the limiting substance can be found out according to the relationships:

$$
\left.
\begin{array}{l}
\dfrac{S_{bd}}{\nu_d MW_d} > \dfrac{S_{ba}}{\nu_a MW_a} \text{ hydrogen acceptor is limiting} \\[3ex]
\dfrac{S_{bd}}{\nu_d MW_d} = \dfrac{S_{ba}}{\nu_a MW_a} \text{ supplies are in balance} \\[3ex]
\dfrac{S_{bd}}{\nu_d MW_d} < \dfrac{S_{ba}}{\nu_a MW_a} \text{ hydrogen donor is limiting}
\end{array}
\right\}
\quad (1.53)
$$

where

S_{bd}, S_{ba} = the bulk liquid concentrations of the hydrogen donor and the hydrogen acceptor, respectively

MW_d, MW_a = the molecular weights of the hydrogen donor and the hydrogen acceptor, respectively

Also for microbial films, the limiting substance can be found out by the same way as the case of suspended biomass under reaction limiting states. In microbial films, however, overall reaction is frequently in the state of diffusion limiting, and then the substance of which supply, or flux N is the smallest relative to its demand will be the limiting substance. For the supplies of D and A to be in balance, the following relation should be kept between fluxes of D and A, N_d and N_a.

$$
\frac{N_d}{\nu_d MW_d} = \frac{N_a}{\nu_a MW_a}
\quad (1.54)
$$

Under this condition, the substances D and A will penetrate to the same depth of microbial film, and so the following equation can be obtained:

$$
N_d / N_a = D_{ed} S_{bd} / D_{ea} S_{ba}
\quad (1.55)
$$

where D_{ed}, D_{ea} = effective diffusion coefficient of D and A, respectively in microbial film.

Moreover, since the assumption $(D_{ed}/D_{ea}) = (D_{wd}/D_{wa})$ is considered to be valid, introducing this relation into Equation (1.54), following equation for the condition of balancing N_d and N_a is obtained.

$$S_{bd} = \frac{D_{ea}\nu_d MW_d}{D_{ed}\nu_a MW_a} S_{ba} = \frac{D_{wa}\nu_d MW_d}{D_{wd}\nu_a MW_a} \qquad (1.56)$$

The following equation, therefore, implies the shortage of supply of D:

$$S_{bd} < \frac{D_{wa}\nu_d MW_d}{D_{wd}\nu_a MW_a} S_{ba} \qquad (1.57)$$

and contrarily, Equation (1.58) represents the shortage of A.

$$S_{bd} > \frac{D_{wa}\nu_d MW_d}{D_{wd}\nu_a MW_a} S_{ba} \qquad (1.58)$$

Examining the relations of Equations (1.57) and (1.58) about two or more substances participating biochemical reaction, the most deficient substance must be the limiting substance. However, hydrogen donor or acceptor is not always a pure substance, but a complex in many cases; then theoretical investigations [45] for such cases is significantly complicated.

Example 1: The aerobic change of glucose is expressed by the following formula.

$$C_6H_{12}O_6 + 2O_2 + 0.8HCO_3^- + 0.8NH_4^+ \rightarrow$$

$$0.8C_5H_7O_2N + 2.8CO_2 + 5.2H_2O$$

where $\nu_d = 1$, $\nu_a = 2$, $MW_d = 180$, $MW_a = 32$, D_{wd}, D_{wa} (diffusion coefficient of glucose and oxygen in water) $= 0.5$ and 1.6 cm^2/sec at 20°C, then substituting these values into Equation (1.57),

$$S_{ba} < \frac{(0.5)(2)(32)}{(1.6)(1)(180)} S_{bd} = 0.11 S_{bd}$$

that is, if the concentration of dissolved oxygen is lowered below 0.11 times that of glucose, oxygen will limit the reaction. On the other hand, it is commonly said that 0.2–0.5 mg/l of dissolved oxygen will do in an aeration tank of activated sludge. In microbial film processes, however, dissolved oxygen concentration required increases in proportion to the concentration of substrate.

Example 2: About 4.5 mg of oxygen is required for nitrification of 1 mg of NH_4-N, then $(\nu_a MW_a / \nu_d MW_d) = 4.5$, $(D_{wd}/D_{wa}) = 0.6$,

$$S_{ba} < (4.5)(0.6)S_{bd} = 2.7\, S_{bd}$$

Dissolved oxygen concentration is desired to be more than 2.7 times that of ammoniacal nitrogen.

Example 3: About 2.5 mg of methanol is consumed for the denitrification if 1 mg of NO_3-N, then $(\nu_a MW_a / \nu_d MW_d) = 2.5$, and (D_{wd}/D_{wa}) is estimated to be ca. 0.5,

$$S_{ba} < \frac{1}{(2.5)}\,(0.5)\,S_{bd} = \frac{1}{5}\,S_{bd}$$

Denitrification rate can be maximized by increasing methanol concentration as high as 5 times that of NO_3-N. But, the stoichiometrically required amount of methanol being only 2.5 times that of NO_2-N, attention must be paid to that excess methanol which will remain in the effluent.

1.4.9 Approach to Reaction Systems Containing Particulate Substrates

While wastewater containing particulate organics is treated with a microbial film process, particulate substances cannot diffuse directly into microbial film. Therefore, application of a theoretical approach is confined, strictly speaking, to systems containing low-molecular weight dissolved substrates alone; so in the analysis of behavior of systems containing large amounts of particulate substrate as municipal sewage, particulate substrates must be distinguished from dissolved substrates. As discussed in section 1.4.5, in microbial film processes, the removal process of dissolved substrate can be approximated by first-order kinetics over a wide range of substrate concentrations; besides, the process of attachment of particulate substrate on microbial films is also considered to be a first-order reaction, because the rate of attachment is proportional to the concentration of particulate substrate. The author et al. [46] simulated the behavior of organic substance in an anaerobic filter for domestic sewage treatment by the following simplified mathematical models.

particulate substrate

$$\frac{dS_o^P}{dt} = \frac{Q}{V}(S_i^P - S_o^P) - \alpha_1 S_o^P$$

particulate substrate captive on microbial film

$$\frac{dS_T^P}{dt} = \alpha_1 S_o^P - \alpha_2 S_T^P \qquad (1.59)$$

dissolved substrate

$$\frac{dS_o^D}{dt} = \frac{Q}{V}(S_i^D - S_o^D) + \alpha_2 S_T^P - \alpha_3 S_o^D$$

where

S^P, S^D = particulate and dissolved substrate concentration, i, o, T are the suffixes for inflow, outflow and captive in filter, respectively.

Q = flow rate

V = volume of the anaerobic filter

$\alpha_1, \alpha_2, \alpha_3$ = rate constants of attachment, dissolution and gasification, respectively.

t = time

Particulate and dissolved BOD in the effluent were simulated for the influent BOD variation as shown in Figure 1.21, and calculated values were fairly coincident with observed values. As seen in Figure 1.21, dissolved BOD in the effluent was sometimes higher than that in the influent owing to the dissolution of particulate BOD, and the proposed model was able to represent such phenomena.

1.4.10 Empirical Formulae for Substrate Removal by Microbial Film

Strict approaches to the substrate removal process by a microbial film must depend on analytical methods as mentioned previously, but it is impossible to obtain the general analytical resolutions of Equation (1.19) or (1.20); so simplification as Equation (1.21) or (1.22) or numerical analyses

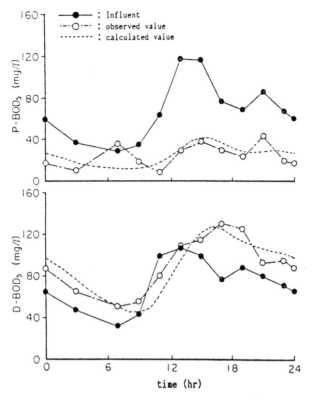

FIGURE 1.21. *Comparison between observed values and calculated ones by Equation (1.59) about effluent BOD of anaerobic filter [$\alpha_1 = 25.4$, $\alpha_2 = 57.9$, $\alpha_3 = 3.84$ (unit day^{-1})].*

of Equation (1.19) or (1.20) are needed. Simplified analyses by empirical formulae, therefore, are useful for some objectives.

The author et al. [47] have found that a relationship similar to Monod's equation can be applied between bulk liquid substrate concentration and overall reaction rate.

$$- \frac{dS_b}{dt} = \frac{KS_b}{K_m + S_b} \tag{1.60}$$

where

S_b = bulk liquid substrate concentration

t = time

K, K_m = constants empirically determined

Equation (1.16) shows a hyperbolic curve of the same type as Equation (1.18), but it implies almost the same matter as Equation (1.35). That is, Equation (1.60) may be used as an approximation of the whole curve in Figure 1.15. Therefore, K_m is the function of various factors, while K_s is a parameter of biological reaction rate alone.

As a general rule, K_m increases with increasing saturation constant K_s, biofilm thickness L, microbial density of biofilm M, and maximum reaction rate in Monod's equation k, and decreases with increasing factors on diffusivity of substrate (k_F, D_e). An example of an empirically determined relationship between k and K_m is shown in Figure 1.22 [47], with K_m corresponding to $k = 0$, i.e., K_m at $L = 0$ is equal to K_s.

Hamoda et al. [48] also pointed out that substrate removal rate could be expressed by a hyperbolic function of substrate concentration in a fixed bed aerobic submerged filter, and it approaches that of a first-order reaction as substrate concentration is decreased.

Haug et al. [49] gave the following equation for nitrification rate of ammoniacal nitrogen in a submerged filter packed with quartz stones:

$$-\frac{dS}{dt} = (0.11T - 0.20)\left(\frac{S}{10}\right)^{1.2} \tag{1.61}$$

where

S = concentration of ammoniacal nitrogen (mg/l)

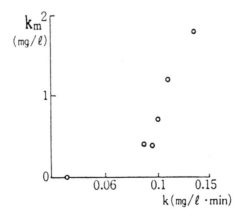

FIGURE 1.22. *Observed values of* k *and* k_m *in Equation (1.60).*

T = water temperature (°C)
t = time (min)

There are many analogous equations with Equation (1.61) which demonstrate the reaction rate by some power of substrate concentration.

Horasawa [50] has given the following formula representing averaged performance of submerged filter:

$$\frac{L_R}{L_o} = 0.845 \ (1.758Q^{-0.313})L_o^{-0.034} \cdot f_T \qquad (1.62)$$

where

L_R/L_o = BOD removal $(-)$
0.845 = BOD removal coefficient (average value for domestic sewage)
Q = hydraulic loading
L_o = BOD loading (g/m³·d)

f_T being the coefficient of temperature, no definite relation could be observed between f_T and temperature, because only little effect was given to f_T in the temperature range of 10–27°C.

Kusumoto et al. [51] has shown a formula of BOD removal in a small scale submerged filter for domestic sewage treatment by a statistical investigation.

$$B_3/(F_1B_1) = K \cdot D^{-0.926} \cdot B_R^{-0.260}(F_1B_1)^{-0.305} \qquad (1.63)$$

where

D = detention time (d)
B_R = BOD loading to unit volume of packed media (kg/m³·d)
F_1 = a parameter indicating biologically treatable ratio of BOD
B_1, B_3 = influent and effluent BOD, respectively
K : constant ($=0.318$)

1.5 REFERENCES

1. Characklis, W. G. 1973. "Attached Microbial Growth–I, II," *Water Res.*, 7(8),(9):1113–1127, 1249–1258.
2. Kornegay, B. H. and J. F. Andrews. 1967. *Proc. 22nd Ind. Waste Conf.*, Purdue Univ.

3. Characklis, W. G. et al. 1982. "Dynamics of Biofilm Process: Methods," *Water Res.*, 16(7):1207–1216.

3-1. Hoehn, R. C. and A. D. Ray. 1973. "Effects of Thickness on Bacterial Film," *J. Wat. Pollut. Control Fed.*, 46:2302–2320.

3-2. Little, B. and D. Lavoie. 1979. "Gulf of Mexico OTEC Biofouling and Corrosion Experiment," *Proceedings OTEC Biofouling. Corrosion and Materials Workshop*, Rosslyn, VA, p. 60.

3-3. Norrman G., W. G. Characklis and J. D. Bryers. 1979. "Control of Microbial Fouling in Circular Tubes with Chlorine," *Devei. Ind. Microbiol.*, 18:581–590.

3-4. Sanders, W. M., 3rd. 1966. "Oxygen Utilization by Slime Organisms in Continuous Culture," *Air Wat. Pollut. Int. J.*, 10:253–276.

3-5. Trulear, M. G. 1980. "Dynamics of Biofilm Processes in an Annular Reactor," M.S. thesis, Rice University, Houston, TX.

3-6. Trulear, M. G. and W. G. Charackils. 1982. "Dynamics of Biofilm Processes," *J. Wat. Pollut. Control Fed.* Accepted for publication.

3-7. Zelver, N. 1979. "Biofilm Development and Associated Energy Losses in Water Conduits," M.S. thesis, Rice University, Houston, TX.

3-8. Picologlou, B. F., N. Zelver and W. G. Characklis. 1980. "Biofilm Growth and Hydraulic Performance," *J. Hydraul. Div. Am. Soc. Civ. Engrs.*, 106(HY5):733–746.

3-9. Bryers, J. D. 1980. "Dynamics of Early Biofilm Formation in a Turbulent Flow System," Ph.D. dissertation, Rice University, Houston, TX.

3-10. Dubois, M., D. A. Giles, J. K. Hamilton, P. A. Rebers and I. Smith. 1956. "Colorimetric Method for Determination of Sugars and Related Substances," *Analyt. Chem.*, 28:350–356.

3-11. Trulear, M. G. 1981. Unpublished results.

3-12. Bryers, J. D. and W. G. Characklis. 1980. "Early Fouling Biofilm Formation in a Turbulent Flow System: Overall Kinetics," *Water Res.*, 15:483–491.

3-13. McCoy, W. 1979. "Immunofluorescence as a Technique to Study Marine Biofouling Bacteria," Directed Research Project-M699, University of Hawaii.

3-14. Corpe, W. A. 1973. "Microfouling: The Role of Primary Film-Forming Marine Bacteria," *Proceedings of the International Congress on Marine Corrosion and Fouling.* Evanston, IL: Northwestern University Press, pp. 598–609.

3-15. Costerton, J. W. and R. R. Colwell, eds. 1979. *Native Aquatic Bacteria: Enumeration, Activity and Ecology.* Philadelphia, PA: ASTM Press.

3-16. Gerchakov, S. M., D. S. Marszalek, F. J. Roth, B. Sallman and L. R. Udey. 1977. "Observation on Microfouling Applicable to OTEC Systems," *Proceeding OTEC Biofouling and Corrosion Symposium*, Seattle, WA, pp. 63–75.

3-17. Geesey, G. G., R. Mutch, J. W. Costerton and R. B. Green. 1978. "Sessile Bacteria: An Important Component of the Microbial Population in Small Mountain Streams," *Limnol. Oceanogr.*, 23:1214–1223.

3-18. Bobbie, R. J., J. S. Nickels, W. M. Davis, et al. 1979. "Measurement of Microfouling Mass and Community Structure during Succession in OTEC Simulators," *Proceedings OTEC Biofouling, Corrosion and Materials Workshop*, Rosslyn, VA, p. 101.

3-19. LaMotta, E. J. 1974. "Evaluation of Diffusional Resistances in Substrate Utilization by Biological Films," Ph.D. dissertation, University of North Carolina at Chapel Hill.

3-20. Dexter, S. C., J. D. Sullivan, J. Williams and S. Watson. 1975. "Influence of Substrate

Wettability on the Attachment of Marine Bacteria to Wetted Surfaces," *Appl. Microbiol.*, 30:298–308.

3-21. Watson, S. W., T. J. Novitsky, H. G. Quinby, F. W. Valois. 1977. "Determination of Bacterial Number and Biomass in the Marine Environment," *Appl. Envir. Microbiol.*, 33:940.

3-22. Fetkovich, J. G., G. N. Granneman, L. M. Mahalingam and D. L. Meier. 1977. "Studies of Biofouling in OTEC Plants," *Proceedings 4th Conference OTEC*, New Orleans, LA, pp. VII15–VII23.

3-23. Knudsen, J. G. 1980. "Apparatus and Techniques for Measurement of Fouling of Heat Transfer Surfaces," *Condenser Biofouling Control*, J. F. Garey et al., eds., Ann Arbor, MI, pp. 143–168.

3-24. Nimmons, M. J. 1979. "Heat Transfer Effects in Turbulent Flow Due to Biofilm Development," M.S. thesis, Rice University, Houston, TX.

3-25. Characklis, W. G., M. J. Nimmons and B. F. Picologlou. 1981. "Influence of Fouling Biofilms on Heat Transfer," *Heat Transfer Engng.*, 3:23–37.

4. Inamori, Y. et al. 1986. *Proc. of 20th Conf. of Japan Soc. of Water Poll. Res.*, S 205, 239–240 (in Japanese).

5. Rouxhet, P. G. and N. Mozes. 1990. "Physical Chemistry of the Interface between Attached Micro-Organisms and Their Support," *Wat. Sci. Tech.*, 22(1/2):1–16.

6. Marshall, K. C. 1967. "Electrophoretic Properties of Fast- and Slow-Growing Species of Rhizobium," *Aust. J. Biol. Sci.*, 20:429.

7. Marshall, K. C. and R. H. Cruickshank. 1973. "Cell Surface Hydrophobicity and the Orientation of Certain Bacteria to Surfaces," *Arch. Microbiol.*, 91:29.

8. Zobell, C. E. 1943. "The Effect of Solid Surface upon Bacterial Activity," *J. Bact.*, 46:39–56.

9. Fattom, A. and M. Shilo. 1984. "Hydrophobicity as an Adhesion Mechanism of Benthic Cyanobacteria," *Appl. Environ. Microbiol.*, 47:135.

10. Mozes, N. and P. G. Rouxhet. 1987. "Method for Measuring Hydrophobicity of Microorganisms," *J. Microbiol. Methods*, 6:99.

11. Heukelekian, H. 1956. "Slime Formation in Polluted Waters. II. Factors Affecting Slime Growth," *Sewage Ind. Wastes*, 28:78–92.

12. Heukelekian, H. 1956. "Slime Formation in Polluted Waters. III. Nature and Composition of Slimes," *Sewage Ind. Wastes*, 28:206–210.

13. Zvyagintsev, D. G. 1958. "Adsorption of Microorganisms by Glass Surfaces," *Microbiology (USSR)*, 28:104–108.

14. Kitao, T. 1980. *Biofilm Process*, S. Iwai and M. Kusumoto, eds., Tokyo: Sangyo-yosui Chosakai (in Japanese).

15. Sanders, W. M. 1966. "Oxygen Utilization by Slime Organisms in Continuous Culture," *Int. J. Air Wat. Pollut.*, 10:253–276.

16. Characklis, W. G. 1967. "Oxygen Transfer through Biological Slimes," M.S. thesis, Univ. of Toledo.

17. Characklis, W. G. 1971. "Effect of Hypochlorite on Microbial Slimes," *Proc. 26th Ind. Waste Conf.*, Purdue Univ.

18. U.S. EPA. 1971. "Application of Rotating Disc Process to Municipal Wastewater Treatment," *Water Poll. Res.*, Series 17050.

19. Hawkes, H. A. 1963. "The Ecology of Waste Water Treatment," in *Waste Treatment*, P. C. G. Isacc, ed., Pergamon Press.

20. Ives, K. J. 1975. "Capture Mechanism in Filtration," in *The Scientific Basis of Filtration*, K. J. Ives, ed., Nato Advanced Study Institutes Series, Series E, Vol. 2.

21. Rittmann, B. E. and C. W. Brunner. 1984. "The Nonsteady-State-Biofilm Process for Advanced Organic Removal," *J. WPCF*, 56(7):874–880.

22. Okey, R. W. and O. E. Albertson. 1989. "Diffusion's Role in Regulating Rate and Masking Temperature Effects in Fixed-Film Nitrification," *J. WPCF*, 61(4):500–509.

23. Stenstrom, M. K. and R. A. Poduska. 1980. "The Effect of Dissolved Oxygen Concentration on Nitrification," *Water Res.*, 14:643–649.

24. Williamson, K. and P. L. McCarty. 1976. "A Model of Substrate Utilization by Bacterial Films," *J. WPCF*, 48(1):9–24.

25. Fujie, K. et al. 1991. "Microbial Growth and Cologgin Factor in Aerated Biofilter," *Japanese J. of Water Poll.*, 8(14):564–573.

26. McHarness, D. D. et al. 1975. "Field Studies of Nitrification with Submerged Filters," *J. WPCF*, 47(2):291.

27. Adachi, T. et al. 1978. "Application of Contact Aeration Process to Secondary and Tertiary Sewage Treatment," *Water and Wastes*, 20(5):521 (in Japanese).

28. Antonie, R. L. 1974. "Nitrification of Activated Sludge Effluent; BIO-SURF Process," *Water & Sewage Works*, 121(11):44–47; 121(12):54–55.

29. Bryers, J. and W. Characklis. 1981. "Early Fouling Biofilm Formation in a Turbulent Flow System: Overall Kinetics," *Water Res.*, 15(4):483–491.

30. Droste, R. L. and J. K. Kennedy. 1986. "Sequential Substrate Utilization and Effectiveness Factor in Fixed Biofilms," *Biotech. Bioeng.*, 28:1713.

31. Harremoes, P. et al. 1980. "Practical Problems Related to Nitrogen Bubble Formation in Fixed Film Reactors," *Prog. Wat. Tech.*, 12:253–269, Toronto.

32. Szwerinski, H. et al. 1986. "pH-Decrease in Nitrifying Biofilms," *Water Research*, 20(8):971–976.

33. Welty, J. R. et al. 1969. *Fundamentals of Momentum, Heat, Mass Transfer.* New York, NY: John Wiley & Sons.

34. Carberry, J. J. 1960. *A.I.Ch.E. Jr.*, 6:460.

35. Wilson, E. J. and C. J. Greankoplis. 1966. *Ind. Eng. Chem. Fundamentals*, 5:9.

36. Kataoka, T. et al. 1972. "Mass Transfer in Laminar Region between Liquid and Packing Material Surface in the Packed Bed," *J. Chem. Eng. of Japan*, 5(2):132–136.

37. DeWalle, F. B. and E. S. K. Chain. 1976. "Kinetics of Substrate Removal in a Completely Mixed Anaerobic Filter," *Biotech. Bioeng.*, 18:1275–1295.

38. Monod, J. 1949. "The Growth of Bacterial Cultures," *Ann. Review of Microbiol.*, 3:371–394.

39. Harremoes, P. 1976. "The Significance of Pore Diffusion to Filter Denitrification," *J. WPCF*, 48(2):377–387.

40. Atkinson, B. and I. J. Davies. 1974. "The Overall Rate of Substrate Uptake (Reaction) by Microbial Films. Part I—A Biological Rate Equation," *Trans. Ins. Chem. Engrs.*, 52:248.

41. Atkinson, B. and S. Y. How. 1974. "The Overall Rate of Substrate Uptake (Reaction) by Microbial Films. Part II—Effect of Concentration and Thickness with Mixed Microbial Films," *Trans. Ins. Chem. Engrs.*, 52:260.

42. Jennings, P. A. et al. 1976. "Theoretical Model for a Submerged Biological Filter," *Biotech. Bioeng.*, 18:1249–1273.

43. Mutharasan, R. 1978. "An Approximate Solution to the Theoretical Model of a Submerged Biological Filter," *Biotech. Bioeng.*, 20:151–156.

44. Chen, Y. S. and H. R. Bungay. 1981. "Microelectrode Studies of Oxygen Transfer in Trickling Filter Slimes," *Biotech. Bioeng.*, 23:781–792.

45. Strand, S. T. and A. J. McDonnell. 1985. "Mathematical Analysis of Oxygen and Nitrare Consumption in Deep Microbial Films," *Water Res.*, 19(3):345–352.

46. Kitao, T., Y. Kiso and K. Kasai. 1991. "A Study on Removal Performances of Anaerobic-Aerobic Bio-Filter System and Removal Mechanism by Numerical Analysis," *Jour. Japan Sewage Works Association Research*, 28(324):17–27.

47. Iwai, S., T. Kitao and S. Teshima. 1975. "Ammonia Nitrogen Removal with Contact Aeration Filter," *Proc. of 30th Ann. Conf. of Japan Soc. of Civil Engrs.*, 2:526 (in Japanese).

48. Hamoda, M. F. 1989. "Kinetic Analysis of Aerated Submerged Fixed-Film (ASFF) Bioreactors," *Wat. Res.*, 23(9):1147–1154.

49. Haug, R. T. and P. L. McCarty. 1972. "Nitrification with Submerged Filters," *J. WPCF*, 44(11):2086–2102.

50. Horasawa, I. 1979. "Performance and Design Criteria of Contact Aerators for Sewage Treatment," *Water and Waste*, 21(8):887–893 (in Japanese).

51. Kusumoto, M. et al. 1980. *Proc. of Ann. Conf. on Microbial-Film Process Res.*, (2):11–23.

CHAPTER 2

Bioreactors Utilizing Microbial Film

2.1 GENERAL CLASSIFICATION OF BIOREACTORS AND CHARACTERISTICS OF EACH REACTOR

Bioreactors can be classified into two large groups—batch or semi-batch reactors and flow reactors (continuous reactor). Most industrial microbial reactors are operated as batch reactors, but biological treatment of waste sludge or wastewater utilizes mostly flow reactors. A batch reactor-activated sludge process has been developed, however, in which a single tank is used both for microbial reaction and for settling, and a batch-wise operation is frequently used for anaerobic digestion.

On the other hand, it can be said that batch operation is never used for a microbial film process, because this process is very adaptable to continuous operation, as microbial mass is immobilized in the reactor. Besides, batch operation shows no merit for microbial film processes, and causes troubles such as peeling off of biofilm brought about by frequent rises and falls of water level.

2.1.1 Batch Reactor

In batch operation, the reactor tank is filled with raw liquid seeded by an appropriate culture, and the reaction is started and continued until expected reaction is completed; then the tank content is withdrawn. Figure 2.1 shows the types of operation and substrate concentration in each type.

2.1.1.1 GROWTH-CYCLE PHASES FOR BATCH CULTIVATION

Typically the number (or amount) of living cells in the culture varies with time, as shown in Figure 2.2 in batch cultivation.

63

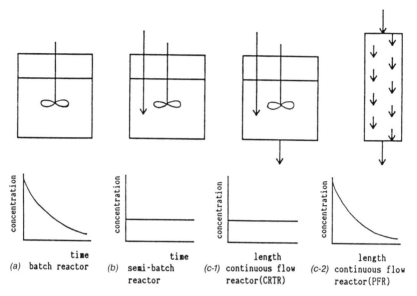

FIGURE 2.1. *Types of operation and concentration change in them.*

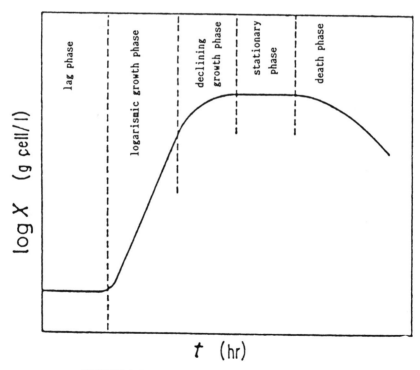

FIGURE 2.2. *Typical growth pattern in batch culture.*

The growth curve consists of (1) lag phase, (2) logarithmic growth phase, (3) declining growth phase, (4) stationary phase and (5) death phase.

Lag Phase

When a fresh medium is inoculated, there is in general a time interval known as the lag phase before steady growth can be started.

During this phase, various intermediates are accumulated once more in the requisite concentration, and the adaption of cells is established in a new environment. Little is known about the factors affecting the length of the lag, so prediction is, in general, difficult. The apparent lag which must be distinguished from the true lag, is caused by non-living cells in the inoculum. Therefore, the apparent lag will continue until the number of non-living cells can be neglected compared with that of living cells [1].

Logarithmic (exponential) Growth Phase

During this phase, the cells can multiply rapidly, and cell mass, or the number of living cells, doubles regularly with time, and the increase of cell number is described by

$$\frac{dn}{dt} = \mu n \quad \text{or} \quad \frac{1}{n} \cdot \frac{dn}{dt} = \mu \tag{2.1}$$

where μ = specific growth rate, and $n = n_0$ at $t = t_{lag}$ (t_{lag} = the length of lag).
Therefore,

$$\ln n/n_0 = \mu(t - t_{lag}) \text{ or } n = n_0 \exp[\mu(t - t_{lag})] \tag{2.2}$$

The time interval t_d required to double the cell population – the doubling time – can be readily obtained by substituting $n = 2n_0$ at $t = t_{lag} + t_d$.

$$t_d = \ln 2/\mu = 0.693/\mu \tag{2.3}$$

Specific growth rates of various microbes are shown in Table 1.2 (this table is in Chapter 1), and as a general rule, the higher a specie of microbe is, the smaller the specific growth rate tends to be.

Specific growth rate μ is affected by various factors as temperature, pH, substrate concentration in the medium, and the relation between μ and substrate concentration is especially important. Monod's equation is most widely used to represent this relation.

$$\mu = \mu_{max} \cdot \frac{S}{K_s + S} \tag{2.4}$$

where

μ_{max} = maximum specific growth rate, i.e., the value of μ when $S \gg K_s$
K_s = saturation constant or half saturation constant, and is equivalent to the substrate concentration which gives $\mu = 1/2 \; \mu_{max}$.

When two or more substrates are growth limiting, combining equations for individual substrate,

$$\mu = \mu_{max} \cdot \frac{S_1}{K_{s1} + S_1} \cdot \frac{S_2}{K_{s2} + S_2} \cdot \; \cdots \qquad (2.5)$$

is used.

Declining Growth Phase

Owing to the deficiency of growth-limiting substrate and/or oxygen, accumulation of toxin as the product of metabolism, and pH decrease etc., the declining growth phase begins where specific growth rate μ is lowered by the deterioration of growing circumstances.

Stationary Phase

In this phase, growth halts completely because of exhaustion of nutrient materials, oxygen and available biological space or unfavorable pH etc., and the population reaches a maximum number.

Supposing the exhaustion of substrate brings about the stationary phase, the maximum concentration of microorganisms can be decided as follows: as a general rule, the increase of microorganisms is proportional to the decrease of substrate, so that

$$dX/dS = - Y_{x/s} \qquad (2.6)$$

where

X = microorganism concentration
S = substrate concentration
$Y_{x/s}$ = proportion coefficient called growth yield

Resolving Equation (2.6) under the conditions $X = X_0$ at $S = S_0$ and $X = X_s$ (X_s = microorganism concentration at the stationary phase, or the maximum concentration of microorganisms) at $S = 0$, X_s is given by the following equation:

$$X_s = X_0 + Y_{x/s}S_0$$

or

$$\frac{X_s - X_0}{S_0} = Y_{x/s} \tag{2.7}$$

Death Phase

The phase after the stationary phase, where living-cell numbers will continue to decrease is called the death phase.

Even in the logarithmic growth phase, a certain portion of microbial population continually decays, so that the process of growth should strictly be expressed by

$$\frac{dX}{dt} = \mu X - k_d X = (\mu - k_d)X \tag{2.8}$$

where k_d = decay constant

The influence of k_d on the total number of microbes can be neglected under the active growth, but once living conditions become bad and μ decreases to zero, living cells die rapidly, and the change of X with time is described by

$$X = X_s e^{-k_d t} \tag{2.9}$$

where

X = the concentration of microorganism
t = the time elapsed since the onset of the death phase
X_s = X in the stationary phase

2.1.1.2 THE LOGISTIC CURVE

The logistic equation is a typical one for batch cultivation, which includes the term of growth inhibition. Assuming that inhibition is proportional to X^2, the logistic equation is represented as

$$\frac{dX}{dt} kX(1 - \beta X) \quad X(0) = X_0 \tag{2.10}$$

where k, β = constants
Equation (2.10) can be easily integrated to give the logistic curve

$$X = \frac{X_0 e^{kt}}{1 - \beta X_0(1 - e^{kt})} \tag{2.11}$$

As illustrated schematically in Figure 2.3, the logistic curve is sigmoidal and lead to a stationary concentration of size $X_s = 1/\beta$, and has an inflection point at $\beta X = 1/2$.

2.1.2 Semi-Batch Reactor

In semi-batch cultivation, the feeding of the medium, including nutrient materials, is continued during the reaction time, but the cultivated liquid is not withdrawn till the end of the cultivation. The substrate concentration is,

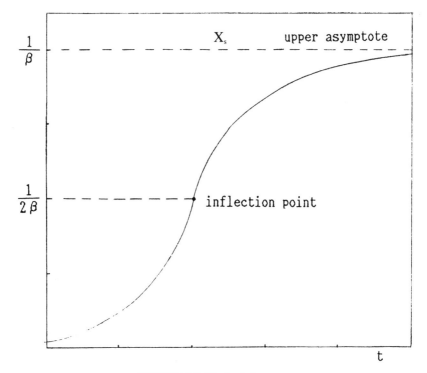

FIGURE 2.3. The logistic curve.

therefore, kept constant as seen in Figure 2.1 in spite of the unsteady operation. The merit of the semi-batch operation is that the substrate concentration in the reactor is controllable to an arbitrary level, and consequently it is more useful than the batch operation when applied to wastewater including an inhibiting substance as substrate. Moreover, the oxygen consumption rate also being constant for aerobic treatment processes, the capacity of the aeration device can be saved. But, in this type of reactor, it should be noted that the volume of the reactor content increases with time. Therefore, vn and vX must be substituted for n and X, respectively into Equation (2.1) as to semi-batch reactor, then

$$d(vn)/dt = \mu(vn) \qquad (2.12)$$

2.1.3 Continuous Reactor (Flow Reactor)

In this type of reactor, the feeding of raw liquid and the withdrawal of effluent stream is continued steadily. Continuous reactors are classified into the ideal flow reactor, the plug flow reactor (PFR), the continuous flow stirred-tank reactor (CSTR), and the non-ideal flow reactor, i.e., the incomplete mixing reactor.

The continuous reactor is usually kept in a steady-state, and chemostats and turbidstats are used for process control, which enables maintenance of a steady-state, with the former being used mostly.

Chemostats maintain a steady-state without any feedback regulation system, only by feeding raw liquid at a constant rate. In turbidstats, the quantity of biomass in the reactor is continuously monitored and maintained at a fixed value with a feedback control of feeding flow-rate. The capacity of the reactor can be raised if the biomass in the reactor effluent is separated and recycled to the reactor. As any chemostat equipped with a recycle system of biomass is susceptible to contamination by various bacteria, the range of its use is confined, but almost all activated sludge processes uses this method. Though the continuous reactor has the advantage of ease of reaction rate control and/or a high efficiency, its practical use in a commercial scale is almost confined to wastewater treatment plants because of its difficulty in keeping the culture pure for a long term.

2.1.3.1 MASS (MATERIALS) BALANCE IN A CONTINUOUS REACTOR

The concept of mass balance is of crucial importance in considering continuous reactors. If the law of conservation of mass is applied to an arbitrary volume of the reactor, then

$$
\begin{bmatrix} \text{net rate of} \\ \text{accumulation} \\ \text{of material } i \end{bmatrix} = \underbrace{\begin{bmatrix} \text{rate of mass} \\ \text{inflow of} \\ \text{material } i \end{bmatrix} - \begin{bmatrix} \text{rate of mass} \\ \text{outflow of} \\ \text{material } i \end{bmatrix}}_{\text{net rate of inflow}} - \begin{bmatrix} \text{rate of} \\ \text{chemical} \\ \text{reaction of} \\ \text{material } i \end{bmatrix}
\tag{2.13}
$$

If the volume in consideration is the whole reactor, the balance is called the overall mass balance, and if a very small element of fluid is considered in which the concentration of each component can be regarded uniform, it is called the differential mass balance.

Considering a volume V in the reactor as shown in Figure 2.4, the rate of accumulation of material i is given by

$$
\begin{bmatrix} \text{net rate of} \\ \text{accumulation} \\ \text{of material } i \end{bmatrix} = \frac{\partial}{\partial t} \left[\int_v c_i dV \right]
\tag{2.14}
$$

where c_i = the concentration of material i

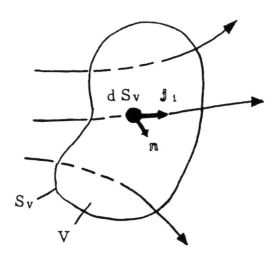

FIGURE 2.4. *Flow through a control volume.*

Letting S_v = the surface area of volume V.

$\mathbf{j}_i$ = the mass flux, i.e., a vector, the magnitude of which represents the mass of i per unit area per unit time crossing an infinitesimal surface dS_v normal to the direction of flow, and the direction of which is parallel to the direction of flow.

$\mathbf{n}$ = a unit vector normal to S_v and directed outward from the surface.

the net rate of mass flow into the volume V is

$$\begin{bmatrix} \text{net rate of} \\ \text{mass inflow} \\ \text{of material } i \end{bmatrix} = -\oint \mathbf{j}_i \, \mathbf{n} dS_v \qquad (2.15)$$

The decay rate of i by reaction is given

$$\begin{bmatrix} \text{decay rate of} \\ \text{material } i \\ \text{within volume } V \end{bmatrix} = \int_v r_i dV \qquad (2.16)$$

where r_i = reaction rate of material i

Substituting Equations (2.14), (2.15) and (2.16) into Equation (2.13), the mass balance equation of i within V is obtained.

$$\frac{\partial}{\partial t}\left[\int_v c_i dV \right] = -\oint_{sv} \mathbf{j}_i \cdot \mathbf{n} dS_v - \int_v r_i dV \qquad (2.17)$$

If the content is completely mixed within V, then c_i and r_i are functions of time only.

$$V\frac{dc_i}{dt} = F_i - Vr_i \qquad (2.18)$$

where F_i = the net inflow flux of i.

When V is the total volume of the reactor (corresponding to a CSTR), Equation (2.18) shows the overall mass balance. Equation (2.15) can be rewritten by employing Green-Gauss' theorem.

$$\oint_{S_v} \mathbf{j}_i \cdot \mathbf{n} dS_v = \int_v \nabla \cdot \mathbf{j}_i dV \qquad (2.19)$$

where ∇ is the del operator, and in rectangular coordinates

$$\nabla = \mathbf{i}_x\frac{\partial}{\partial x} + \mathbf{i}_y\frac{\partial}{\partial y} + \mathbf{i}_z\frac{\partial}{\partial z}$$

where $\mathbf{i}_x$, $\mathbf{i}_y$, and $\mathbf{i}_z$ are the unit vectors in the x, y and z directions. The left side of Equation (2.17) can be transformed according to Leibnitz's formula,

$$\frac{\partial}{\partial t}\int_v c_i dV = \int_v \frac{\partial c_i}{\partial t} dV \qquad (2.20)$$

Substituting Equations (2.19) and (2.20) into Equation (2.17), and transposing all quantities to the left side,

$$\int_v \left[\frac{\partial c_i}{\partial t} + \nabla \cdot \mathbf{j}_i + r_i\right] dV = 0 \qquad (2.21)$$

or, because V is arbitrary,

$$\frac{\partial c_i}{\partial t} + \nabla \cdot \mathbf{j}_i + r_i = 0 \qquad (2.22)$$

which is the fundamental equation of mass balance. To solve Equation (2.22) and obtain the special and temporal distribution of i, it is necessary to relate the flux $\mathbf{j}_i$ and the reaction rate r_i to the concentration c_i.

The flux $\mathbf{j}_i$ of material i is related to diffusion and flow, so that $\mathbf{j}_i$ is represented as

$$\mathbf{j}_i = -(D_{wi} + \epsilon_i)\nabla c_i - \nabla c_i \qquad (2.23)$$

where

D_{wi} = molecular diffusion coefficient of material i in the liquid
ϵ_i = the eddy diffusion coefficient of material i
∇ = flow velocity in vector

In completely turbulent flows, ϵ_i is likely to be several orders of magnitude greater than D_{wi}.

2.1.3.2 MASS BALANCE IN CSTR (OR THE COMPLETELY MIXED FLOW REACTOR, CMF)

Because of complete mixing, the concentrations of c_i and r_i etc. are, of course, the same throughout the complete mixed reactor, so Equation (2.18) can be employed.

$$V\frac{dc_i}{dt} = F_i - Vr_i \tag{2.18}$$

Letting

Q = inflow or outflow rate in the reactor
S_i = substrate concentration of feeding liquid
S = the substrate concentration within the reactor

We have following two formulae.

$$F_i = QS_i - QS = Q(S_i - S) \tag{2.24}$$

$$r_i = r(S) \tag{2.25}$$

Introducing above two equations

$$V\frac{dS}{dt} = Q(S_i - S) - Vr(S) \tag{2.26}$$

or

$$\frac{dS}{dt} = D(S_i - S) - r(S) \qquad \text{where } D = \frac{Q}{V} \tag{2.27}$$

D is the reciprocal of detention time, and called the dilution rate. For a first-order reaction, r is represented

$$r = \alpha S \tag{2.28}$$

and the left side is zero under a steady-state, so

$$D(S_i - S) - \alpha S = 0, \quad S = \frac{DS_i}{D + \alpha} = \frac{1}{1 + \alpha/D}S_i \tag{2.29}$$

For a CMF reactor series of m tanks, the concentration of reactant remaining in the effluent from the final reactor is easily given.

$$S = \left(\frac{1}{1 + \alpha/D}\right)^m S_i \qquad (2.30)$$

On the other hand, applying Equation (2.18) to the microorganism concentration in the reactor,

$$V\frac{dX}{dt} = QX_0 - QX + V\left(\frac{dX}{dt}\right)_G \qquad (2.31)$$

where $(dX/dt)_G$ implies the increase of microorganism concentration by growth. Neglecting the decrease by death, the increase can be expressed,

$$\left(\frac{dX}{dt}\right)_G = \mu X \qquad (2.32)$$

If $X_0 = 0$, Equation (2.31) is

$$\frac{dX}{dt} = -DX + \mu X = (\mu - D)X \qquad (2.33)$$

In a steady-state, the left side of Equation (2.33) is zero,

$$D = \mu = \mu_{max}g(S) \qquad g(S) \leq 1 \qquad (2.34)$$

where $g(S) = $ some function of S and S can be obtained by employing Equation (2.34).
    If

$$D > \mu_{max}g(S_i) \qquad (2.35)$$

a steady-state can never be retained, and the microbes in the reactor are drawn off entirely. This state is called wash out.
    Equations (2.29) and (2.30) were obtained assuming a first-order reaction, but if a Monod's type equation is employed to Equation (2.35),

$$g(S) = \frac{S}{K_s + S} \qquad (2.36)$$

and the substrate concentration in the effluent is given by

$$D = \mu_{max} \frac{S}{K_s + S} \qquad S = \frac{K_s D}{\mu_{max} - D} \qquad (2.37)$$

The microorganism concentration in the reactor is given by

$$X = Y_{x/s}(S_i - S) = Y_{x/s} \left(S_i - \frac{K_s D}{\mu_{max} - D} \right) \qquad (2.38)$$

2.1.3.3 MASS BALANCE IN PFR

The microorganism concentration is inevitably zero in a plug flow reactor unlike in a completely mixed reactor, if the microorganism concentration in the feeding liquid is zero, and so the duration of reaction cannot be expected. We assume, therefore, that the concentration of microorganism in the reactor is in a steady-state by some way.

Supposing a reactor with one-dimensional flow as shown in Figure 2.5,

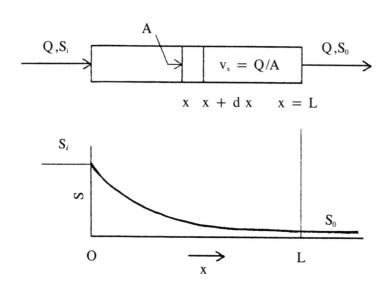

FIGURE 2.5. *Differential mass balance between cross sections at* x = x *and* x = x + dx.

Equation (2.17) is applied to a thin slice of the reactor with a thickness dx and an area A. The balance of the substrate is given by

$$A \cdot dx\left(\frac{\partial S}{\partial t}\right) = Av_x S - Av_x\left(S + \frac{\partial S}{\partial x}dx\right) - A \cdot dxr(S) \quad (2.39)$$

where v_x = the x component of velocity.
Rearranging Equation (2.39),

$$\frac{\partial S}{\partial t} = -v_x\frac{\partial S}{\partial x} - r(S) \quad (2.40)$$

In a steady-state, $\partial S/\partial t = 0$, and partial differentials can be rewritten to ordinary differentials.

$$\frac{dx}{v_x} = -\frac{dS}{r(S)} \quad (2.41)$$

Solving above equation under the boundary conditions; $S = S_i$ at $x = 0$, and $S = S$ at $x = x$,

$$\int_0^x \frac{dx}{v_x} = -\int_{S_i}^{S} \frac{dS}{r(S)} \quad (2.42)$$

Considering a first-order reaction,

$$\frac{x}{v_x} = -\int_{S_i}^{S} \frac{dS}{\alpha S} = \frac{1}{\alpha} \ln S_i/S \quad (2.43)$$

Or

$$S = S_i \exp\left(-\frac{\alpha x}{v_x}\right) \quad (2.44)$$

represents the concentration profile of the substrate in the x direction. Substituting $x = L$, the substrate concentration in the effluent is given by

$$S = S_i \exp\left(-\frac{\alpha L}{v_x}\right) = S_i \exp(-\alpha t_d) \quad (2.45)$$

where t_a = the detention time

2.1.3.4 INCOMPLETELY MIXED REACTOR

Most practical full-scale reactors do not have fully idealized flow patterns as completely mixed flow or plug flow. The deviations from ideal flow are caused by short-circuiting flows, recycling flow and/or the presence of stagnant zones in the reactor.

Non ideal reactor models are as follows:

(1) A completely mixed reactor with a bypassing flow [Figure 2.6(a)]
(2) A completely mixed reactor with a stagnant zone [Figure 2.6(b)]
(3) Completely mixed reactors in series [Figure 2.6(c)]
   - simple CMF series [Figure 2.6(c1)]
   - CMF series with a recycling flow [Figure 2.6(c2)]
   - CMF series with back-mix flows [Figure 2.6(c3)]
(4) A plug flow reactor with longitudinal dispersion [diffusion model, Figure 2.6(d)]

The diffusion model is most widely utilized, and is a model where flow velocity and eddy diffusion coefficient in the longitudinal direction are uniform throughout the reactor. The substrate concentration variation along the length of the reactor is expressed by the following equation, considering one-dimensional flow and diffusion in Equations (2.22) and (2.23):

$$\frac{\partial S}{\partial t} = E_x \frac{\partial^2 S}{\partial x^2} - v_x \frac{\partial S}{\partial x} - r(S) \tag{2.46}$$

where E_x = longitudinal eddy diffusion coefficient

Assuming that $r(S) = 0$, the residence time distribution function $E(\phi)$ is represented by the following equation when some conservable substance is added into the reactor as a tracer (2).

$$E(\phi) = \frac{Vc}{V_0 c_0} = 4 \sum_{n=1}^{\infty} \frac{(-1)^{n+1} \mu_n \exp(PeB/2)}{(PeB)^2 + PeB + \mu_n^2} e^{-((PeB/2 + \mu_n)/PeB)\phi} \tag{2.47}$$

where

V = the volume of the reactor
V_0 = the volume of tracer solution added
c = the tracer concentration in the output flow at $t = t$
c_0 = the concentration of tracer solution added
$\phi = t/t_d$ ($t_d = L/v_x$, the nominal residence time)
$Pe = v_x d/E_x$ (Peclet number)

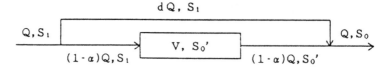

$$dQ, S_1$$

Q, S_1 | Q, S_0

$$V, S_0'$$

$$(1-\alpha)Q, S_1 \qquad (1-\alpha)Q, S_0'$$

(a) Model with a bypass flow.

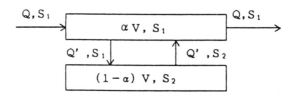

Q, S_1 Q, S_1

$$\alpha V, S_1$$

$$Q', S_1 \qquad Q', S_2$$

$$(1-\alpha) V, S_2$$

(b) Model with a dead zone.

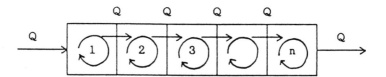

$$Q \qquad Q \qquad Q \qquad Q$$

Q 1   2   3   n Q

(c-1) Model with a completely mixed tank series.

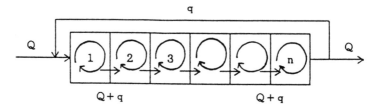

$$q$$

Q 1   2   3   n Q

$$Q + q \qquad\qquad Q + q$$

(c-2) Model with a completely mixed tank series
having a recycle flow.

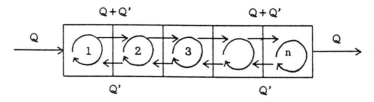

$$Q + Q' \qquad\qquad Q + Q'$$

Q 1   2   3   n Q

$$Q' \qquad\qquad Q'$$

(c-3) Model with a completely mixed tank series
having backmix flows.

FIGURE 2.6. Models for a incompletely mixed CSTR.

78

$B = L/d$

d = the characteristic length

μ_n = the nth positive solution of $\cot \mu = (2\mu/PeB - PeB/2\mu)/2$ ($\mu_1 <$ $\mu_2 < \ldots < \mu_n < \mu_{n+1} \ldots$

$E(\phi)$ is determined only by a dimensionless number, $PeB(v_x L/E_x)$ as shown in Figure 2.7. Since the flow in consideration is one-dimensional, utilizing the reactor length, L as the characteristic length d, PeB is equal to

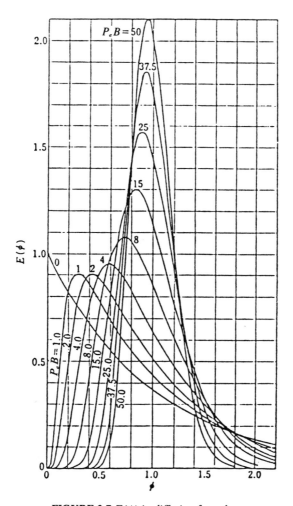

FIGURE 2.7. E(ϕ) *in diffusion formula.*

Pe which is equivalent to the second term in the right-side of Equation (2.46) divided by the 1st term and means the relative magnitude of transport by flow to diffusion by mixing. When *Pe* = 0, therefore, transport by flow is negligible compared with diffusion by mixing, and the reactor is regarded as a completely mixed flow reactor. The effect of the term of transport by flow increases with increment of *Pe* or *PeB*, and it means a plug flow in the strictest sense, in that *Pe* = ∞. Hence, the peak height of the tracer concentration becomes higher and the peak width becomes narrow as the value of *Pe* increases. Under certain operating conditions, the larger the length of the reactor, the larger the value *Pe* (*PeB*) becomes, and the nearer the flow approaches plug flow.

Equation (2.41) is rewritten into the following equation under a steady-state if the reaction is of the first-order:

$$E_x \frac{d^2S}{dx^2} - v_x \frac{dS}{dx} - \alpha S = 0 \tag{2.48}$$

Wehner and Wilhelm [3] have given the following equation as the solution of Equation (2.48).

$$\frac{S_0}{S_i} = \frac{4\bar{\beta} \exp[1/2(v_x L/E_x)]}{(1 + \bar{\beta})_2 \exp[(\bar{\beta}/2)(v_x L/E_x)] - (1 - \bar{\beta})^2 \exp[-(\bar{\beta}/2)(v_x L/E_x)]} \tag{2.49}$$

where $\bar{\beta} = [1 + (4\alpha L/v_x)(E_x/v_x L)]^{0.5}$

And, if we use L as the characteristic length, $v_x L/E_x = Pe$. Then, Equation (2.44) is rewritten into

$$\frac{S_0}{S_i} = \frac{4\bar{\beta} \exp(Pe/2)}{(1 + \bar{\beta})^2 \exp(\bar{\beta}Pe/2) - (1 - \bar{\beta})^2 \exp(-\bar{\beta}Pe/2)} \tag{2.50}$$

The characteristics of the reactor can be fully represented by Pe, v_x and L. Setting a dimensionless number

$$Nr \equiv \alpha L/v_x \tag{2.51}$$

$\bar{\beta}$ is given by

$$\bar{\beta} = \sqrt{1 + 4Nr/Pe} \tag{2.52}$$

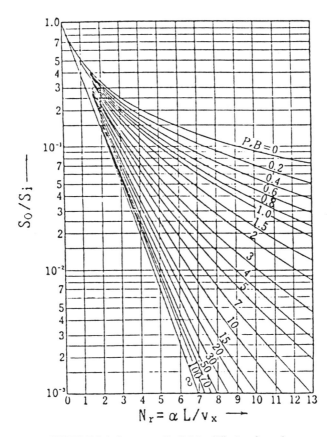

FIGURE 2.8. *Recovery (S_0/S_i) in diffusion formula.*

and then, it is understood that S_0/S_i is governed by two dimensionless numbers, *Pe* and *Nr*.

The relation between *Pe* (*PeB*), *Nr* and S_0/S_i are shown in Figure 2.8.

2.2 THE FACILITIES AND THE CONFIGURATIONS OF MICROBIAL FILM REACTORS

It was already stated in Chapter 1 that although microbial film processes are largely classified into three types – submerged biological filters, rotating biological contactors and trickling filters – the configuration, the operation, the performance and the range of use are extremely diverse.

Also, the difference between microbial film processes and other biological treatment processes is not always distinct. For example, as seen in Figure 2.9(a), if the surface area of attached microbial growth is far larger than its thickness (usually less than several millimeters), it will clearly be recognized as a microbial film. But the term "film" would not be able to be applied to the attached matter as in Figure 2.9(b), whose thickness is considerably larger than the size of carrier particle. Besides, pelleted or gelled particles of biomass without carrier [Figure 2.9(c)] can hardly be recognized as microbial films. In Figures 2.9(a)–(c), however, when the thickness or the particle size of biological solid is not less than 1 mm, the substrate removal characteristic is intensely diffusion-limiting, which is the distinguishing feature of a microbial process. Hence, it can be concluded that the biomasses as in Figure 2.9(b) and (c) have a common nature as biofilms. In this text, therefore, all processes which utilize biological solids having common substrate removal characteristic are included in microbial

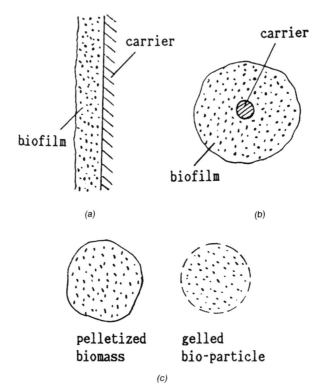

FIGURE 2.9. Various types of biomass.

film processes. On the contrary, even pelleted biomass cannot be regarded as microbial films if the particle size is small and the diffusion-limiting nature is not observed concerning the substrate removal characteristic. Though such classification would not be coincident with the practical one in a common sense, it is impossible to distinguish microbial film processes from the other biological treatment processes based on the shape and/or size of biomass. The above mentioned interpretation of microbial film processes enlarges significantly the range of submerged filters. Various types of submerged filters are described in the following paragraphs.

2.2.1 The Submerged Filter

The submerged filter can be defined as a treatment system which can purify wastewater by bringing into contact with microbial film grown on the surface of some solids, contact media, filter media, packing media or carrier etc., submerged in the wastewater. In many cases, biological solids are stuck on the surface of media, but they are sometimes deposited on it. Only a one time contact of wastewater with microbial film will do sometimes, but repeated contacts by recycling of wastewater are required when an excellent efficiency or performance of treatment device is desired. Especially in an aerobic microbial film process, it is indispensable to pass repeatedly the aerated wastewater along the microbial film surface for the supply of oxygen.

2.2.1.1 THE AEROBIC SUBMERGED FILTER

The Fixed Bed Submerged Filter

This type of submerged filter is most popular, in which the media supporting microbial film are fixed without any movement. Typical fixed submerged filters are packed with corrugated plastic plates or plastic tubings, etc., with specific surfaces ranging from several tens to hundreds m^2/m^3, so their overviews are similar to trickling filters submerged in wastewater. But, unlike a trickling filter, artificial aeration with air diffusers or mechanical aeration devices are indispensable to a fixed bed submerged filter in order to supply oxygen to biofilms. In a diffused air system, some vacant (not packed with media) space is set in the reactor as depicted in Figure 2.10(a)–(e), and air diffusers are installed just above the level of the bottom of packed media. Diffusing air from the diffusers, oxygen supply and recirculation flow are accomplished to bring aerated wastewater into contact with biofilms. The classification of reactors based upon the aeration method is as follows: the central aeration method in which air is diffused into the

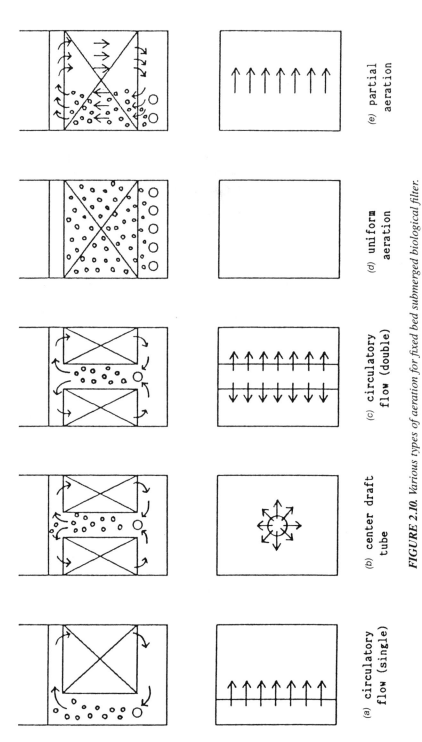

FIGURE 2.10. Various types of aeration for fixed bed submerged biological filter.

(a) circulatory flow (single)

(b) center draft tube

(c) circulatory flow (double)

(d) uniform aeration

(e) partial aeration

draft tube set vertically at the center of the reactor and radial recirculation flow is formed [Figure 2.10(b)]; the one-directional recirculation method in which air diffusers are installed along a side wall of the reactor [Figure 2.10(a)]; and the two-directional recirculation method in which air diffusers are set along the centerline of the reactor and two-directional recirculation flow is formed [Figure 2.10(c)]. It is also possible to introduce air bubbles directly into the packed media (bubble oxygenation system); in one type, air is diffused into a portion of the filter and recirculation flow is formed in the reactor [Figure 2.10(e)], and in the other type, diffused air is introduced uniformly over the whole filter. The oxygen dissolution efficiency is higher for bubble oxygenation system than for recirculation systems as shown in Figure 2.10(a)–(c), but it is necessary to remove suspended solids peeled off by air bubbles in order to maintain a high quality of effluent. Mechanical aeration systems can also be used for oxygen supply and recirculation flow formation, but mechanical aeration systems are much less popular than diffused air systems (Figure 2.11). In some types of submerged filters, oxygen is dissolved into water using some oxygen dissolution device located outside the reactor (preoxygenation system), and feedwater with adequate dissolved oxygen is introduced into the reactor.

Figure 2.12 shows schematically an example of preoxygenation system. In such systems, the dissolved oxygen concentration in the liquid within the reactor is sufficiently high as not to limit the reaction rate in microbial films, so the efficiency of biological oxidation is significantly high.

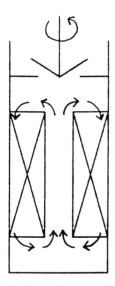

FIGURE 2.11. *Mechanically aerated biological filter.*

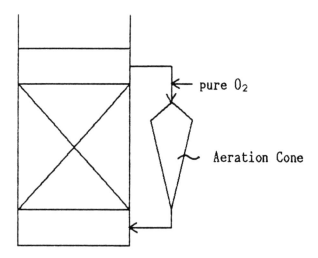

FIGURE 2.12. Preoxygenation system.

Filter media used in various types of submerged filters are classified as Figure 2.13 depending upon the state of the surface supporting microbial film, and also classified as follows:

(1) Shapeless granulated media—sand, quartz stone, volcanic rock, coke, coal cinder, oystershell, plastic piece, cork piece, wood piece, flock, etc.

(2) Definite formed granulated media—Intalox saddles, Rachig rings, plastic tubing, Pall-rings, etc.

(3) Pole-like, lace-like media—wood pole, branch of tree, lace with numerous radial fiber rings (D.B. lace or Ring lace), etc.

(4) Plate-like media—wood plate, plastic net, corrugated plastic plate, etc.

(5) Porous block-like media—porous plastic tubing, honeycomb tubing, etc.

(6) Mat-like media—saran mat, etc.

As a general rule, granulated media catch suspended solids not only more effectively but also in greater amounts than the other shaped media. On the other hand, they have larger resistances to water flow and are apt to be more easily blockaded. Especially, shapeless granulated media being random in their shapes and/or sizes and having low porosities, have the greater disadvantages of granulated media, not being used so widely. On the contrary, flat-plate, corrugated plates, and nets are poorer in suspended solids-catching ability, though they are resistant to clogging only if arranged properly.

General requirements for the packing media for an aerobic fixed bed submerged filter are as follows:

a. Proper degree of attachability of microbial film

b. Large specific surface area

c. High porosity of themselves and columns packed with them

d. Low resistance to water flow

e. Chemical and biological stability and high resistibility to the change of quality

f. Sufficient mechanical resistibility to buckling, destruction, and abrasion and enough durability

g. High capability to catch suspended solids

h. Uniform granular size or uniform spacing of plates, which enables uniform flow through the filter

i. No elution of toxicant as heavy metals, etc.

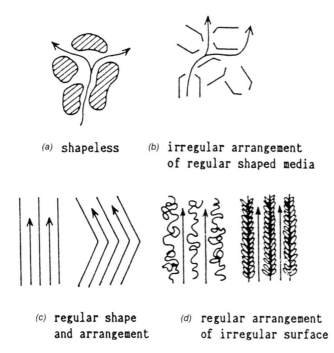

(a) shapeless   (b) irregular arrangement
                     of regular shaped media

(c) regular shape   (d) regular arrangement
    and arrangement   of irregular surface

FIGURE 2.13. *Various types of media.*

j. Small difference in specific gravity from water, giving no heavy load on submerged structures or on tank bottoms

k. Low price and stable supply

l. Ease of transportation, fabrication and construction

There is no medium which meets all of these requirements, because any two or more of the above requirements are antinomic to each other. For example, the requirements b. and d. are obviously antinomic, and to thicken plates in order to satisfy the requirement f. will be disadvantageous to c., k., etc. It is primarily important, therefore, to decide the priority among requirements, depending on the purpose of treatment, and on the restrictions of design and maintenance of the treatment plant etc., and to select the most appropriate shape and size of medium on a synthetic judgment. Obviously, media with a larger specific surface area enables a higher efficiency in the reactor, but the speed of filter clogging will be greater at the same time. Hence, the specific surface area of media used for a fixed bed submerged filter is usually not more than 100 m^2/m^3 (at most several hundreds m^2/m^3 or less), and BOD volumetric loading is restricted in order not to exceed 1 kg·BOD/m^3·media (0.5 kg·BOD/m^3·media in many cases). This minimizes the speed of clogging of the filter by the surplus thickening of microbial films. Yet, an extreme deterioration of effluent quality is inevitable, caused by clogging of the filter bed or by peeling off of overaccumulated biofilms while a fixed bed submerged filter is used for a long period. So overaccumulated biofilms must be stripped off and be withdrawn from the unit by periodic or timely back-washing, usually using air bubbles. The air diffusers are arranged underneath the filter beds with appropriate spacings, and diffused air bubbles wash off overaccumulated biofilm by shearing force.

Backwashing is continued for about 10–20 min, then pieces of microbial film peeled off from the media are settled and drawn off from the bottom of the filter unit. Subsequently, the unit is returned to the normal state of operation, but early in the normal operation, the water in the unit contains a considerable concentration of suspended solids not able to be removed with settlement. Such suspended solids, however, are rapidly caught by the filter media, and most of them will usually be removed in 30 min or less.

A conventional fixed bed submerged filter system includes a settling basin (the final settling basin) after the filter unit to remove suspended solids from the filter effluent and to improve the quality of the final effluent. It is useless to say that the settled sludge in the final settling basin must not be recycled to the filter unit. Nevertheless, an improved submerged filter system, as it were, a combined system of submerged filter and activated sludge process, has been developed in which settled sludge is recycled to the filter unit. The presence of a high concentration of suspended solids in a submerged filter unit spoils some of its advantages, but new advantages will be produced at the same time. Making efficient use of both microbial film and suspended biomass not only heightens treatment efficiency but also brings about the following effects from the viewpoint of the improvement of the activated sludge process:

First, since the biota of microbial film is much more diverse than that of activated sludge, and especially abounds in the higher ranks of predators, the coexistence of film microbes and activated sludge elongate the food chain of the ecosystem and reduce excess sludge production. Among existing predators, many ingest living or dead bacterial materials, contributing to clarification of the effluent.

Secondly, the existence of microbial film is fairly conducive to improving settleability and condensability of activated sludge. The reason for such improvements is not yet clear, but possibly the microzoa on the filter media prey upon filamentous bacteria or fungi, the growth of which frequently cause the increase of SVI or bulking – an extreme state of SVI increase – of activated sludge.

Thirdly, the combined process is significantly faster in start-up than conventional activated sludge process, and this fact implies that the combined process is suitable to treatment of wastewaters from buildings discharging highly fluctuating effluent as schools, assembly halls, etc.

On the other hand, the following advantages can be pointed out from the viewpoint of the improvement of microbial film processes:

In general, the amount of diffused air flow required to form recirculation flow with enough velocity to bring bulk liquid in effective contact with microbial films is larger than that required for the sufficient supply of oxygen. Therefore, the supply of oxygen is in excess in many cases, so the supply and the consumption of oxygen can be equilibrated by augmenting biological treatment capacity of the submerged filter unit with recycling biomass from the final settling basin and by increasing volumetric organic loading to the filter bed. Moreover it also is a great advantage that, since suspended microbes compete with film microbes for food materials, the rate of increment of biofilm thickness will be reduced. It should be noted, however, that, in order to recycle sludge to a submerged filter, such media must be used as do not clog even under such operating conditions. The above mentioned combined process is often called activated biofilter (ABF; refer to Chapter 3, Paragraph 5).

Another improved process of fixed bed submerged filter is called the biofilm filtration process, and has the configuration as depicted in Figure 2.14. Filter depth is 2–3 m, and relatively small-sized (about 3–7 mm in diameter) particles are used for packing media such as ceramic, volcanic

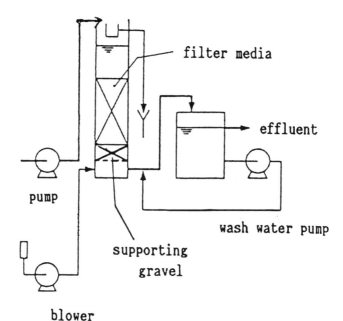

FIGURE 2.14. *Schematic configuration of biofilm filter.*

stone, anthracite, etc. Air is diffused from beneath the filter bed, and air bubbles rise through the bed. On the other hand, the influent enters above the top of the bed and flows down through the bed; then treated water is withdrawn from the bottom. Hence, this process is highly energy-saving, because water and air are contacted in a counter flow, which enables high efficiency of oxygen dissolution of 15% or more.

Most of the biomass produced in the bed and/or the suspended solids in the influent are caught by packed particles and collected in the bed, and consequently the suspended solid concentration level of the effluent is very low. The filter bed is considered, therefore, to have functions of both biological treatment and filtration, and so the name biofilm filtration process has been given to this process. Since the clogging rate of biofilm filters is generally high as might be anticipated from its mechanism of purification, influent BOD of this process should not exceed 100 mg/l, and a typical example of its application is the polishing-up operation of the secondary effluent of municipal sewage. Even under such operation conditions, the filter will be clogged within several days of operation, and so backwashing of the filter is necessary with a frequency of once per 1–2 days. Backwashing is practiced by passing air and water concurrently upward through the filter.

The wastewater produced by backwashing is wasted through the gutters installed above the filter. An example of the results of treatment obtained with biofilm filtration is shown in Table 2.1. Biofilm filtration plants have deficiencies that the filtering operation must be stopped during backwashing, that the head loss through the filter varies very widely, and that wastewater by backwashing is discharged concentrated in a short time. To remedy these deficiencies, a moving bed biofilm filtration system has been developed as outlined in Figure 2.15. As is shown in this figure, the media in the bottom of the filter bed are lifted upward to the top with air lifting, and the media and wash wastewater are separated by the screen around the top of the draft tube. Thus, the filter media are able to be washed at a constant rate, and consequently, a constant-rate filtering is available with a constant head loss.

TABLE 2.1. Treatment Results with Biofilm Filtration.

	Wastewater	Treatment/Condition		
		L/V (m/day)	BOD Loading (kg/m³·day)	Air to Water Volume Ratio
A	Raw sewage (S city)	41–46	3.1–4.8 (4.0)	2–3
B	Primary effluent of sewage (U city)	43–67	2.9–6.7 (4.3)	0.8–2.3 (2.0)
C	Secondary effluent of sewage (S city)	180	0.5–0.7 (0.6)	0.15
D	Secondary effluent of sewage (H city)	120	0.7–2.6 (1.6)	0.5
E	Secondary effluent of sewage (K city)	180	0.2–0.6 (0.5)	0.2
F	Chemical industry	43–75	2.1–4.0	0.7
G	Activated sludge effluent of beer brewery wastewater	120	0.2–1.2 (0.4)	0.1
H	Flotation effluent of papermill wastewater	120	0.6–1.6 (1.0)	0.3
I	River water	240	0.2–0.4	0.01

( ) Means average.

TABLE 2.1a. Treatment Results with Biofilm Filtration.

Sample		Temperature (°C)	Transparency (cm)	S/S (mg/ℓ)
A	Influent	27–31(29)	3–5(4)	96–270(170)
	Effluent	27–33(30)	>100	<1–3.0(1.8)
	Removal (%)			98–100(99)
B	Influent	18–26(22)	3.5–6.5(4.9)	73–220(120)
	Effluent	17–27(22)	20–>100	3–18(8)
	Removal (%)			90–96(93)
C	Influent	25	16–34(25)	13–21(17)
	Effluent	25	>100	1.0–1.4(1.2)
	Removal (%)			89–95(92)
D	Influent	16–18(17)	11–35(18)	11–40(25)
	Effluent	15–18(17)	58–>100	<1–6(3)
	Removal (%)			85–99(88)
E	Influent	18–20(19)	17–33(23)	5.3–8.6(7.1)
	Effluent	19–20(19)	92–>100	<1–1.9(1.3)
	Removal (%)			73–92(81)
F	Influent	18–25(23)		52–59(55)
	Effluent	14–22(18)		3–5(4)
	Removal (%)			92–94(93)
G	Influent	25–31(28)	14–18(19)	13–83(28)
	Effluent	25–30(27)	59–>100	<1–12(5)
	Removal (%)			75–95(82)
H	Influent	25–30(28)		5.6–32(13)
	Effluent	25–30(27)		<1–5.1(1.9)
	Removal (%)			75–92(84)
I	Influent	6–8(7)	40–>100	2.4–12(6.1)
	Effluent	6–8(7)	>100	<1–4.6(2.6)
	Removal (%)			40–78(60)

( ) Means average.

The Expanded Bed (Submerged) Filter

When wastewater is passed upward through a filter packed with fine particles with diameters similar to those of sand used for rapid filtration, the filter media are lifted by the friction between the grains of the media and the rising water. If rising flow velocity were too high, the filter bed would be fluidized. An appropriate rising velocity will result in about 20 to 30% expansion of the filter media. Unlike a fluidized bed, the filter media grains do not change their relative position in an expanded bed. Since a porosity not less than a certain value is maintained in an expanded bed owing to the expansion of the bed, the trouble of filter clogging can be conquered, which is inevitable in fixed bed filters. To aerate an expanded bed directly for oxygen supply will destroy it, so preoxygenation of influent is usually practiced. This type of process is used, therefore, much more as anaerobic process than as aerobic process. Quartz sand, plastic pellets, granular activated carbon or anthracite grain, etc., is used frequently as the filter medium (the carrier). Self-pelletizing of biological sludge is also available, which implies the formation of pellets from suspended sludge under appropriate conditions of flow velocity, organic loading, etc., without usage of carrier particles. Self-pelleted particles cannot be regarded as a kind of microbial film, strictly speaking, because it contains no carrier particles, but also it is use-

TABLE 2.1b. Treatment Results with Biofilm Filtration.

Sample		COD_{Mn} (mg/t)	BOD (mg/t)	Sol-BOD (mg/t)
A	Influent	59–90(71)	150–220(180)	54–120(85)
	Effluent	7.7–14(11)	1.4–6.3(4.2)	<1–4.0(2.6)
	Removal (%)	80–87(83)	96–99(98)	96–99(97)
B	Influent	58–90(67)	77–160(120)	
	Effluent	18–31(20)	5–21(12)	
	Removal (%)	66–73(70)	87–94(90)	
C	Influent	13–16(15)	5.6–7.2(6.6)	2.8
	Effluent	8.6–9.3(8.9)	2.0–2.8(2.5)	<2
	Removal (%)	31–45(38)	59–64(62)	>28
D	Influent	16–38(27)	10–37(22)	
	Effluent	10–20(14)	1.8–9.7(4.1)	
	Removal (%)	38–52 (48)	74–85 (81)	
E	Influent	10–13(11)	2.8–6.6(5.1)	<1–1.8(1.1)
	Effluent	7.3–9.3(8.5)	<1–2.3(1.4)	<1–1(<1)
	Removal (%)	18–29(24)	62–85(73)	0–83(37)
F	Influent	27–91(50)	45–130(74)	
	Effluent	4–20(11)	5–17(10)	
	Removal (%)	73–89(78)	82–93(86)	
G	Influent	14–44(23)	4–20(7)	
	Effluent	11–17(13)	1–4(2)	
	Removal (%)	21–61(43)	65–80(71)	
H	Influent	17–37(23)	9.4–26(17)	7.5–23(15)
	Effluent	7.9–17(11)	1.0–5.0(2.9)	1.0–4.1(2.3)
	Removal (%)	44–64(54)	69–91(82)	65–94(83)
I	Influent	2.1–5.6(3.7)	1.4–3.7(2.4)	
	Effluent	2.1–4.7(3.4)	1.1–2.9(1.8)	
	Removal (%)	0–18(8)	7–57(25)	

( ) Means average.
Sol-BOD was determined for filtrate by 1 μ filter.

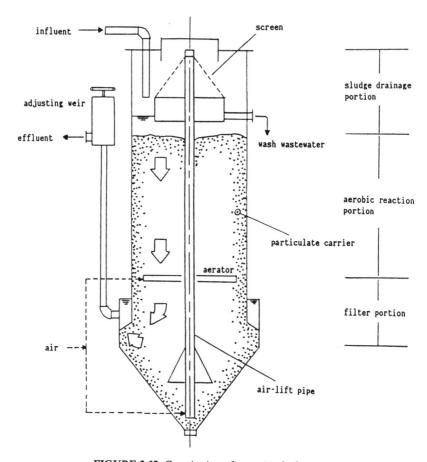

FIGURE 2.15. *Constitution of a moving bed reactor.*

less to say that it has the same substrate ingestion characteristic as a particle containing a sand grain as the nucleus. Utilization of fine carriers or self-pelleted particles offers the specific area of biofilm surface as much as several thousands of square meters per cubic meter of filter bed, which enables several tens of times the efficiency of an expanded bed compared with a conventional fixed bed submerged filter.

The Fluidized Bed (Submerged) Filter

Passing water upward with a higher velocity than in an expanded bed through a bed packed with sand or granular activated carbon, etc., will cause the fluidization of the bed, as in Figure 2.16. A fluidized bed submerged filter has an extremely large specific surface area as well as an expanded bed filter, and is able to accomplish such a treatment within several minutes to several tens of minutes, compared to several hours using a conventional biological treatment process [4,5]. Moreover, the settling velocity of biomass particles, being much higher than that of activated sludge, means they can be removed by a simplified settling facility or a screen, thus achieving a very compact treatment plant.

The most typical fluidized bed reactor consists of a long reactor bed, as in Figure 2.16, with water being passed from beneath the bed at a high enough velocity to fluidize the bed. Generally the velocity required for fluidization of the bed is much higher than that determined by the hydraulic retention time needed for the biological reaction desired. Therefore, for the coincidence of both velocities, the effluent from the bed is usually recycled to the inlet in order to increase the rising flow velocity.

Several modified types of fluidized bed reactors have been developed. Figure 2.17(a) shows the type of fluidized bed process in which a settling basin is equipped to recover the particles of immobilized microbes from the effluent. There is a packaged type of reactor and settling basin [Figure 2.17(b)], and a type with a screen at the outlet of the reactor to prevent the outflow of biological particles [Figure 2.17(c)]. In any of these three types,

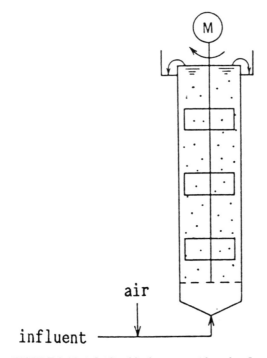

FIGURE 2.16. *A fluidized bed reactor with a plug-flow.*

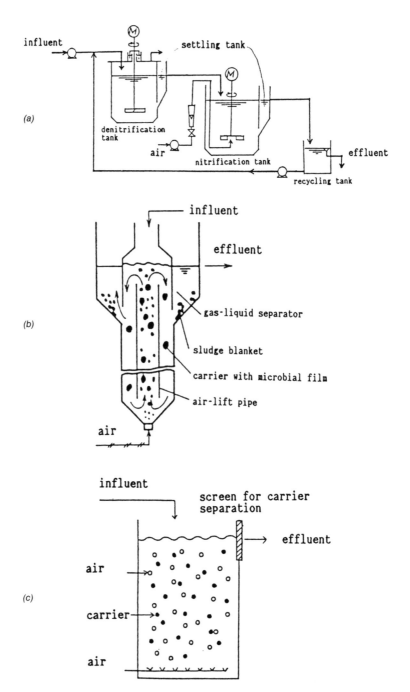

FIGURE 2.17. *Various fluidized bed reactors: (a) fluidized bed reactors with settling basin; (b) fluidized bed reactor combined with a settling basin and (c) fluidized bed reactor with a screen for carrier separation.*

the fluidization of bioparticles is accomplished by aeration, enabled by the use of carrier particles with a specific gravity near water.

To keep the fluidized state in a longitudinal reactor (Figure 2.16), an appropriate rising flow velocity is required. Let us consider the velocity for fluidization.

If the flow velocity upward is higher than that of the free settling of a single particle, then particles will be carried away from the reactor tower. Therefore, the flow velocity upward available for the operation of a fluidized bed reactor is confined to a narrow range between the minimum velocity for fluidization and the free settling velocity of the particles included in the bed.

The minimum velocity for fluidization is given by the following empirical formula:

$$V_{mf} = 0.0381 \frac{d_{60}^{1.82}[\gamma(\gamma_s - \gamma)]^{0.94}}{\mu^{0.88}} \qquad (2.53)$$

where

V_{mf} = minimum fluidization velocity (gpm/ft²)
d_{60} = 60% finer size of the filter medium (mm)
γ_s and γ = specific weight of the filter medium and of water, respectively (lb/ft³)
μ = water viscosity (centipoise)

Or, V_{mf} (cm/hr) is shown by the following equation.

$$V_{mf} = \frac{d^2(\varrho_s - \varrho)g}{1650\,\mu} \qquad (2.54)$$

where

d = diameter of the filter medium (m)
ϱ_s and ϱ = specific gravity of the filter medium and of water, respectively (g/m³)
μ = water viscosity (g/m·hr)
g = gravitational acceleration (m/hr²)

Accordingly, as the biomass becomes attached to the filter medium, the apparent density of the filter medium will decrease. The apparent density can be calculated from the attached amount of biomass on the assumption that the microbial film density is 1.005 g/cm³.

Head loss through a fluidized bed is given by the following equation.

$$H_L = \frac{H(\gamma_s - \gamma)(1 - \epsilon)}{\gamma} \qquad (2.55)$$

where

H_L = head loss (m)
H = fluidized bed height (m)
ϵ = porosity of fluidized bed ($-$)

The treatment processes with an expanded or a fluidized bed can be considered to have an efficiency close enough to the potential upper limit of biological treatment process, taking into consideration the extremely high microbial densities and the high effectiveness factors of biofilm in these processes.

However, needless to say, the maintenance of these processes is not as easy as the fixed bed reactors, because it is difficult to maintain the reactor bed in a stably expanded state or in a stably fluidized state.

2.2.1.2 THE ANAEROBIC SUBMERGED FILTER (THE ANAEROBIC FILTER)

Anaerobic wastewater treatment processes are defined as treatment processes which utilize the metabolic reactions of microbes under molecular oxygen free conditions. The three main processes are decomposition of organic substances by gasification (methane fermentation), reductive decomposition of nitrite or nitrate nitrogen (denitrification), and biological phosphorus removal by the repetition of aerobic and anaerobic conditions (luxury ingestion of phosphorus). Among these three processes, biological phosphorus removal is difficult, if not impossible, to apply to microbial film reactors, so hereafter in this paragraph only methane fermentation and denitrification are related. Though they are equally anaerobic processes, there is a wide difference in several points. Accordingly, the outlines and the applications as microbial film processes are described separately in the following section.

Methane Fermentation and the Anaerobic Filter for It

By methane fermentation, various organic matters are decomposed and gasified into methane and carbon dioxide through the actions of anaerobic bacteria, and the decomposition process consists of the following three processes:

(1) *First step:* hydrolysis of suspended organic solids or soluble organic macromolecules

(2) *Second step:* degradation of low molecular weight organics into volatile fatty acids, finally into acetic acid

(3) *Third step:* methane formation from acetic acid or from hydrogen and carbon dioxide

Sometimes the first and second steps are collectively called liquefaction, with the third step being called gasification.

The group of anaerobic bacteria responsible for liquefaction is called the Acid producing bacteria or Acid formers, and that for gasification is called the Methane producing bacteria or Methane formers. The course of anaerobic decomposition is schematically shown as Figure 2.18.

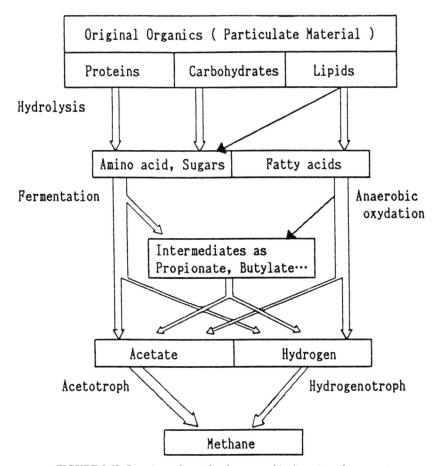

FIGURE 2.18. *Reaction scheme for the anaerobic digestion of wastes.*

Concerning these two groups of anaerobic bacteria and the mixed culture of them, Henze and Harremoës [6] have shown the growth parameters as in Table 2.2. The marked characteristic of anaerobic bacteria, especially of methane bacteria, is that they have small yield coefficients [refer to Equation (2.6)], and consequently have fairly small specific growth rates, μ. Hence, when anaerobic bacteria are cultivated using a cultivation tank of the chemostat type, methane bacteria will be washed out, as is expected from Equation (2.35), unless the dilution rate is controlled in a very low range. That is, a steady-state cannot be kept in this type of anaerobic reactor, provided the hydraulic retention is not very long, so that the application of this type of reactor is restricted to wastewaters with BOD not less than 10,000 mg/l or to organic sludges. To overcome such a restriction on the application, it is required to maintain the retention time of methane bacteria longer than a certain value independently of the hydraulic retention time. To realize such a condition, various types of reactors can be used as shown in Figure 2.19. A settling basin and a membrane separation unit are equipped after the bioreactor in Figures 2.19(a) and 2.19(b), respectively, to recycle

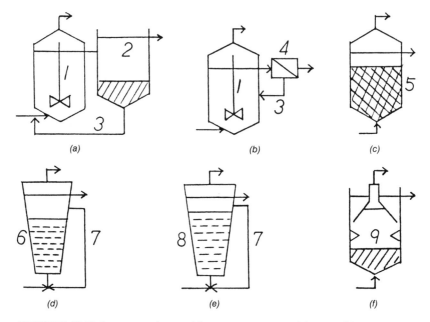

FIGURE 2.19. Various types of anaerobic treatment process: (a) anaerobic contact process; (b) membrane separation process; (c) anaerobic filter process; (d) expanded bed process; (e) fluidized bed process and (f) sludge blanket process. 1. continuously stirred tank reactor (CSTR); 2. clarifier; 3. sludge recycle; 4. membrane separator; 5. fixed bed; 6. expanded bed; 7. liquid recirculation; 8. fluidized bed; and 9. sludge blanket reactor.

TABLE 2.2. *Growth Constants of Anaerobic Cultures.*

Parameter Culture	μ_{max} Maximum Specific Growth Rate, 35°C (d^{-1})	Y_{max} Maximum Yield Coefficient (kg VSS/kg COD)	$r_X = \mu_{max}/Y_{max}$, Maximum Substrate Removal Rate, 35°C [kg COD/(kg VSS·d)] 100% Active (VSS)	50% Active (VSS)	K_S, Half Velocity Constant (kg COD/m³)
Acetic acid producing bacteria	2.0	0.15	13	7	0.2
Methane producing bacteria	0.4	0.03	13	7	0.05
Combined culture	0.4	0.18	2	1	—

101

the biomass and to operate the processes as a kind of turbidstat for the satisfaction of the above condition. On the other hand, the processes in Figure 2.19(c)–(f) enable the confinement of biomass within the bioreactors by holding as a fixed bed, expanded bed, fluidized bed, and a sludge blanket, respectively, irrespective of the flow rate of the wastewater; thus the same required condition is satisfied. Among these processes, the fixed bed is able to maintain the biomass retention time longer than a certain length with the simplest facility and with the most simplified operational procedure. Moreover, the expanded or fluidized bed reactor is also very effective in holding a large amount of biomass and maintaining sludge retention time (SRT) fairly long, though they are inferior to the fixed bed in the ease of operation because of their requirement to regulate the upward flow velocity through the bed. Besides, the diffusion-limiting nature of microbial film processes has been repeatedly pointed out so far, but the application of methane fermentation which has optimum temperatures at about 53°C (thermophilic fermentation) or at about 37°C (mesophilic fermentation), at ambient temperature (not more than 25°C) weakens the nature of microbial film processes, so the rationality of various anaerobic filter processes is increased. For the above mentioned reasons, the application of anaerobic filters to methane fermentation has extended the range of application of methane fermentation, even to middle or low organic concentration of wastewaters and the range of operational conditions to ambient temperatures, and the hydraulic retention time of several hours or so. In addition to the general merits of submerged filters, anaerobic submerged filters have the following merits:

a. Operation without or with a little energy consumption

b. Ability of the collection of valuable material, methane gas

c. Low clogging rate of filter and small amounts of excess sludge production because of low growth rate of microbial film

d. No rate limitation by dissolved oxygen unlike aerobic treatment processes

e. Ability of obtaining almost suspended solid free effluent

On the other hand, there are disadvantages:

f. It is often impossible to gain the desired quality of effluent with an anaerobic filter alone, and some additional treatment is required, e.g., secondary treatment.

g. Not applicable to those wastewaters which contain such substances which can be converted into inhibitory compounds such as hydrogen sulfide, through anaerobic reaction.

The most basic type of anaerobic submerged filter consists of a fixed bed

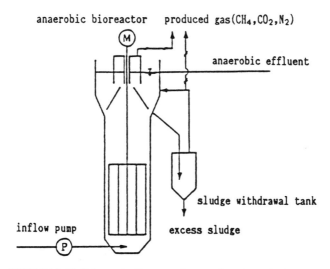

anaerobic bioreactor   produced gas(CH_4,CO_2,N_2)

anaerobic effluent

sludge withdrawal tank

inflow pump   excess sludge

FIGURE 2.20. Schematic diagram of anaerobic sludge bed reactor.

reactor as in Figure 2.19(c), and, unlike an aerobic submerged filter, oxygen supply to the filter bed being unnecessary, wastewater should only be passed through the filter bed once without recirculation within it (Once-through System). The filter medium packed in a usual anaerobic filter is similar to that used for aerobic filters. In anaerobic fluidized bed reactors, as shown in Figure 2.19(f), the filter media are fluidized by a high upward velocity caused by a high degree of recycled filter effluent, and biofilm is formed on the media surface. It is also able to use self-pelletized biomass for fluidized bed reactors which can be formed only by a slow mixing of filter content, as in Figure 2.20, without usage of any carrier. The results of anaerobic submerged filters for organic removal are listed in Table 2.3.

Denitrification and the Anaerobic Filter for It

Denitrification is widely employed for removal of various forms of nitrogen from wastewaters, combined with a kind of aerobic reaction, nitrification. Organic nitrogen in wastewaters is converted into ammoniacal nitrogen when organics are degraded either aerobically or anaerobically. Then, ammoniacal nitrogen is oxidized by aerobic bacteria to be converted into nitrate nitrogen via nitrite nitrogen.
Nitrosomonas sp.:

$$NH_4^+ + 3/2O_2 \rightarrow NO_2^- + H_2O + 2H^+$$

TABLE 2.3. Results of Wastewater Treatment with Anaerobic Submerged Filters.

Wastewater	Type of Filter	COD (mg/l)		COD Removal (%)	HRT (days)	COD Loading (kg/m³·day)	Reference
		Influent	Effluent				
Waste molasse	fixed	10,000–50,000	—	57–79	2.5–5.0	2–12	[7]
Food processing	fluidized	7,530	1,900	75	0.31	24.1	[8]
Whey	fluidized	27,300	5,000	82	5.3–7.4	—	[9]
Black liquor from kraft mill	⎧ fixed	—	—	≥80	—	1–2	⎫
	⎨ expanded	—	—	≥80	—	–10	⎬ [10]
	⎩ fluidized	415 ± 50	—	≥80	—	2.5–13	⎭
Sewage	expanded	88–306	—	51–90	0.17–3.0 hr	—	[11]
Sewage	fixed	238*	—	50–90*	1.04–1.54	0.16–0.23*	[12]

*BOD.

104

Nitrobactor sp.:

$$NO_2^- + 1/2O_2 \rightarrow NO_3^-$$

The above formulated changes are called nitrification.

Nitrate nitrogen produced by nitrification is utilized by anaerobic bacteria as the hydrogen acceptor. The reaction is formulated as follows when methanol is used as the hydrogen donor.

The first step:

$$6NO_3^- + 2CH_3OH \rightarrow 6NO_2^- + 2CO_2 + 4H_2O$$

The second step:

$$6NO_2^- + 3CH_3OH \rightarrow 5CO_2 + 3N_2\uparrow + 7H_2O + 60H^-$$

That is, nitrate nitrogen is converted into harmless nitrogen gas via nitrite nitrogen. This process, called denitrification, is the most important and widely used anaerobic reaction similar to methane fermentation. Though the organic compounds originally contained in wastewater are available for denitrification, many authors have agreed that methanol is most effective as a hydrogen donor, taking into account many conditions, if the addition of some organic compound is necessary.

Anaerobes responsible for denitrification are common soil bacteria or fungi, or various wastewater-borne bacteria, the yield of which is a little smaller than that of aerobic bacteria. Therefore, an anaerobic filter for denitrification is more easily clogged and produces larger amounts of excess sludge than one for methane fermentation. Any of the fixed, expanded, or fluidized beds are available for denitrification, but influent water into the filter must contain not only nitrite-or-nitrate nitrogen, but also organic compounds for the electron donor (organic carbon). When raw wastewater quality meets such requirements, only passing wastewater through the bed will do. When raw water contains organic nitrogen or ammoniacal nitrogen, nitrification is indispensable prior to denitrification. Two methods exist for this requirement.

In the first method, two submerged filters are used in series, the former and the latter of which are used for denitrification and for nitrification, respectively, and the effluent from the filter for nitrification is recycled to the filter for denitrification (Figure 2.21).

In this case, letting α = denitrification ratio in the denitrification filter, β = nitrification ratio in the nitrification filter, and n = the ratio of

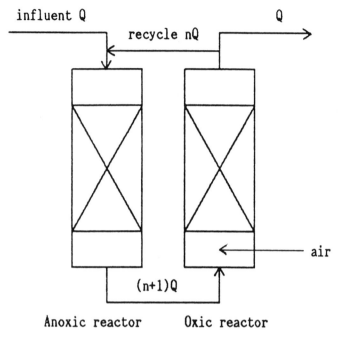

FIGURE 2.21. *Schematic diagram of nitrogen removal using two steps of submerged biological filters.*

recycled flow rate to treated flow rate, the nitrogen removal by this combined filter system is given by

$$E = \frac{n(n + 1)\alpha\beta}{(n\alpha + 1)(n\beta + 1)} \qquad (2.56)$$

where it should be noted that α and β are not mere constants, but are dependent on the operation conditions including n.

The concentrations of ammoniacal and nitrate nitrogen remaining in the treated water are shown by

$$c_{e1} = (1 - \beta)c_0/(n\beta + 1) \qquad (2.57)$$

$$c_{e2} = \frac{\beta(n + 1)c_0}{(n\alpha + 1)(n\beta + 1)} \qquad (2.58)$$

where

c_{e1} = ammoniacal nitrogen concentration in the effluent

c_{e2} = nitrate nitrogen concentration in the effluent

c_0 = kjeldahl nitrogen concentration in the influent

In the above formulae, the influent is supposed to contain no nitrate nitrogen. Moreover, in the introduction of Equations (2.56)–(2.58), the nitrogen ingestion accompanied by the biological growth is neglected.

The second way to remove nitrogen with submerged filters is using a single filter with intermittent aeration by which nitrification and denitrification are practiced one after the other. In order to bring the bulk liquid in effective contact with biofilm, it is desirable to recirculate mechanically the water in the unit during the time periods where aeration is stopped, but it is not essential. In microbial film processes, all biomass in the film is not always aerobic even in the time period of aeration, so intermittently-aerated microbial film processes are much more complicated in their behavior than intermittently-aerated activated sludge processes. Therefore, the optimum lengths of aeration period and aeration-stopping period, or the optimum ratio between them must be determined empirically.

2.2.1.3 THE SUBSTRATE CONCENTRATION CHANGE ALONG THE AXIS OF A SUBMERGED FILTER

Since the flow velocity in the axial direction is generally very high, the substrate concentration change in the direction is very small in an aerated submerged filter. That is, an aerated submerged filter can be looked upon as a CSTR. However, in the cases where axial flow velocity is not so high, or where wastewater is passed only once through the filter, the water quality change along the filter axis is of much importance. Mathematical modeling of substrate concentration in these cases is described in the following.

In a submerged filter schematically shown in Figure 2.22, z axis is set vertically and the mass balance of substrate is considered for the hatched area put between $z = z$ and $z = z + dz$.

The inflow of substrate through the surface, $z = z$ within unit time is:

$$Av_z S_z - AE_z \frac{\partial S_z}{\partial z}\bigg|_{z=z}$$

The outflow of substrate through the surface, $z = z + dz$ within unit time is:

$$Av_z S_{z+dz} - AE_{z+dz} \frac{\partial S_z}{\partial z}\bigg|_{z=z+dz}$$

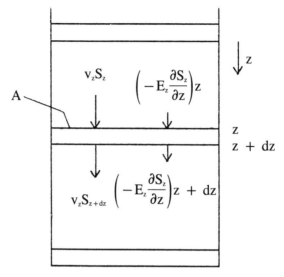

FIGURE 2.22. *Differential mass balance between cross sections* $z = z$ *and* $z = z + dz$ *in a submerged biological filter.*

The accumulation rate of substrate in the sliced portion of filter in consideration is:

$$\epsilon Adz \frac{\partial S_z}{\partial t} - Adz \cdot a_w N$$

where

A = cross-sectional area of the filter
v_z = superficial velocity through the filter
S_z = substrate concentration at $z = z$
E_z = diffusion (dispersion) coefficient in the direction of z axis
ϵ = porosity of the filter
t = time
a_w = biofilm area per unit volume of filter bed
N = substrate removal flux per unit area of biofilm

General relation among these items is shown as [Inflow] − [Outflow] = [Accumulation] + [Consumption], so the following equation is obtained.

$$\epsilon Adz \frac{\partial S_z}{\partial t} + Adz \cdot a_w N = Av_z S_z - AE_z \frac{\partial S_z}{\partial z}\bigg|_{z=z}$$

$$- \left(Av_z S_{z+dz} - AE_{z+dz} \frac{\partial S_z}{\partial z}\bigg|_{z=z+dz} \right) \qquad (2.59)$$

Dividing Equation (2.59) by $A dz$, and rearranging

$$\epsilon \frac{\partial S_z}{\partial t} + a_w N = -v_x \frac{\partial S_z}{\partial z} + \frac{\partial}{\partial z}\left(E_z \frac{\partial S_z}{\partial z}\right) \qquad (2.60)$$

is obtained. E_z is regarded as a constant independent of z, so Equation (2.60) can be rewritten into

$$\epsilon \frac{\partial S_z}{\partial t} + a_w N = -v_x \frac{\partial S_z}{\partial z} + E_z \frac{\partial^2 S_z}{\partial z^2} \qquad (2.61)$$

Equation (2.61) is the fundamental equation of substrate concentration change in the axial direction. Under a steady-state, $\partial S_z / \partial t = 0$, so that from Equation (2.61)

$$E_z \frac{d^2 S_z}{dz^2} - v_z \frac{dS_z}{dz} - a_w N = 0 \qquad (2.62)$$

is obtained. This equation is of the same form as Equation (2.46). That is to say, the behavior of substrate concentration in the axial direction of a submerged filter is similar to that in the flow direction within an empty tank reactor. From Equation (1.41),

$$N = \eta N_b = \eta LMQ(S_z) \qquad (2.63)$$

is substituted into Equation (2.62)

$$E_z \frac{d^2 S_z}{dz^2} - v_z \frac{dS_z}{dz} - a_w \eta LMQ(S_z) = 0 \qquad (2.64)$$

where

η = effectiveness factor
L = thickness of biofilm
M = microbial density within biofilm
$Q(S_z)$ = metabolic rate of substrate at the concentration, S_z per unit biomass

In Equation (2.64), if the term of reaction is of the first-order in S_z, i.e., if $\eta Q(S_z)$ is equal to $k_1 S_z$ (k_1 being a constant), then Equation (2.64) is of the same form as Equation (2.48), and the solution of Equation (2.64) is given by Equation (2.50), substituting $\alpha = a_w k_1 LM$.

On the other hand, if the term of reaction is independent of substrate concentration (zero-order reaction), Equation (2.64) is changed into

$$E_z \frac{d^2 S_z}{dz^2} - v_z \frac{dS_z}{dz} - K_0 = 0 \tag{2.65}$$

where $K_0 = a_w \eta LMQ(S_z)$ = constant and the boundary condition at the inlet is given by

$$S_z = S_{z0+} \tag{2.66}$$

S_{z0+} implies the substrate concentration at the inlet, and the following equation is available

$$-E_z \left(\frac{dS_z}{dz} \right)_{z=0+} = v_z (S_{z0} - S_{z0+}) \tag{2.67}$$

where S_{z0} = substrate concentration in the influent.

Equation (2.67) signifies that a discontinuous change of substrate concentration will occur immediately after the inflow. But, when E_z is much smaller than v_z, S_{z0+} is equal to S_{z0}.

The boundary condition at the outlet is

$$-E_z \left(\frac{dS_z}{dz} \right)_{z=Z-0} = v_z (S_{zZ} - S_{zZ-0}) \tag{2.68}$$

where Z = the depth of the submerged filter.

Under the boundary conditions of Equations (2.67), Equation (2.65) is solved and the following equation on substrate concentration distribution is obtained.

$$S_z = S_{z0+} + \frac{E_z K_0}{v_z^2} \exp\left(\frac{v_z}{E_z}(z - Z) \right) - \frac{K_0}{v_z} z \tag{2.69}$$

where S_{z0+} is decided from Equations (2.67) and (2.69), and represented as follows.

$$S_{z0+} = S_{z0} - \frac{E_z K_0}{v_z^2} \left(\exp\left(-\frac{v_z}{E_z} Z \right) - 1 \right) \tag{2.70}$$

The term of diffusion in Equation (2.64) is often negligible compared with the term of bulk flow transfer in a submerged filter, except for a fluidized bed. That is, it is assumed that $Pe \rightarrow \infty$, then Equation (2.64) is simplified into

$$\frac{dS_z}{dz} + k_z \eta Q(S_z) = 0, \qquad k_z = a_w LM \qquad (2.71)$$

and when $\eta Q(S_z)$ is expressed as follows depending upon the order of biological reaction,

$$\eta Q(S_z) = \begin{cases} k_0 \text{ (zero-order reaction)} \\ k_{1/2} S_z^{1/2} \text{ (half-order reaction)} \\ k_1 S_z \text{ (first-order reaction)} \end{cases}$$

then the solutions of Equation (2.65) are shown by Equation (2.72).

$$\left. \begin{array}{l} S_z = S_{z0} - k_z k_0 z \text{ (zero-order reaction)} \\ S_z^{1/2} = S_{z0}^{1/2} - k_z k_{1/2} z \text{ (half-order reaction)} \\ S_z = S_{z0} \exp(-k_z k_1 z) \text{ (first-order reaction)} \end{array} \right\} \qquad (2.72)$$

These equations coincide with the substrate concentration distributions in a PFR.

2.2.1.4 THE DESIGN OF SUBMERGED FILTER

In spite of a large number of theoretical and empirical investigations already carried out concerning submerged filters, a specific design approach has not yet been established. In most cases, therefore, the process design of a submerged filter is practiced depending upon the results of experimental treatment using a pilot-plant, existing data, or similar examples.

The reason for the above mentioned is as follows:

a. Not only the purification capacity but also unquantifiable factors as the arrangement for maintenance, etc., must be included in the main factors for the process design.

b. The mechanism of purification of submerged filter is much more complicated than that of processes using suspended biomass, so that theoretical design has its limit.

c. The condition for an optimum design has not yet been made clear.

d. Although there are many configurations of microbial film processes, its history of research and development is much shorter than that of the activated sludge process.

Among these reasons, b. and d. seem to need no explanation. However, a. and c. require additional explanation.

As an example, when treatment of domestic sewage with conventional fixed bed submerged filter is planned to attain BOD removal of 90% or more, the volumetric organic loading of as much as 1.0 kg-BOD/m^3-media·day or more is applicable judging from only the purification performance in a short period. But, microbial film will rapidly become thicker and the filter will be clogged within one or two weeks, if such a high loading is continued. Hence, frequent backwashing of the filter bed is required to maintain the performance of the process, resulting in net operational time loss of the filter and increase in labor for maintenance and excess sludge production. High efficiency does not always mean a good process design. It is rather, the unquantifiable factors that govern the design of submerged filters.

From above, the description of the design of submerged filters based on explicit theory is impossible, but how to design or what factors should be in consideration for design will be described in the following.

Though there are a large number of factors associated in the design of submerged filters, many of them are common to the design of general biological treatment processes. The following items are among them:

(1) The quality of wastewater and the effect of pretreatment on it
(2) The properties of filter media
(3) Organic loading on filter media
(4) Aeration device and recirculation flow velocity
(5) Installations for backwashing and for sludge withdrawal
(6) The others

These items are intermingled, and cannot be correctly discussed separately.

The Quality of Wastewater and the Effect of Pretreatment on It

Water quality indices associated in the design of submerged filters are suspended solids, nutrient salts, inhibitory substances, pH and temperature, etc.

Generally, suspended solids exert a bad effect on the result of treatment, and particularly inorganic suspended solids have various bad effects, because of reduction of microbial density of biofilm, obstruction to the contact between bulk water and biofilm by covering the surface of biofilm, and acceleration of filter clogging. Hence suspended solids, especially inorganic suspended solids ought to be removed sufficiently by a pretreatment, such as settling.

It should be noted that some dissolved metals may be converted into hydroxides within the filter and may behave like suspended solids. The effect of organic suspended solids is not so bad as inorganic ones, because they are dissolved and degraded sooner or later. Moreover, the removal rate of organic suspended solids is smaller because they cannot enter directly into biofilm by diffusion and must be hydrolyzed before they can enter into the same path as dissolved organic solids. Therefore, the higher the ratio of dissolved organic matter in the influent water is, the easier it can be treated. For these reasons, the removal of suspended solids by pretreatment is of much importance, and particularly when a filter media with a large specific surface area is used, sufficient pretreatment is effective to prevent filter clogging. However, the removal of organic suspended solids by a physical process such as pretreatment causes a disadvantage that the amount of sludge produced by the whole treatment system will increase compared with the system where suspended solids are fed into the filter without any pretreatment. To decrease sludge production, it is worthy of consideration to send physically-collected sludge into the submerged filter after anaerobic hydrolyzation. In addition to this, the process combination as a septic tank−an aerobic submerged filter or an anaerobic filter−an aerobic submerged filter, will serve for reduction of sludge production.

Concerning the desired amount of nutrient salts (*N* and *P*) in the influent flow, attention should be paid so that the supply of nutrient salts should meet not only stoichiometric demand for cell synthesis but also sufficient flux into biofilm as described in Chapter 1, section 1.4. If any of the fluxes of nutrient salts are relatively deficient as compared with the flux of substrate, biological reaction by biofilm will be retarded. Moreover, it must be noted that it is not the concentration in the influent but that in the reactor which controls the flux of substrate or nutrient salts. Generally the net yield of biomass being small in microbial film processes, the demand of nutrient salts is rather small.

When inhibitory substances are contained in the raw wastewater, attention should be paid to their effect just as in the activated sludge process. But, the degree of inhibitory effect on biofilm is lower than on suspended biomass, because biofilm does not make contact directly with inhibitory substances. Particularly when the inhibitory substance in the wastewater are cyanides, phenol formaldehyde or hydrogen sulfide, etc., which are biologically decomposable; the concentration of such substance within biofilm is always lower than that in the bulk water. Moreover, many of these compounds are refractory compounds, the removal of which microbial film processes are fitted. For the same reason as their tolerance to inhibitory substances, microbial film processes are thought to be tolerable to abrupt and wide changes in pH. It has already been described in Chapter 1, section 1.3, that microbial film processes which have a rather diffusion-limiting nature, are not susceptible to temperature fluctuations. Since low water temperature is usually accompanied by low atmospheric temperature, the trickling filter process and the rotating biological contactor process are particularly apt to cause temperature decreases among microbial film processes, because they utilize microbial film exposed entirely or partly to the atmosphere. Therefore, it can be concluded that the submerged filter process which is the microbial film process utilizing entirely submerged biofilm, is the process least susceptible to temperature fluctuation.

The Properties of Filter Media

Various carriers are being investigated for expanded beds or fluidized beds. These include materials such as quartz sand, zeolite, polypropylene pellets (PP pellets), fine coke particles, small pieces of synthetic plastic foam, and porous ceramics or activated carbon, etc. Such materials are suitable to carriers which are easy to be fluidized and to gather biofilm on themselves, and not easily carried over from the bed.

High attachability of microbial film on the media is desirable in principle, though it sometimes makes trouble by clogging of filter beds by over-attachment of microbial film, or in exhibiting difficulty in sloughing off microbial film during backwashing. Factors affecting the attachability include physical factors as shape and macroscopic surface roughness, physicochemical factors such as microscopic structure, elementary composition, electrical charge, hydrophilicity and material of media surface, and water in contact with the media. The outline of the effects of these factors were mentioned in Chapter 1. Studies on suitable materials of filter medium from the viewpoint of biofilm attachability so far are very few, and this problem remains an important one to be solved in the future.

Among factors associated with hydraulic characteristics of filter medium are porosity, specific surface area, shape and size, and packing ratio. The porosity affects directly the actual retention time and the ultimate amount of biomass held in the reactor, and the head loss through the filter decreases with increasing porosity of the medium. Moreover, higher porosity reduces the quantity of material used for the manufacturing of unit volume of filter medium, resulting in decrease in cost, so that the higher the porosity of a medium, the better its quality as far as mechanical strength and specific surface area are desirable.

The specific surface area controls the amount of microbial film which is able to be held in a unit volume of filter bed, and so it has a direct and most important influence on the function of the filter medium. Harremoës [13] pointed out that the half-order reaction kinetics were able to be applied to the denitrification rate in anaerobic filters, depending upon the data obtained by a number of experimental investigations using filter media of various shapes and sizes; he also determined half-order rate constants based on volume of filter media, $k_{1/2v}$ and ones based on surface area of filter media, $k_{1/2a}$. The result is shown in Table 2.4. The values of $k_{1/2v}$ showed a wide spread distribution but the distribution of $k_{1/2a}$ was in a much more narrow range, suggesting that the controlling factor of filter medium function was specific surface area. A larger specific surface area probably causes a correspondingly larger frequency of collision between the filter medium and the suspended particles in the bulk water, resulting in a higher efficiency of removing suspended solids by the medium.

On the other hand, the larger the specific surface area of a filter medium, the larger will be the head loss and thus the energy required to pass water at a certain velocity through the filter bed.

Besides, a filter bed packed with a medium possessing a larger specific surface area is more susceptible to clogging, and increasing the frequency of backwashing to prevent filter clogging is apt to cause the washout of protozoa and micro-metazoa, resulting in the shortening of the food chain and an increment of excess sludge production. Thus, the most important merit of microbial processes will be diminished.

The head loss through a submerged filter seems to be simulated as the frictional loss through round pipes, and is given by the following formula by Darcy and Weisbach

$$h_L = f \cdot \frac{L}{D} \cdot \frac{v^2}{2g} \qquad (2.73)$$

where

h_L = frictional loss of head
f = coefficient of frictional loss
D = piezometric depth
L = length of pipes
v = flow velocity through pipes
g = gravitational acceleration

When the Reynolds number (Re) is not more than 2000, the flow through the filter bed is a laminar flow, and the relation between f and Re is given by the following equation, independent of the roughness of pipe wall.

$$f = 64/Re \qquad R = vD/v \qquad (2.74)$$

where v = kinematic viscosity

TABLE 2.4. Data on Half-Order Volume and Surface Removal Rates for Denitrification Filters.

Type of Medium	Medium Size (mm)	Porosity	Filter Diam. (cm)	$k_{1/2}v$ mg$^{1/2}$/1$^{1/2}$ min × 10^{-3}	ω dm^2/dm^3	$k_{1/2}a$ mg$^{1/2}$/dm$^{1/2}$ min × 10^{-3}	Temperature (°C)
Activated carbon	0.65	–	7.5	1,400	–	–	26
Sand	0.60	–	46	12,000	–	–	23
Gravel, crushed	4.0	0.45	18.6	250–900	112	2.5–8.0	6–19
Gravel, round	3.5	0.39	18.6	250–1,000	130	1.8–7.4	6–19
Ceramic saddle	13	0.73	12.5	80–140	62	1.3–2.2	20
Sand	3–4	0.40	180	300	82.5	3.7	26
	15 × 15	0.92	300	90	34.6	2.6	27
Stones	25	0.40	45	22–57	13.9	1.6–4.1	12–22
Flexrings	25 × 25	0.96	45	6.9–37	21.4	0.3–1.7	10–26
Gravel	3.4	0.28	10	290	80	10.4	27
Stones	5.9	0.34	10	520	58	9.0	27
Stones	14.5	0.37	10	570	28	7.1	27
Glass beads	3	–	3	300	67	>4	5–30
Plexiglass cylinders	16–28	0.80	10	90–230	26	4–9	25

Substituting Equation (2.74) into Equation (2.73)

$$h_L = \frac{32\nu Lv}{gD^2} \tag{2.75}$$

is obtained. Because the hydraulic radius, D is equal to the cross-sectional area divided by the wetted perimeter, D can be shown as

$$D = \frac{\text{the cross-sectional area of flow}}{\text{the wetted perimeter}} \cdot \frac{\text{the flow length}}{\text{the flow length}}$$

$$= \frac{\text{total volume of pore within the filter bed}}{\text{total surface area within the filter bed}}$$

and if we consider unit volume of the filter bed, D is represented as

$$D = P_s/S_s \tag{2.76}$$

where

P_s = porosity of filter medium
S_s = specific surface area of filter media

Substituting Equation (2.76) into Equation (2.75), we obtain

$$h_L = \frac{32\nu LvS_s^2}{gP_s^2} \tag{2.77}$$

Energy loss ϵ_L per unit time by passing water through a filter is proportional to the product of v and h_L, so that

$$\epsilon_L \propto vh_L = \frac{32\nu L(vS_s)^2}{gP_s^2} \tag{2.78}$$

Equation (2.78) is applicable to a laminar flow, and the flow through a submerged filter is usually laminar flow. The flow velocity through a submerged filter, therefore, is inversely proportional to the specific surface area, and consequently Reynolds number, *Re,* is inversely proportional to the square of the specific surface area, under a fixed ϵ_L, i.e., under a fixed driving force. That is, the efficiency of the contact between water and biofilm abruptly decrease when the specific surface area increases (refer to Chapter 1, section 1.4).

Hence, the specific surface area of filter medium used for a conventional fixed bed submerged filter does not usually exceed 100 m²/m³, and those for a biofilm filtration and a expanded (or a fluidized) bed should be some hundreds m²/m³ and some thousands m²/m³, respectively.

Two or three stages of submerged filters are often used in series. In such case, the volume of the first stage should be biggest, and the volume should be reduced as the stage number increases to equalize volumetric loading of pollutants on each stage as well as possible. Even so, the former stage tends to be clogged more easily, and the coarser medium should be used for the former stage.

No investigation recognized definitely the effect of medium shape on the purification function.

For all media packed in a filter to work effectively, the flow velocity throughout the filter needs to be as uniform as possible. Practically, the flow velocity distribution is shown in Figure 2.23 and the flow direction is often upward in the neighborhood of the aeration space, indicating that a recirculation flow is formed within the cross section of the filter bed. Accordingly, the flow velocity in the central zone is extremely small, and a dead

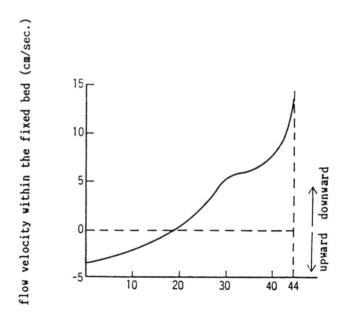

FIGURE 2.23. Flow velocity profile in an AFBSF with spiral flow packed with honeycomb tubing (filter width 44 cm).

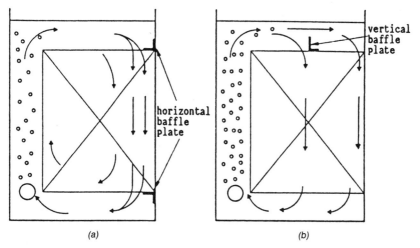

FIGURE 2.24. *Improvement of flow pattern by baffle plates (equalization of flow velocity): (a) horizontal baffle plate and (b) vertical baffle plate.*

zone is formed in the neighborhood in most cases. To improve flow velocity distribution, a parabolic shape of the top of the filter bed was proposed as a trial, but it is not so practical. A somewhat widespread method is to provide baffle walls on the top of the filter as shown in Figure 2.24.

Organic Loading

The capacity of a submerged filter is considered to increase by increasing the specific surface area of filter media, not only theoretically but also by empirical evidence associated with "the properties of filter media" in this paragraph. Therefore, the organic loading on a filter should fundamentally be determined based on a valid value of organic surface loading on the medium, but actually it is determined based on the volumetric loading in most cases. Though such design procedure appears to be invalid theoretically, it is a practical way indeed. The reason: If a high volumetric loading is applied on the bases of a high specific surface area of the medium employed, the filter bed will be clogged rapidly; and to maintain the designed value of volumetric loading will be impossible in a short period of time because of overaccumulation of biomass. Thus, in the determination of filter loading, such conditions should be maintained that will enable the minimizing of the clogging rate of the filter, with priority to the original capacity of the filter.

For example, when a conventional fixed bed submerged filter is applied to domestic sewage, it is safe to set the volumetric organic loading not more than 0.5 kg-BOD/m³-medium·day, and a specific surface area of 100 m²/m³ is sufficient under such a moderate value of loading.

It is the biofilm filtration process which has been developed in order to enable the application of several times larger volumetric loading than the conventional fixed submerged filter. For this purpose, filter media with a specific surface area of 800–1000 m²/m³ are used, corresponding with the higher value of volumetric loading, and the filter media are backwashed with air and water at the frequency of once in one to three days. For expanded or fluidized bed reactors, media with more than several times of specific surface area are used, and the filter constitution itself is considered not to be easily clogged.

As mentioned above, the optimum organic loading is determined in relationship to the filter bed clogging, and filter media in correspondence with the loading is selected. Therefore, if such a device were developed which could wash a filter bed easily, it would contribute very much to the compactness of submerged filters. Moreover, in the treatment of wastewaters containing organics with small yield coefficient, larger loading than usual is available, making submerged filters especially suitable for removing such organics.

The guideline for volumetric loadings for submerged filters are listed on Table 2.5.

Aeration Device

The concentration of dissolved oxygen in a bioreactor using suspended biomass like activated sludge is thought commonly to be sufficient only if a fairly low level of dissolved oxygen is detected in the reactor, because microbial oxygen ingestion rate is almost independent of dissolved oxygen,

TABLE 2.5. BOD Loading Range for
Submerged Biofilter.

Type of Submerged Bio-Filter	Volumetric BOD Loading (kg/m³·day)
Conventional fixed bed (coarse medium)	0.1–1.0
Biofilm filtration (fine medium)	0.1–5.0
Fluidized bed (fine medium)	1.0–10

and the presence of a trace level of dissolved oxygen indicates that the rate of oxygen supply is equal to its consumption.

On the other hand, concerning microbial film processes, the concentration of dissolved oxygen in the bulk water affects the depth of aerobic layer, i.e., the depth of the effective layer, and consequently the effectiveness factor of microbial film (Chapter 1, section 1.4). That is, the concentration of dissolved oxygen does have an effect on the apparent activity of microbes. By way of example, Haug et al. [14] reported that fairly high nitrification is attainable if the influent into a submerged filter is preoxygenated with pure oxygen to dissolve a high concentration of oxygen. On the contrary, the dissolved oxygen concentration should be as low as possible to keep the power efficiency high for aeration, because the efficiency of oxygen dissolution is proportional to the oxygen deficit in the bulk water. Taking into consideration both apparent microbial activity and power efficiency, the optimum dissolved oxygen concentration is estimated to be in the range 2–3 mg/l at ambient temperature.

There are two methods of supplying oxygen into submerged filters. One method is recycling preoxygenated water to the filter (Preoxygenation), and the other diffuses air bubbles directly into the filter (Bubble Oxygenation). Preoxygenation is much more popular, but Bubble Oxygenation has the advantages of high efficiencies of oxygen dissolution, liquid-solid contact on biofilm surfaces, and higher tolerability to filter clogging; though there is a disadvantage, in that a higher suspended solids concentration is apparent in the filter effluent. Particularly, the advantage of higher tolerability to filter clogging is of high importance, since filter clogging is prevented or at least delayed, by a kind of constant backwashing caused by diffused air bubbles, or that, even if the filter is clogged, the recovery of filter function is easy. McHarness et al. [15] compared Preoxygenation with Bubble Oxygenation in the nitrification of secondary effluent of municipal sewage, using a fixed bed submerged filter packed with quartz stone of 1–1.5 inch in diameter.

In the Preoxygenation system, only periodic drainage of the filter enables continuous operation for a long term with the retention time of 60 minutes, but additional backwashing (once in 4 to 20 days) is necessary when hydraulic retention time is lowered to 30–40 minutes, or the suspended solids concentration is higher than normal condition.

In the Bubble Oxygenation system, suspended solids were continuously discharged with the effluent in the turbulence of the rising air bubbles, and continued operation was possible for 6 months with drainage of once or twice a week.

The biofilm filtration process, too, is regarded as a Bubble Oxygenation system, and the remarkable merit of this process is its high efficiency of oxygen dissolution, owing to its long period of gas-liquid contact. The efficiency of oxygen dissolution—the ratio of absorbed oxygen to the amount contained in the air diffused into the filter—reaches as high as 15–20%, so that this process is highly energy-saving. Although no systematic investigation was found concerning the amount of oxygen required in a submerged bed, the required amount would probably seem to be fundamentally the same as that of the activated sludge process and represented by the following equation:

$$O_2 = a'L_r + b'\eta S_v \qquad (2.79)$$

where

O_2 = the amount of oxygen required per day (kg/day)
a' = oxidized ratio of removed BOD ($-$)
L_r = removed amount of BOD per day (kg/day)
b' = coefficient of endogeneous respiration (1/day)
η = effectiveness factor, the value defined by Equation (1.41), ($-$)
S_v = total amount of biomass retained in the bioreactor as microbial film (kg)

It should be noted that η in Equation (2.79) is not a fixed value, but varies, depending on the dissolved oxygen concentration in the bulk water by which the depth of aerobic layer in the microbial film is controlled. Moreover, it is not as easy as it is with activated sludge to determine the total amount of biomass, S_v retained in a submerged filter.

When it is difficult to estimate S_v or parameters in Equation (2.79), it is practical to determine the amount of oxygen which should be supplied. The mass-balance of oxygen in a submerged filter is expressed as follows.

$$\frac{dO}{dt} = \left(\frac{dO}{dt}\right)_A - \left(\frac{dO}{dt}\right)_R \qquad (2.80)$$

where

O = dissolved oxygen concentration
t = time elapsed
$(dO/dt)_A$ = dissolved oxygen concentration change with time by aeration
$(dO/dt)_R$ = dissolved oxygen concentration change with time by microbial reaction

In a steady-state, dO/dt being equal to zero,

$$\left(\frac{dO}{dt}\right)_A = \left(\frac{dO}{dt}\right)_R \qquad (2.81)$$

Using Equation (2.81), $(dO/dt)_R$ is determinable from the empirically determined value of $(dO/dt)_A$. $(dO/dt)_A$ is determined empirically from the difference of oxygen content between fresh air and exhausted gas, and the volume of air diffused per unit time.

If the overall mass transfer coefficient k_La is already known, $(dO/dt)_A$ is obtained by the following equation:

$$\left(\frac{dO}{dt}\right)_A = k_La\,(O_s - O) \tag{2.82}$$

where

O_s = saturated value of dissolved oxygen
O = concentration of dissolved oxygen in a steady-state

Note that k_La with microbial film in a submerged filter is not always equal to one without microbial film.

On the other hand, letting $(dO/dt)_A = 0$ in Equation (2.80), we obtain

$$\frac{dO}{dt} = -\left(\frac{dO}{dt}\right)_R \tag{2.83}$$

Therefore, from the dissolved oxygen concentration change in a short period after the stoppage of aeration, $(dO/dt)_R$ can be determined directly. The fault in this method is that $(dO/dt)_R$ varies depending upon the flow pattern in the filter which is affected by aeration.

Sludge Production

It has been repeatedly pointed out that microbial film processes, including submerged filter processes, produce relatively small amounts of sludge because of the lengthened food chains in them. This is commonly recognized to be a very important advantage in a biological treatment process. For example, Iwai et al. [16] gained 0.18 kg-VSS/kg-BOD removed (1/3–1/4 of activated sludge process) as the ratio of volatile suspended solids produced per unit BOD removal at a volumetric loading of 0.47 kg/m³·day, in an experimental treatment of municipal sewage using a three-staged fixed bed aerated submerged filter. Moreover, the volatile suspended solids content of sludge withdrawn from the first, the second and the third stage of filter was 53.3, 38.2 and 34.4%, respectively, and these low values distinctly exhibit an extremely high degree of mineralization of sludge which seems to be higher in the latter stages. Though it has been pointed out that the higher the organic loading is on a filter, the larger the sludge production ratio will be. No quantitable discussion is possible because of the lack of data accumulation.

2.2.2 The Rotating Biological Contactor

A rotating biological disk (RBD), or a rotating biological contactor means a number of slowly rotating thin disks fixed to the center shaft at regular intervals and positioned so as to be submerged in a semi-cylindrical tank about 40%, as depicted in Figure 2.25. The rotating disks come into contact with water and air alternately, and absorptions of pollutant and oxygen are repeated correspondingly by microbial films on both sides of the disks. The disks are usually rotated mechanically, but there are some types of rotating disks which utilize the buoyancy of air, as in Figure 2.26. Generally, they are rotated with a slow peripheral velocity not exceeding 20 m/min.

2.2.2.1 THE ROTATING DISKS

The diameter of the disk ranges from 0.5 to 5 m, mostly from 2.0 to 3.6. The thickness varies widely with the material of the disk and ranges from 1 to 25 mm. The materials widely used are polystyrene foam, polyvinyl chloride and polyethylene, etc. The surface of the disk is either flat or corrugated. From some tens to some hundreds of such disks are fixed to one center shaft in regular intervals of 20–30 mm to make a rotating unit. The standard of submergence ratio of disks is 40%, which should be increased as the concentration of wastewater decreases. Rotating biological contactors can also be used for anaerobic treatment (anaerobic digestion or denitrification) with the entire disk unit submerged. The standard of peripheral velocity is 10–20 m, which corresponds to the revolution speed of 3–6 rpm

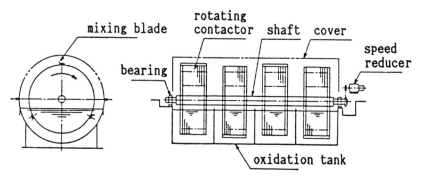

FIGURE 2.25. *An example of constitution of rotating biological contactor.*

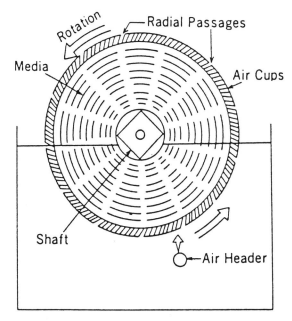

FIGURE 2.26. *Air driven rotating biological contactor.*

for disks of 1 m in diameter, and 1-2 rpm for one of 3 m in diameter, respectively. The speed of revolution should be adjusted to the strength of the wastewater to be treated.

2.2.2.2 THE CONTACT TANK

The adequate space between the perimeter of disks and the bottom of the tank is about 10% of the disk diameter. The tank volume can also be determined based on the liquid volume-disk area ratio, G, i.e., the volume of the tank divided by the total area of disk surface.

Generally, G should be larger than 5 l/m², and G of 5-9 l/m² is widely used.

2.2.2.3 THE DESIGN FACTORS OF THE ROTATING BIOLOGICAL CONTACTOR

BOD Surface Loading

BOD surface loading is equal to grams of BOD input into the unit per day divided by the total surface area of the disks. The smaller this design factor is, the higher the BOD removal will be.

Hydraulic Loading

Hydraulic loading means the flow rate of wastewater per unit surface area, and is expressed by

$$L_H = (Q/A) \cdot 10^3 \qquad (2.84)$$

where

L_H = hydraulic loading (l/m² · day)
A = total surface area of disks (m²)

Disk Intervals

The disk interval should be increased as the strength of inflow wastewater raises, because of the increment of the thickness of microbial film. In multistaged rotating biological contactor units, the center-to-center spacing between disks should be 25–35 mm for the first stage, and 10–20 mm for latter stages.

Stage Number

Because each stage of rotating biological contactors can be regarded as a CSTR, the flow pattern through a multistage unit approaches that of a plug-flow as the stage number increases.

For conventional secondary treatment of municipal sewage, four staged units are often used, and a similar number of stages are added for nitrification or denitrification.

Other Factors

Other factors, such as submergence ratio of disks, rpm, direction of inflow (parallel or perpendicular to disks) and temperature, etc., are the factors affecting the efficiency of rotating biological contactors.

2.2.2.4 THE DESIGN FORMULAS OF THE ROTATING BIOLOGICAL CONTACTOR

Pöpel's Formula

Pöpel [17] has proposed the following formula for determining the required surface area of rotating disks:

$$F = f(F/F_w) \cdot f(\eta) \cdot f(t) \cdot f(T) \cdot Q \cdot O_{bz} \qquad (2.85)$$

where

F = required surface area of disks (m²)
$f(F/F_w)$ = ratio of total area to submerged area (−)
$f(\eta)$ = a function of BOD removal, given by Equation (2.86)
$f(t)$ = a function of contact time t (hr), given by Equation (2.87)
$f(T)$ = a function of water temperature T (°C), given by Equation (2.88)
Q = treated flow rate (m³/day)
O_{bz} = concentration of influent BOD (mg/l)

$$f(\eta) = 0.01673\ \eta^{1.4}/(1 - \eta)^{0.4} \qquad (2.86)$$

$$f(t) = 1 - 1.24 \cdot 10^{-0.114t} \qquad (2.87)$$

$$f(T) = 10^{-0.02T} + 1.713 \cdot 10^{-0.1T} + 0.35 \cdot 10^{-4} \cdot 10^{0.1T} \quad (2.88)$$

Ishiguro's Formula [18]

$$F_w = \frac{4.63 \cdot 10^{-3} \cdot L_0 \cdot q \cdot \eta^{2.19}}{(1 - \eta)^{1.19}} \qquad (2.89)$$

where

F_w = required submerged area of disks (m²)
L_0 = influent BOD (mg/l)
q = flow rate of wastewater (m³/day)

In addition to process design based on formulas as shown above, the methods based on hydraulic loading [19], or on microbial film reaction theory [20], and the like have also been proposed.

2.2.2.5 SLUDGE PRODUCTION

For municipal sewage treatment, it is reported that sludge production by rotating biological contactors is only 0.04–0.08% of treated sewage or 0.06 kg/m³ on a dry basis, while the activated sludge process is generally thought to produce a little less than 2% of sludge. The U.S. EPA has shown, based upon the report by Antonie et al. (Figure 2.27) the relation between BOD removal and sludge production per unit BOD removed in rotating biological contactors. As might be expected, sludge production had a tendency to decrease as BOD removal was raised, and was mostly lower than 0.7 kg-SS/kg-BOD removed, where BOD removal was 80% or more.

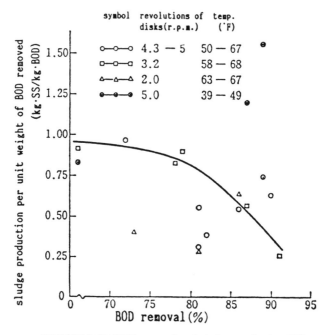

FIGURE 2.27. *BOD removal and sludge production [21].*

2.2.3 The Trickling Filter

2.2.3.1 CLASSIFICATION

Trickling filters are classified into four groups depending on loading, as in Table 2.6. The low rate filter (conventional filter) does not incorporate effluent recycling, and is used at the lowest loading, offering a stable quality of effluent nitrified to a high degree. Figure 2.28 shows the cross section of a conventional trickling filter and loading rate increases ranging from moderate to high to ultrahigh, with the stability of the effluent decreasing in the same order, with less nitrification. In all of these three types of trickling filter, the effluent is recycled to raise the performance.

2.2.3.2 CONFIGURATION

Various kinds of flow diagrams are shown in Figure 2.29 for municipal sewage treatment using trickling filters. Any of the trickling filters in these flow diagrams are high rate filters employing effluent recycling. The benefits of recycling are generally thought to be flow equalization, improved dis-

TABLE 2.6. *Classification of Trickling Filters and the Ranges of Loading.*

	Low Rate Trickling Filter	Moderate Rate Trickling Filter	High Rate Trickling Filter	Super-High Rate Trickling Filter
Hydraulic loading* ($m^3/m^2 \cdot day$)	1–4	4–10	10–30	30–80
BOD loading** ($kg/m^3 \cdot day$)	0.1–0.4	0.2–0.5	0.5–1.0	0.5–2

*Including recycle flow.
**Not including BOD of recycle flow.

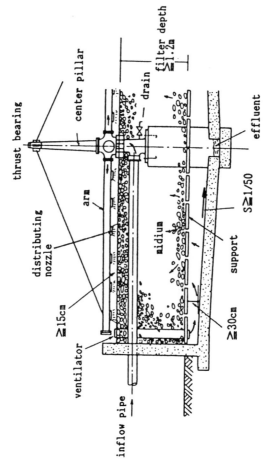

FIGURE 2.28. Cross section of a conventional trickling filter.

thrust bearing

center pillar

distributing nozzle

≧15cm

arm

ventilator

inflow pipe

drain

medium

support

≧30cm

filter depth ≧1.2m

effluent

S≧1/50

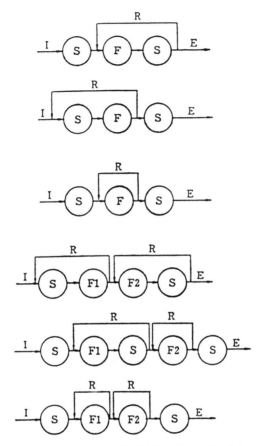

FIGURE 2.29. *Typical constitutions of trickling filter: I = influent, E = effluent, R = recycled water, S = settling basin, F = trickling filter, F_1 and F_2 = 1st and 2nd stages of trickling filter.*

tribution of flow over the media, improved filter maintenance, prevention of the growth of pychoda files, and improved organic removals. Recycling ratios are in the range of 1 to 5, and it should be noted that recycling effluent effects the settling basins.

2.2.3.3 DESIGN FORMULAS

BOD change with time through a trickling filter is commonly approximated by first-order reaction kinetics.

$$L_e/L_0 = \exp(-Kt) \qquad (2.90)$$

where

L_e = effluent BOD
L_0 = influent BOD
t = contact time
K = specific BOD removal rate

Contact time is often represented in the following formula:

$$t = \frac{c'Z^n}{(Q/A)^m} \tag{2.91}$$

where

m,n,c' = constants depending upon the properties of microbial film and wastewater
Z = filter depth
Q = flow rate
A = the cross-sectional area of the trickling filter

Eckenfelder [22] has suggested using Equation (2.92)

$$L/L_0 = \exp(-KA_v^m D/q^n) \tag{2.92}$$

where

L = BOD at the depth D
A_v = the specific surface area of media
q = the hydraulic loading, i.e., flow rate per unit cross-sectional area

BOD change is given by Equation (2.92) for trickling filters employing effluent recycling.

$$\frac{L}{L_0} = \frac{\exp[-KA_v D/(1 + N)^n q^n]}{(1 + N) + N \exp[-KA_v D(1 + N)^n q^n]} \tag{2.93}$$

where

N = the ratio of recycling
q = hydraulic loading without recycling

The National Research Council (NRC) has suggested the following empirical formula for BOD removal by the first stage of trickling filter [23].

$$E_1 = \frac{1}{1 + 0.0085 \sqrt{W/VF}} \tag{2.94}$$

where

E_1 = BOD removal by single or first stage of trickling filter, including the effects of recycling and settling
W = BOD loading on the filter (lb/day)
V = volume of media (acre/ft)
F = coefficient of recycling

The coefficient of recycling can be determined by the following formula

$$F = \frac{1 + R}{(1 + R/10)^2} \tag{2.95}$$

where R is the ratio of recycling Q_r/Q (Q_r is recycling flow rate).
BOD removal by second stage, E_2 is

$$E_2 = \frac{1}{1 + \dfrac{0.0085}{1 - E_1} \sqrt{\dfrac{W'}{VF}}} \tag{2.96}$$

where W' is BOD loading on second stage (lb/day)
The influence of water temperature on BOD removal rate is given by Reference [24]

$$K_T = K_{20}(1.035)^{T-20} \tag{2.97}$$

where K_T, K_{20} = specific BOD removal rates at temperature $T\,°C$ and $20°C$, respectively.
The reason for the small temperature coefficient for BOD removals by microbial processes was already explained in Chapter 1.

2.3 REFERENCES

1. Hinshelwood, C. N. 1946. "Chemical Kinetics of the Bacterial Cell," Oxford: Clarendon Press.

2. Danckwerts, P. V. 1953. *Chem. Eng. Sci.*, 2:1.

3. Wenner, J. E. and R. H. Wilhelm. 1956. "Boundary Conditions of Flow Reactors," *Chem. Eng. Sci.*, 6:89.

4. Cooper, P. F. and S. C. Williams. 1990. "High-Rate Nitrification in a Biological Fluidized Bed," *Wat. Sci. Tech.*, 22(1/2):431–442.

5. Jeris, J. S. et al. 1977. "Biological Fluidized-Bed Treatment for BOD & Nitrogen Removal," *Jr. WPCF*, 49(5):816–831.

6. Henze, M. and P. Harremoes. 1983. "Anaerobic Treatment of Wastewater in Fixed Film Reactors—A Literature Review," *Wat. Sci. Tech.*, 15(8/9):1–101.

7. Carrondo, M. J. T. et al. 1983. "Anaerobic Filter Treatment of Molasses Fermentation Wastewater," *Wat. Sci. Tech.*, 15:117.

8. Jeris, J. S. 1983. "Industrial Wastewater Treatment Using Anaerobic Fluidized Bed Reactors," *Wat. Sci. Tech.*, 15:169.

9. Sutton, P. M. et al. 1981. "Anitron and Oxitron Systems: High-Rate Anaerobic and Aerobic Biological Treatment System for Industry," *Proc. 36th Ind. Waste Conf.*, Purdue Univ., p. 665.

10. Norrman, J. 1983. "Anaerobic Treatment of a Black Liquor Evaporator Condensate from a Kraft Mill in Three Types of Fixed-Film Reactors," *Wat. Sci. Tech.*, 15:247.

11. Jewell, W. J. et al. 1981. "Municipal Wastewater Treatment with the Anaerobic Attached Microbial Film Expanded Bed Process," *J. Wat. Pollut. Control Fed.*, 53:482.

12. Kitao, T. et al. 1986. "Anaerobic Submerged Biofilter for the Treatment of Domestic Wastewater," *Proc. EWPCA Wat. Trea't. Conf.*, p. 736.

13. Harremoes, P. 1976. "The Significance of Pore Diffusion to Filter Denitrification," *Jr. WPCF*, 48(2):377–388.

14. Haug, R. T. and P. L. McCarty. 1972. "Nitrification with Submerged Filters," *Jr. WPCF*, 44(11):2086–2102.

15. McHarness, D. D. et al. 1975. "Field Studies of Nitrification with Submerged Filters," *Jr. WPCF*, 47(2):291–309.

16. Iwai, S., H. Ohmori and T. Tanaka. 1977. "An Advanced Sewage Treatment Process-Combination of Submerged Biological Filtration and Ultra-Filtration with Pulverized Activated Carbon," *Proc. Int'l. Cong. on Desalination and Water Reuse, Vol. II*, Tokyo, pp. 29–36.

17. Pöpel, F. 1964. "Leistung, Berechnung und Gestaltung von Tauchtropfkorperanlagen," *Stuttgarter Berichte zur Siedlungabwasserwritschaft*, p. 11.

18. Ishiguro, M. et al. 1975. "A Study on the Advanced Treatment of Sewage by Rotating Bio-Disk Unit, Part 1—Evaluation of Treatment Characteristics," *Jr. Japan Sewage Wks. Assoc.*, 12(129):46–54.

19. Antonie, R. L. 1975. *Fixed Biological Surfaces—Wastewater Treatment.* CRC Press.

20. Watanabe, Y. 1985. "Mathematical Modelling of Nitrification and Denitrification in Rotating Biological Contactors," in *Mathematical Models in Biological Waste Water Treatment*, S. E. Jorgensen and M. J. Gromiec, eds., Elsevier, pp. 419–471.

21. U.S. EPA. 1971. "Application of Rotating Disc Process to Municipal Wastewater Treatment," *Water Poll. Res.*, Series 17050.

22. Eckenfelder, W. W., Jr. 1966. *Biotech. Bioeng.*, 8:389.

23. National Research Council, "Sewage Treatment at Military Installations," *Sewage Wks. Jr.*, 8(5):787.

24. Howland, W. E. 1953. *Sewage and Ind. Wastes*, 25(2):161.

CHAPTER 3

Wastewater Treatment Systems with Microbial Film Processes

3.1 INTRODUCTION

The characteristics and the purification kinetics of microbial film processes were discussed, and their effective usages were pointed out based on such discussion in Chapter 1. In Chapter 2, a wide variety of constitutions and configurations of microbial film processes were presented. This wide variety of microbial film treatment facilities have their advantages and disadvantages, so that it is very important to apply microbial film processes to those uses in which the advantages might be used, and the disadvantages minimized. Therefore, it will be able to offer a very profitable treatment system for a variety of purposes of wastewater treatment, if the optimum process or the optimum process combination is successfully selected from a wide variety of microbial film processes on the basis of a profound knowledge of the fundamental characteristics of them.

The purposes of this chapter are to show the essentials and the design data of microbial film process application and to cultivate a better understanding of the fundamentals of the processes, at the same time, through the presentation of many examples of appropriate applications of them.

3.2 THE PROCESSES USING AEROBIC FIXED BED FILTERS

3.2.1 The Processes Using Aerobic Fixed Bed Submerged Filters

Aerobic fixed bed submerged filters (AFBSFs) are the most basic type of submerged filters. The range of use of AFBSFs, therefore, is widespread, including both the same uses as the activated sludge process and further purification processes for wastewaters pretreated with some processes.

That is, the lower the concentration of the wastewater, the easier it can be treated with a microbial film process, because the main factor of microbial film process design is organic loading. AFBSFs are frequently used solely or added to anaerobic processes or activated sludge process for polishing operations or advanced treatment.

3.2.1.1 MUNICIPAL SEWAGE TREATMENT WITH AFBSFs

AFBSFs are now widely used for comparatively small municipal sewage treatment plants because they meet the three main demands of small scale treatment plants: high tolerability to influent quantity and quality fluctuations, ease of maintenance and small sludge production. Moreover, they enable higher quality of effluent as compared with trickling filters. For example, about 99% of domestic sewage treatment plants with flow rates not more than 50 m³/day and about 70% of those with somewhat larger flow rates, now employ this process in Japan. They are also the most effective method for polishing operations for secondary effluents of domestic sewage.

Secondary Treatment of Municipal Sewage

Hoshino et al. [1], the members of the Japan Sewage Works Agency, carried out experimental investigations on secondary treatment of municipal sewage using pilot plants of two-staged AFBSF, using as filter media honeycomb tubings of polyvinyl chloride with a cell-size of 20 mm (HC) and laces with numerous radial fiber rings (D.B. lace or Ring lace). The laces were stretched vertically in regular intervals of 40 mm and 50 mm in the first stage and the second stage of filter, respectively. The experimental treatments were conducted evenly in high and low water temperature periods. A part of the conditions and the results of treatment are shown in Tables 3.1(a)–(c), showing that BOD of effluents from both the first and the second stage of AFBSF were lower than 10 mg/l with either of the two kinds of media, and that the results obtained using HC were a little better, though the difference of performance between two kinds of media was little. Concerning the maintenance, however, it was necessary to periodically brush the upper portion of packed media to remove algae and the like attached when HCs were used as packing media, while no maintenance was necessary for the media itself for the units packed with laces, for as long as eighteen weeks of experiment. Thus, volumetric loading must be restricted on those filter media which are susceptible to clogging, no matter how excellent the performance is, so that they cannot be regarded as useful media.

TABLE 3.1a. Results of Sewage Treatment Using Two-Stages of AFBSFs—Packing Media/Honeycomb Tube HC (Run A).

Run	A – 1			A – 2			A – 3		
Sample	Influent	Effluent		Influent	Effluent		Influent	Effluent	
Index		1st Stage	2nd Stage		1st Stage	2nd Stage		1st Stage	2nd Stage
Temp., °C	–	24–25	–	–	25–8	–	–	20–27	–
T-BOD mg/l	69.5	9.0	3.0	87.6	4.2	2.6	86.5	8.0	4.3
S-BOD mg/l	19.5	5.0	1.5	13.6	2.0	1.0	17.1	2.9	1.5
T-COD$_{Mn}$ mg/l	43.0	14.5	9.0	53.6	11.4	9.4	58.5	16.8	12.1
S-COD$_{Mn}$ mg/l	15.5	11.5	8.0	15.8	10.2	8.4	18.4	12.9	10.0
SS mg/l	92.0	6.5	2.0	121.2	2.8	2.2	108.4	10.8	5.6
T-K-N mg/l	21.1	16.8	1.6	23.4	4.1	1.4	24.1	8.7	2.1
S-K-N mg/l	13.6	15.0	1.2	14.5	3.7	1.1	16.3	8.0	1.7
NH$_4^+$-N mg/l	12.3	14.4	0.4	13.1	2.3	0.07	14.7	6.5	0.28
NO$_2$-N mg/l	0.01	0.22	0.32	0.04	0.39	0.05	0.08	0.42	0.06
NO$_3$-N mg/l	0.09	0.41	12.9	0.12	7.0	8.6	0.28	9.3	15.8
T-N mg/l	21.20	17.43	14.82	23.56	11.49	10.05	24.46	18.42	17.96
T-P mg/l	3.4	2.5	2.4	5.2	3.0	3.3	5.2	3.5	3.4
Alkalinity mg/l	135	142	46	142	72	55	149	82	36
Transparency	–	58–66	100	–	58–98	83–100	–	19–46	37–96
pH	7.3–7.4	7.7	7.3–7.5	7.0–7.5	7.1–7.5	7.1–7.6	7.2–7.5	7.1–7.7	6.9–7.3

137

TABLE 3.1b. Results of Sewage Treatment Using Two-Stages of AFBSFs—Packing Media/Rope-Wise Material (Ring Lace or D.B. Lace, Run B).

Run	B – 1			B – 2			B – 3		
Sample	Influent	Effluent		Influent	Effluent		Influent	Effluent	
Index		1st Stage	2nd Stage		1st Stage	2nd Stage		1st Stage	2nd Stage
Temp., °C	—	24–25	—	—	25–28	—	—	20–27	—
T-BOD mg/l	69.5	8.0	5.5	87.6	6.8	4.6	86.5	10.5	5.4
S-BOD mg/l	19.5	3.5	2.0	13.6	4.6	1.2	17.1	5.1	2.0
T-COD$_{Mn}$ mg/l	43.0	13.0	11.0	53.6	14.4	12.6	58.5	19.3	14.1
S-COD$_{Mn}$ mg/l	15.5	10.5	9.5	15.8	11.8	10.8	18.4	15.1	11.8
SS mg/l	92.0	6.5	3.5	121.2	5.8	3.6	108.4	10.8	5.4
T-K-N mg/l	21.1	13.5	3.1	23.4	12.2	2.0	24.1	18.3	7.0
S-K-N mg/l	13.6	12.2	2.3	14.5	11.3	1.5	16.3	17.1	6.3
NH$_4^+$-N mg/l	12.3	11.3	1.4	13.1	10.5	0.3	14.7	15.1	4.7
NO$_2$-N mg/l	0.01	0.41	1.15	0.04	0.35	0.80	0.08	0.40	0.36
NO$_3$-N mg/l	0.09	0.87	7.8	0.12	1.4	13.3	0.28	1.26	11.1
T-N mg/l	21.20	14.78	12.05	23.56	13.95	16.10	24.46	19.96	18.46
T-P mg/l	3.4	2.1	2.0	5.2	3.1	3.0	5.2	3.4	3.2
Alkalinity mg/l	135	126	59	142	121	34	149	148	71
Transparency	—	62–64	89–90	—	57–73	—	—	25–46	37–60
pH	7.3–7.4	7.6	7.2–7.4	7.0–7.5	7.3–7.6	7.0–7.4	7.2–7.5	7.3–7.6	7.0–7.4

TABLE 3.1c. Results of Sewage Treatment Using Two-Stages of AFBSFs—Packing Media/Rope-Wise Material, Low Water Temperature.

Run	C			D		
Sample	Influent	Effluent		Influent	Effluent	
Index		1st Stage	2nd Stage		1st Stage	2nd Stage
Temp. °C	—	11–16	—	—	11–16	—
T-BOD mg/l	109.3	14.1	5.6	109.3	23.4	8.9
S-BOD mg/l	22.0	8.3	3.8	22.0	12.3	5.9
T-COD$_{Mn}$ mg/l	60.0	19.4	14.7	60.0	22.8	16.2
S-COD$_{Mn}$ mg/l	19.1	15.3	12.7	19.1	17.2	14.5
SS mg/l	122.0	10.2	3.2	122.0	15.7	3.8
T-K-N mg/l	27.2	20.8	14.2	27.2	31.3	19.0
S-K-N mg/l	18.4	19.4	13.5	18.4	19.7	18.1
NH$_4^+$-N mg/l	16.3	17.5	12.0	16.3	18.0	16.1
NO$_2$-N mg/l	0.08	0.15	0.45	0.08	0.18	0.54
NO$_3$-N mg/l	0.20	0.52	5.31	0.20	0.21	1.80
T-N mg/l	27.48	21.47	19.96	27.48	21.69	21.34
T-P mg/l	4.0	2.2	2.0	4.0	2.4	2.2
Alkalinity mg/l	153	155	119	153	159	149
Transparency	—	15–70	39–100	—	15–41	36–100
pH	6.9–7.4	7.1–7.6	6.8–7.7	6.9–7.4	7.0–7.6	7.1–7.7

When laces were used, the sludge production percentages determined as the ratios of produced weight of sludge to total weight of suspended solids inflow were a little higher than 50% and 60–70%, during the periods of high and low water temperature, respectively. It is possible that the reduction degree of produced sludge is affected by the biota in microbial film, and so by ambient temperatures.

Iwai et al. [2] used three stages of AFBSFs, each of them being packed with 4 m³ of modified shaped plastic tubings as the media for experimental treatment of 1.2 m³/hr (28.8 m³/day) of municipal sewage with BODs of 110–367 (average 197) mg/l. Accordingly, the volumetric loadings on overall three stages and on the first stage were 0.47 kg-BOD/m³·day and 1.4 kg-BOD/m³·day, respectively, and the total retention time was 10 hr. The effluent from AFBSFs was further treated with ultra-filtration after the addition of 1.9–2.0 wt % of pulverized activated carbon (activated carbon was used repeatedly). The result obtained is shown in Table 3.2, which is representative of excellent BOD removal and nitrification, suggesting some denitrification; this result is thought to highlight the feature of microbial film processes, that is, the effects of the coexistence of aerobic-anaerobic actions.

The quantity and quality of sludge production by this experiment were already touched upon in Chapter 2, section 2.2.1, the quantity being only 1/3–1/4 of that of activated processes, but further reduction of sludge production is expected if organic loading were smaller.

Because of the ease of maintenance and decreased production of excess sludge, AFBSF is particularly suitable to small scale plants for domestic sewage, and so almost 100% of on-site domestic sewage treatment facilities for individual houses in Japan, the smallest scale treatment facilities, are based on this process. As an example of investigation for an actual proof of the excellent performance of AFBSF as a small scale facility, Hisakawa et al. [3] surveyed three kinds of flow diagrams of on-site domestic sewage treatment facilities (Figure 3.1). The results are detailed in Table 3.3, and reflect the fact that the facilities using microbial film process were superior to those using the activated sludge process in both performance and ease of operation and maintenance.

For maintenance of on-site domestic sewage treatment facilities for individual houses, it is known that the required operations for maintenance are no more than backwashing of filters and sending sludge sloughed off from the filter to a primary settling tank once in several months, and withdrawal of scum and deposited sludge from a primary settling tank once a year. Another kind of small-scale facility using an anaerobic filter instead of a primary settling tank is also coming into wide use throughout Japan [4].

TABLE 3.2. *Qualities of Raw Sewage and Effluent from Each Treatment Step.*

	BOD$_5$ (mg/l)	Kjeldahl-N' (mg/l)	NO$_2$-N (mg/l)	NO$_3$-N (mg/l)	O-P (mg/l)	T-P (mg/l)
Sewage	110–370	20–30	Tr	Tr	—	4
Effluent from no. 2 ASB filter	8–10	8–11	0.2	3–4	1.6	1.8
Effluent from no. 3 ASB filter	3≥	3–5	Tr	7–9	1.4	1.7
Final effluent from U-filter	—	2–4	Tr	5–9	1.9	2.2

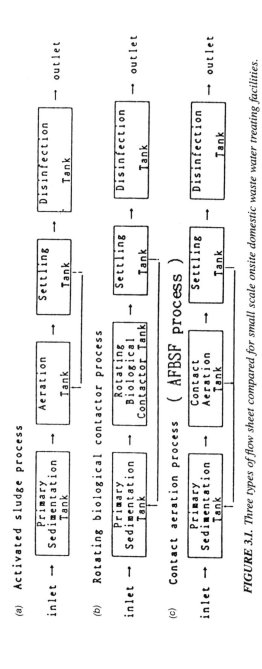

(a) Activated sludge process

inlet → Primary Sedimentation Tank → Aeration Tank → Settling Tank → Disinfection Tank → outlet

(b) Rotating biological contactor process

inlet → Primary Sedimentation Tank → Rotating Biological Contactor Tank → Settling Tank → Disinfection Tank → outlet

(c) Contact aeration process (AFBSF process)

inlet → Primary Sedimentation Tank → Contact Aeration Tank → Settling Tank → Disinfection Tank → outlet

FIGURE 3.1. Three types of flow sheet compared for small scale onsite domestic waste water treating facilities.

142

TABLE 3.3. Performance of Small Scale Domestic Wastewater Plants.

(a) Activated Sludge Process

Plant No.	Effective Volume of Tank			Number of Users (capita)	Quantity of Water (l/c·d)	BOD Loading (kg/m³·d)	Effluent BOD (mg/l)
	Primary Sedimentation	Aeration (m³)	Settling				
1	1.5	1.18	0.38	3	600	0.322	58 (2–208)
2	1.5	1.61	0.40	5	152	0.060	17 (2–73)

(continued)

143

TABLE 3.3. (continued).

(b) Rotating Biological Contactor Process

| Plant No. | Effective Volume of Tank | | | | Disk Area (m²) | Number of Users (capita) | Quantity of Water (l/c·d) | BOD Loading (g/m²·d) | Effluent BOD (mg/l) | Interval of Cleaning (days) |
| | Primary Sedimentation | | Contactor (m³) | Settling | | | | | | |
	1st	2nd								
1	1.16	0.56	0.13	0.65	23	3	222	2.6	14 (3–37)	514
2	1.66	0.56	0.13	0.65	23	3	192	3.1	10 (3–54)	548
3	1.66	0.56	0.13	0.65	23	4	150	3.2	13 (4–69)	612
4	2.27	0.73	0.34	0.88	53	6	188	3.8	11 (3–15)	465
5	2.27	0.73	0.34	0.88	53	5	388	5.6	14 (6–31)	450
6	2.27	0.73	0.34	0.88	53	4	233	1.0	7 (3–15)	442

TABLE 3.3. (continued).

(c) Contact Aeration (AFBSF) Process

Plant No.	Effective Volume of Tank				Number of Users (capita)	Quantity of Water (l/c·d)	BOD Loading (kg/m³·d)			Effluent BOD (mg/l)		Interval of	
	Primary Sedimen-tation	Aeration (m³)		Settling >			1st	2nd	Total	1st	2nd	Backwash (days)	Cleaning (days)
		1st	2nd										
1	2.00	1.02	—	—	5	223	0.106	—	—	10 (8–11)	—	43	365
2	1.62	0.72	—	0.20	6	248	0.408	—	—	13 (9–18)	—	43	270
3	1.62	0.72	—	0.20	5	246	0.398	—	—	17 (12–26)	—	43	270
4	2.12	1.57	1.06	0.34	5	232	0.043	.006	.025	7 (3–14)	5 (2–7)	511	536
5	2.05	1.57	1.11	0.27	5	230	0.065	.012	.038	20 (7–41)	13 (4–18)	—	411
6	2.05	1.57	1.11	0.27	7	406	0.164	.045	.097	16 (11–19)	7 (5–9)	950	216
7	2.01	1.50	0.99	0.29	6	203	0.118	.031	.071	25 (8–72)	8 (2–14)	840	426

145

Treatment of Gray Water for Reclamation

The authors et al. [5], experimentally investigated the treatment of office building drainage with a pilot-plant treatment system including AFBSFs, to produce reclamation water for toilet flushing, car washing, air conditioning and the like. The flow diagram of the treatment system consists of AFBSFs, moving bed filtration with chemical coagulation and activated carbon adsorption as in Figure 3.2, with design treating capacity of 24 m^3/day. The reasons for the utilization of AFBSF as a biological treatment process are the ease of maintenance, small sludge production, the ability to tolerate both quantity and quality fluctuation of raw wastewaters, and the capability of efficient treatment of wastewaters with relatively low strength with a compact facility as mentioned before.

The effective volumes of the 1st and 2nd stage of AFBSF were 4 m^3 each, and those of the 3rd and 4th, were 3 m^3 each. The pilot plant was operated at four steps of hydraulic loading, and the result as shown in Figure 3.3 was obtained as the BOD profiles in the treatment system. In spite of wide fluctuations of quantity and quality of raw wastewater, the qualitative variation range of the effluent from the 4th stage of AFBSF was extremely narrow, and the stability of the effluent quality increased further through filtration with chemical coagulation and carbon adsorption. The characteristic in organic removal by this treatment system was investigated with gel-chromatography. Figure 3.4 shows gel-chromatograms of effluents from each stage of the system, which shows that both TOC and color are composed of high molecular weight (small fraction No.) and low molecular weight (large fraction No.) compounds, and that about 40% of the high molecular weight organics and most of the low molecular weight organics were removed with AFBSFs. Color originating in both high molecular and low molecular weight organics was somewhat removed with AFBSFs, but it is reasonable that a significant removal efficiency cannot be expected with any biological treatment. High molecular weight fractions of both TOC and color were removed effectively with the coagulation-filtration process, and additional activated carbon adsorption removed middle and low molecular weight organic compounds almost completely. Consequently, AFBSF, a kind of biological process and coagulation filtration, a kind of physicochemical process worked by compensating each other, and thus the rationality of combining these processes was confirmed. In addition, Coliphages ϕx 174 were added into raw wastewater to investigate their behavior in the treatment system. The removal efficiency of them with 4 stages of AFBSF was about 90%, and much higher removal was obtained with the coagulation-filtration process. The result corresponds to the above mentioned results of organic compound removal characteristics, because viruses too, are a sort of high molecular weight organic compound.

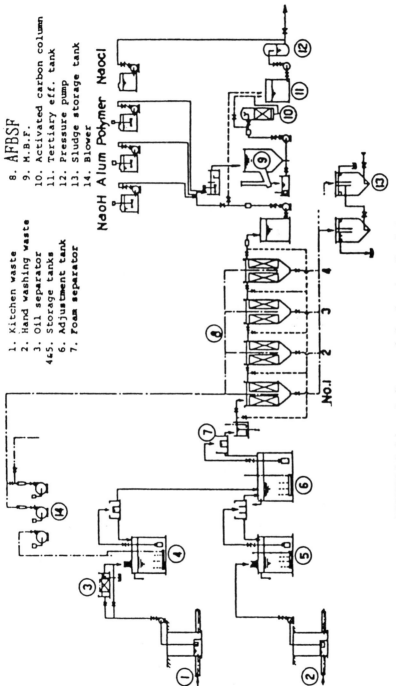

1. Kitchen waste
2. Hand washing waste
3. Oil separator
4&5. Storage tanks
6. Adjustment tank
7. Foam separator

8. AFBSF
9. M.B.F.
10. Activated carbon column
11. Tertiary eff. tank
12. Pressure pump
13. Sludge storage tank
14. Blower

NaoH Alum Polymer Naocl

FIGURE 3.2. Schematic diagram of pilot plant for office drainage reuse.

147

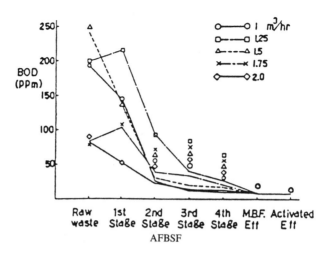

FIGURE 3.3. BOD profile in the office drainage treatment system.

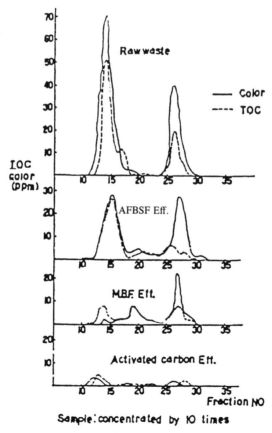

FIGURE 3.4. Gel-chromatogram of samples from the treatment system.

Polishing Operation or Tertiary Process for Municipal Sewage Treatment

Microbial film processes, aerobic submerged filters in particular, are capable of effective treatment of diluted organic wastewaters, and so AFBSF is useful in removing organic compounds and/or suspended solids in secondary effluent of municipal sewage to lower levels.

The authors et al. [6] treated the secondary effluent of domestic sewage from an extended aeration process, using two stages of AFBSF. Each filter was packed with modified shaped plastic tubing, and the total hydraulic retention times were 50 and 100 minutes. The results are shown in Table 3.4, representing significant degrees of organic compound removal and nitrification, but somewhat poor removal of suspended solids. pH decreased markedly as nitrification proceeded. The relatively poor performance on the whole at hydraulic retention time of 50 min was probably due to seasonal factors.

McHarness et al. [7] carried out an experimental research on the treatment of secondary effluent of municipal sewage using two AFBSFs of Plexiglas column (inner diameter 13.97 cm, height 1.13 m) and packed with quartz stones with diameter of 2.54–3.18 cm (filter depth 91.4 cm). Pure oxygen was used for the oxygen supply in order to elevate dissolved oxygen in the bulk water to increase the depth of effective layer of microbial film. The other systems used for comparing oxygen supply were preoxygenation in which the influent was oxygenated with pure oxygen before entering into AFBSF (Figure 3.5), and Bubble oxygenation where oxygen bubbles were directly introduced into the filter bed. The preoxygenation system required the recycling of the effluent when NH_3-N concentration in the influent was higher than 10 mg/l, and in addition, periodical drainage and backwashing of the filter were indispensable to prevent clogging. In the AFBSF unit using Bubble oxygenation system, suspended solids were continuously discharged in the effluent because of turbulence from rising bubbles, so that continuous operation for six months was possible with drainages only once or twice a week. The results of nitrification and removals of BOD, COD and SS are shown in Table 3.5(a)–(d), representative of an excellent performance at high loadings owing to pure oxygen aeration.

Adachi et al. [8] applied an AFBSF unit packed with honeycomb tubings with a specific surface area of 200 m^2/m^3, to secondary and tertiary treatments of municipal sewage, and obtained empirical formulas between BOD removal rates (kg-BOD/m^3·day) and BOD in the effluent, c (mg/l).

$$\left. \begin{array}{l} \text{secondary treatment : BOD removal rate } = 8.67 \ c^{1.28} \\ \text{tertiary treatment : BOD removal rate } = 2.44 \ c^{1.98} \end{array} \right\} \quad (3.1)$$

It may be a reasonable assumption that BOD in the filter can be approx-

TABLE 3.4. *AFBSF Performance for Secondary Effluent of Domestic Wastewater (total retention time 50 min BOD loading 0.84 kg/m³·d).*

	BOD (ppm)			COD$_{Mn}$ (ppm)			Total-N (ppm)			NH$_3$-N (ppm)		
	Avg.	Max.-Min.	Removal (%)	Avg.	Max.-Min.	Removal (%)	Avg.	Max.-Min.	Removal (%)	Avg.	Max.-Min.	Removal (%)
Influent	29			15.1	21–11		30.0	38.5–26.1		19.3	26.8–11.7	
1st stage	9.8	15.5–5.1	66.7	11.3	13–10	25	8.6	13.7–5.1	71.3	5.1	9.6–2.1	73.6
2nd stage	7.1	10.0–3.3	26.5 / 75.5	12.8	17–8	16.9	9.9	11.7–8.9	67	6.3	8.2–2.3	67.4

	NO$_2$-N + NO$_3$-N (ppm)		SS (ppm)		pH		Cl⁻(ppm)	Transparency	
Influent	4.0	11–1.4	24	31–13	7.1	7.3–6.8	59.4–49.4	11.8	21–6
1st stage	15.4	20.9–2.1	27	88–3	6.6	6.9–5.2	57.8–52.9	17.1	30–8
2nd stage	18.9	20.7–15.8	13	38–1	5.8	6.2–5.2	57.8–52.9	22.1	30–12

150

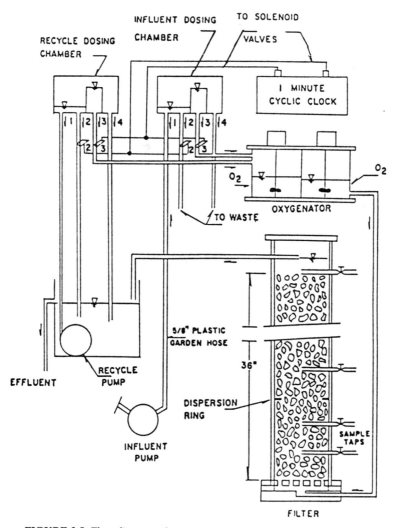

FIGURE 3.5. *Flow-diagram of preoxygenation recirculation AFBSF system.*

imated by BOD in the effluent. Therefore, it is reasonable, considered from the viewpoint of reaction kinetics in microbial film, that a higher dependency of BOD removal rate was observed on BOD concentration, c in tertiary treatment than in secondary treatment, or that BOD removal rate has a higher power of c in the formula of tertiary treatment than that of secondary treatment.

TABLE 3.5. Performance in Experimental Treatment of Secondary Effluent of Municipal Sewage Using AFBSF with Pure Oxygen Aeration.

(a) Nitrogen Compound Concentrations in Influent and Effluent

Detention Time (min)	Organic N (mg/l)		NH_3-N (mg/l)		NO_2^--N (mg/l)		NO_3^--N (mg/l)		Total N (mg/1)	
	Influent	Effluent	Influent	Effluent	Influent	Effluent	Influent	Effluent	Influent	Effluent
60	3.6	1.5	14.3	1.0	0.05	0.6	0.1	15.9	18.0	19.0
40	–	–	19.6	5.6	0.05	4.1	–	6.9	–	–
30	5.7	2.5	15.2	6.9	0.04	1.7	0.1	6.0	21.0	17.1

(b) BOD Removal

Detention Time (min)	Number of Samples	Influent BOD (mg/l)	Effluent BOD (mg/l)	BOD Removal (%)
60	4	35 ± 6	5 ± 3	86 ± 10
40	2	38	3.4	91
30	3	37	9.6	74

(c) COD Removal

Detention Time (min)	Number of Samples	Influent COD (mg/l)	Effluent COD (mg/l)	COD Removal (%)
60	4	104 ± 15	51 ± 12	51 ± 6
40	2	113	61	46
30	4	124 ± 10	54 ± 11	56 ± 2

(d) SS Removal

Detention Time (min)	Number of Samples	Influent SS (mg/l)	Effluent SS (mg/l)	SS Removal (%)
60	4	27 ± 3	3.5 ± 4	87 ± 6
40	3	38	6.6	83
30	6	25 ± 5	6.5 ± 3	74 ± 4

3.2.1.2 TREATMENT OF INDUSTRIAL WASTEWATER

Some industrial wastewaters contain pollutants which can be removed more profitably with microbial film processes than with the activated sludge process. The reason for the disadvantages of the activated sludge process can be summarized in the difficulty of accumulating high concentration of activated sludge with good settleability and a high content of microbes capable of decomposing such pollutants. The reasons for the difficulties are (1) small growth rate of the microbes because of low uptake rate and/or low growth yield of substrate, (2) inability of microbes responsible for removing substrate to form flocs with good settleability, (3) excessive extent of low substrate concentration in the raw wastewater. Such conditions occur more frequently in industrial wastewater treatment than in municipal sewage treatment, and hence microbial film processes have various uses for industrial waste treatment.

The authors et al. [9] investigated the decomposition characteristics of phenol, cyanide and thiocyanate which are the main components of coke plant wastewater, using a bench scale AFBSF packed with Intalox saddles. The relationship between the volumetric phenol loading and the removal of phenol are shown in Figure 3.6. Phenol was removed completely when the volumetric loading of phenol was less than 0.9 kg/m³·day (2.88 kg/m³·day as BOD loading). At a fixed phenolic volumetric loading of 0.9 kg/m³·day with the influent phenol concentration range of 28.5–1000 mg/l, the relationship between hydraulic retention time and removals of phenol, COD_{Mn} and BOD is shown in Figure 3.7. This represents a constant removal when influent phenol concentration was higher than 40 mg/l (corresponding to the hydraulic retention time of 10 hr or more) independent of influent phenol concentration.

Concerning the tests on cyanide, two different substrates were compared. One is the case where the influent contained cyanide alone as the carbon source, and in the other, cyanide and, as a supplementary carbon source, fructose or acetonitrile were contained together. The result is shown in Figure 3.8. It was revealed that supplemental carbon sources, in particular fructose, were effective in accelerating decomposition of the cyanide. Decomposition of thiocyanate was carried out using microbial film grown on phenol, and thiocyanate removals of 92% and 60.8% were obtained at volumetric loadings of 0.04 kg/m³·day and 0.2 kg/m³·day, respectively, suggesting that thiocyanate could not be removed sufficiently unless its loading was fairly small.

Gasanov et al. [10] investigated the biodegradation of hydrocarbons using aerated biofilters. They found that biodegradation of oil and its associated products was intensified by adding a microbial culture consisting of bacteria isolated from oil-bearing soil. Generally speaking, microbial film pro-

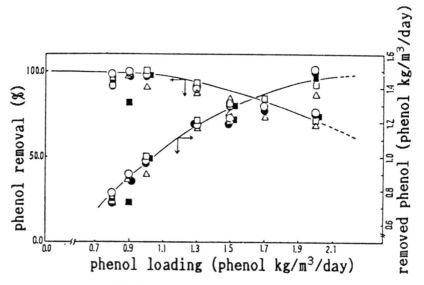

FIGURE 3.6. *The relation between volumetric phenol loading and removal percentage or removed amount. Temp. 20°C, HRT {△ :1 hr, ○ :3 hr, ● :6 hr, □ :12 hr, ■ :24 hr.*

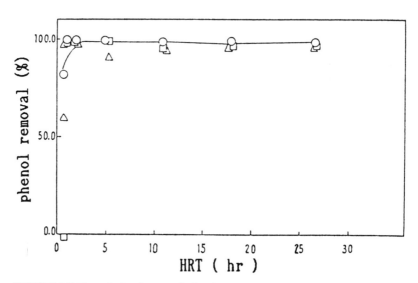

FIGURE 3.7. *The relation between hydraulic retention time and removals of phenol, COD_{Mn} and BOD at volumetric phenol loading of 0.9 kg/m³·day at 20°C. ○ :phenol, □ : COD_{Mn}, △ : BOD.*

154

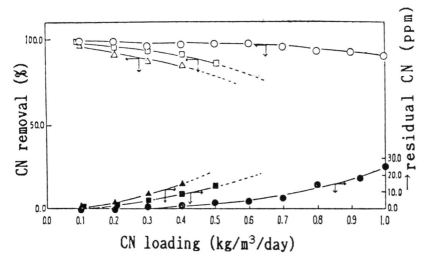

FIGURE 3.8. *Volumetric cyanide loadings and cyanide removals (HRT 6 hr, Temperature 20°C)* – ○ ● *: mixture of cyanide and fructose;* □ ■ *: mixture of cyanide and acetonitrile;* △ ▲ *: cyanide only.*

cesses are convenient for accumulating a certain specific species of microbe.

Costa Reis et al. [11] treated diluted stillage of alcohol distilleries presenting COD content of 3000–3500 mg/l, using a bench scale AFBSF of Bubble Oxygenation type, to assess the performance of aerobic submerged bed reactors applied to the treatment of a concentrated influent. The reactor was continuously operated at five hydraulic retention times varying from 23 to 4.5 hr, and the results showed that 60–80% of COD removal could be obtained with 10–16 hr of hydraulic retention time. Such operating conditions correspond to an extremely high COD loading.

3.2.2 The Processes Using Biofilm Filtration Processes (Biofilm Filters)

Biofilm filtration processes can be regarded as a kind of AFBSF with Bubble Oxygenation. But, biofilm filters utilize much smaller media than typical AFBSFs, possess a high capability of filtration owing to the fine medium utilized, and require frequent backwashings to recover their performance, so that it would be valid to regard biofilm filters as another kind of process.

As this process has the effect of both biological oxidation and physical filtration, no settling basin is needed for suspended solids separation from the effluent, but backwashing of the filter bed is indispensable once or several times a day, and a thickening device must be installed to condensate wastewater from backwashing.

3.2.2.1 SECONDARY OR TERTIARY TREATMENT OF MUNICIPAL SEWAGE

Linear velocity and volumetric loading of pollutants are usually employed as the operational indexes of biofilm filters, and appropriate values of linear velocity and volumetric BOD loading are 20–40 m and 1.5–4.0 kg-BOD/ m^3·day, respectively, for raw or primarily settled sewage. The volume of air diffused per unit volume of sewage is 2–3 m^3, only about a half or one third of the air volume required for conventional activated sludge process, clearly showing that this process is very energy-saving.

Fuchu [12] pointed out the relationship between volumetric loading and removal of BOD (Figure 3.9), which implies that the removal of pollutants by biofilm filters has a tendency to increase with the reduction of loading, and reaches an upper limit below a certain value of loading (about 5 kg-BOD/m^3·day in Figure 3.9). When this process is used for tertiary treatment of municipal sewage, a larger linear velocity can be applied than the one for secondary treatment. It can be operated at the linear velocity of 50–120 m/day, and volumetric BOD loading of 0.5–1.2 kg/m^3·day, with

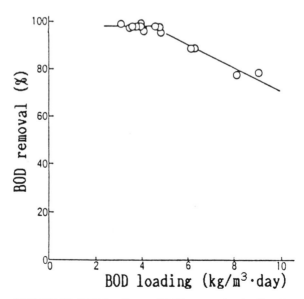

FIGURE 3.9. BOD loading and BOD removal in biofilm filter.

BOD removal more than 60%. Most of NH_3-N can be removed at volumetric NH_3-N loading not more than 0.5 kg/m³·day. Moreover, 95% or more of NH_3-N is removable at NH_3-N loading not exceeding 0.3 kg/m³·day. In biofilm filters, in addition to a filtering action, biological effects are expected, and so a higher degree of purification can be gained than with a rapid sand filter, if operated at the same linear velocity. As to the problems of using biofilm filters for secondary treatment, the frequent need of back-washing caused by a rapid progress of filter clogging, owing to a large amount of suspended solids trapped, must be recognized. Hence, it is not an advisable way to treat municipal sewage with this process, directly or after a primary settling whose removal of suspended solid is not so high. Primary treatment processes which enable high degrees of suspended solids removal as filtration, chemical coagulation and the like, and the combined processes of biofilm filters with such primary treatment are now under development.

3.2.2.2 TREATMENT OF INDUSTRIAL WASTEWATER

Biofilm filtration processes can be used to treat wastewaters containing low concentration of suspended solids and relatively low concentration of organic compounds, so that applications in food processing and the chemical industries may be in order.

3.3 THE PROCESSES USING ANAEROBIC FIXED BED SUBMERGED FILTERS (ANAEROBIC FILTERS)

It was previously described in Chapter 2, section 2.2 that anaerobic filters are divided into those for nitrogen removal by denitrification and those for organic compound removal by methane fermentation, and that the contents of anaerobic filters differ widely depending on their purposes. They may also be categorized into those under anoxic conditions and those under anaerobic conditions. In the case where the purpose of treatment is denitrification, an anoxic-oxic treatment system must be planned in combination with some nitrification process, a kind of aerobic process. Such a complex system shall be referred to in a later portion of this chapter, and some descriptions of examples of application of anaerobic filters will also be made, but, restricted to organic compound removal.

3.3.1 Secondary Treatment of Municipal Sewage

The authors et al. [13] treated municipal sewage using a pilot plant consisting of three stages of anaerobic filters packed with modified shaped plastic tubings, and obtained the relationship shown in Figure 3.10 between volumetric loading and removal of BOD. The following approximation was recognized between influent and effluent BOD:

$$c = c_0 e^{-0.0747T} \qquad (3.2)$$

where

c_0, c = BOD of influent and effluent, respectively (mg/l)
T = superficial retention period in anaerobic filter

That is, the BOD reduction process is represented by first-order reaction kinetics.

Kitao et al. [14] applied mass-balance equations as Equation (1.59) to the result of treatment of domestic sewage with a pilot scale anaerobic filter using vertically stretched laces as media, paying special attention to the difference of behaviors between dissolved and particulate substrates. They also compared numerically the output—effluent quality—from the filter between the two filter models—a single CSTR and two stages of CSTR. As a result, it was revealed that the number of stages had little effect on the output, and that Equation (1.59), in which first-order reaction kinetics are applied to both anaerobic reactions and trapping of suspended solids, could simulate water quality changes in the filter fairly well. A portion of the result of the experiment is shown in Figure 1.20.

Domestic sewage usually has 30–40 mg/l of SO_4-S, which is reduced to

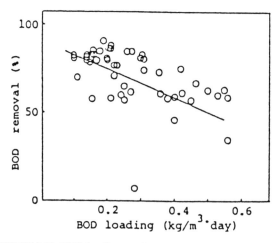

FIGURE 3.10. BOD loading vs. BOD removal in anaerobic filter.

S^{2-} in anaerobic filters. S^{2-} in the effluent from anaerobic filters is oxidized in turn, as $S^{2-} \rightarrow S \rightarrow SO_4^{2-}$, in stages of aerobic treatment. This, in turn, causes a large amount of oxygen consumption, and delayed degradation of organic compounds or nitrification.

Kitao and coworkers experimentally found two effective methods for removal of S^{2-}. One is flush aeration, and the other is coagulation with addition of $FeCl_3$ to form FeS. Other research has been carried out on treatment of municipal sewage with anaerobic filters [15–19].

3.3.1.1 TREATMENT OF INDUSTRIAL WASTEWATERS

Various industrial wastewaters contain higher concentrations of organic compounds than domestic sewage, and are suitable for anaerobic treatment, so that a lot of research has been focused on their treatment with anaerobic filters.

Carrondo et al. [20] investigated the treatment of waste molasses, a typical concentrated organic wastewater, with two kinds of upflow anaerobic filters packed with polyethylene tubing (6 × 7.8 mm) or raschig rings. COD reduction at ambient temperature was 57–79% with gas production of 4.8 m³/m³·day, under the conditions of COD in influent liquid 10,000–50,000 mg/l (volumetric COD loading 2–12 kg/m³·day) and of HRT 2.5–5 days. The methane content was, however, not more than 40% and a high concentration of hydrogen sulfide was detected in the gas produced. The filters were able to be operated for seven months with no withdrawal of sludge.

Wheatley et al. [21] made a comparison between the characteristics of aerobic and anaerobic filters, which is shown in Figure 3.11. BOD removal with aerobic filters is substantially higher, but the difference decreases rapidly with increment of BOD loading. At BOD loadings higher than 4 kg/m³·day, the efficiency of anaerobic filters is even higher than that of aerobic filters.

Van den Berg et al. [22] treated bean-bleaching wastewater with a downflow anaerobic filter packed with vertical polymeric parallel plates, and compared the results with experiments obtained with those of other anaerobic processes. The results of the experiments and comparisons are summarized in Table 3.6, which shows that an anaerobic filter is able to remove more COD with much shorter hydraulic retention times than an anaerobic contact process or a completely mixed anaerobic digester. As shown in this example, in spite of its simple operation, an anaerobic filter is able to be operated not only with much higher efficiencies than those of processes using suspended cultures, but also with a COD loading of biological treatment as high as 10–20 kg/m³·day. Moreover, the suspended solids

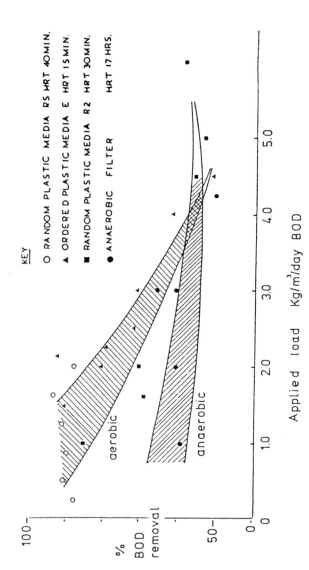

FIGURE 3.II. *Comparison between aerobic and anaerobic filter performance.*

TABLE 3.6. *Performance Data for Anaerobic Fixed Film Reactors,
Contact Process Digesters and Completely Mixed Digesters
Fed Bean Blanching Waste.*

	Fixed Film Reactor		Contact Process Digester[c]	Completely Mixed Digester
	0.7 Liter[a]	35 Liter[b]		
Minimum hydraulic retention time, days	0.5	1	3	8.5
Maximum COD loading rate (M-COD-LR), kg/m³/day	20[d]	10	7	1.2[e]
COD removal efficiency, %, at M-COD-LR	86	86	80	75
Suspended COD of fermenter liquid %			0.9	
Suspended COD of effluent				
1) %	0.09	0.09	0.16	0.20
2) % of total effluent COD	65	65	45	80

[a]Surface to volume ratio, 140 m²/m³.
[b]Surface to volume ratio, 120 m²/m³—reactor may not have reached maximum loading rate in test run (160 days).
[c]Fed Waste with 2% COD; loading rates would have been less with lower waste strength. Other reactors fed 0.95% COD waste.
[d]Independent of waste strength (0.5–2.0% COD).
[e]Fed waste with 0.95% COD—loading rates would increase about proportional with waste strength.

concentration in the filter effluent was 900 mg/l, which is smaller than in the raw wastewater (1500 mg/l), suggesting a very small yield coefficient of biomass from COD.

3.4 THE PROCESSES USING FLUIDIZED BEDS

In fluidized bed reactors, both the amount of biomass held and total microbial film area in the reactor is large because of the fine carrier particles supporting the biomass. Fluidized bed reactors, therefore, are able to treat wastewaters with extremely short hydraulic retention time or at very large volumetric loadings, corresponding to their large microbial film areas. For example, fluidized bed reactors can be operated with a superficial retention time of 10 minutes or more, or at a volumetric BOD loading of 10 kg/m³·day or so. On the other hand, they require much closer and finer operations for maintenance than fixed bed reactors to form and maintain stable fluidized beds.

3.4.1 Processes Using Aerobic Fluidized Bed Reactors

Aerobic fluidized bed reactors are classified into four types shown in Figures 2.16 and 2.17.

Jeris et al. [23] reported the results of the investigation on nitrification, denitrification and BOD removal with fluidized bed units with specific surface areas larger than 3300 m²/m³. The apparatus used for nitrification consisted of a fluidized bed, aeration cone for dissolving pure oxygen in the influent into the fluidized bed, the instrument to separate biofilm from carriers (sweco), the sand separation tank to recover carrier particles from the reactor effluent, and the tank for recycling the effluent as shown in Figure 3.12. Table 3.7 shows the results of nitrification and denitrification. Nitrification and denitrification of efficiencies more than 99% were accomplished.

Kim et al. [24] treated night soil at dilution rates of 2–10 with a fluidized bed, and concluded that this process is more profitable than others because of its smaller requirements of reactor volume and dilution water.

Removal of heavy metals with fluidized bed reactors from industrial wastewaters was described by Remacle and Houba [25] who pointed out that more than 90% of Zn, Pb, Cu and Cd were removed and concentrated

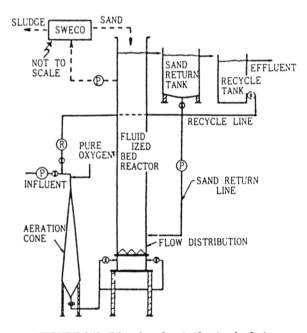

FIGURE 3.12. Pilot plant for nitrification by Jeris.

TABLE 3.7. Performance Data for
Fluidized Bed Reactor for Nitrification and Denitrification.

(1) Summary of Nitrification (8/11/75–9/9/75)	
Secondary effluent flow (1/min)	52
Recycle flow (1/min)	122
Detention time (min)	10.6
Temperature (°C)	24
Secondary effluent NH_3-N (mg/l)	19.1
Fluidized bed effluent NH_3-N (mg/l)	0.2
% NH_3-N removal	99
Secondary effluent organic-N (mg/l)	3.1
Fluidized bed effluent organic-N (mg/l)	2.3
(2) Summary of Denitrification (10/3/73–12/3/73)	
Nitrified effluent flow (1/min)	98.4
Detention time (min)	6.5
Temperature (°C)	20
Nitrified effluent $(NO_3 + NO_2)$-N (mg/l)	21.5
Fluidized bed effluent $(NO_3 + NO_2)$-N (mg/l)	0.2
% $(NO_3 + NO_2)$-N removed	99

in particles coated by microbial film. Omura et al. [26] carried out an experimental investigation on biological oxidation of ferrous ion in high acid mine drainage with a fluidized bed reactor, and gained ferrous ion removals higher than 90%.

A corn starch wastewater was treated with a fluidized bed reactor using shell sand as the medium, at food-to-microorganism (F/M) ratios ranging 0.42–1.61 g-BOD/g-TVS·day [27]. Removal behaviors of BOD and TKN (total Kjeldahl nitrogen) as shown in Figure 3.13 were obtained. A good combined BOD removal and nitrification, with removals larger than 90%, could be achieved provided that F/M ratio and mean cell residence time were maintained at less than 1.0 g-BOD/g-TVS·day and longer than five days, respectively. The biomass concentration in the reactor ranged from 5,000 to 16,000 mg-TVS/l.

3.4.2 Processes Using Anaerobic Fluidized Bed Reactors

Anaerobic fluidized beds for degradation of organic compounds are divided into those in which combined liquefaction and gasification is performed in a single reactor, and those where two phases of anaerobic reaction proceed separately in two reactors.

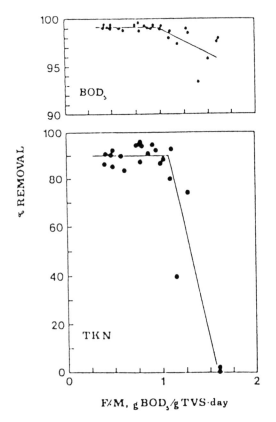

FIGURE 3.13. *Effects of F/M ratio on removal of BOD$_s$ and TKN in experimental fluidized bed unit.*

Jewell et al. [28] treated primary effluent of municipal sewage at 20°C with an anaerobic fluidized bed reactor, as in Figure 3.14, utilizing pulverized polyvinyl chloride (particle size not more than 1 mm) as the carrier. The influent, with COD and SS ranging from 88–306 (avg. 186) mg/l and 40–186 (avg. 86) mg/l, respectively, was treated at superficial retention times and volumetric COD loadings of 5–180 min and 0.65–35 kg/m³·day, respectively. As shown in Figure 3.15, the maximum COD removal was about 80%, and COD removal was maintained higher than 60% at superficial retention times longer than 20 minutes.

Many other experimental treatments were fulfilled using actual municipal sewages [29–31].

Many investigations have been reported on the application of anaerobic fluidized bed reactors to food processing wastewaters such as cheese whey or to wastewaters from soft drink manufacturing [32–36].

Binot et al. [37] reported results of treatment of citric acid fermentation wastewater with a bench-scale anaerobic fluidized bed reactor having glass beads or glass particles as media. It was shown that volumetric COD loadings up to 42 kg/m³·day with more than 70% conversion to gas were achieved under steady state conditions and volumetric loading increases of 10% per day. They also compared the performances of this reactor and conventional CSTR, the acceptable maximum organic loadings for the former (32 kg-VS/m³·day or more) being 16 times or more larger than that for CSTR corresponding to the amount of active biomass in each reactor.

Jeris showed in Figure 3.16 the relationship between volumetric COD loadings and COD reductions of various high strength wastewaters. High efficiencies of COD removal were attained at very high COD loadings for all of these wastewaters.

Anaerobic fluidized bed reactors are feasible not only for wastewaters containing natural organic compounds, but also for ones mainly containing chemically-synthesized organic compounds, as from the chemical processing industries.

Wang et al. [39] used an anaerobic expanded bed reactor packed with granular activated carbon as a carrier for the degradation of polycyclic

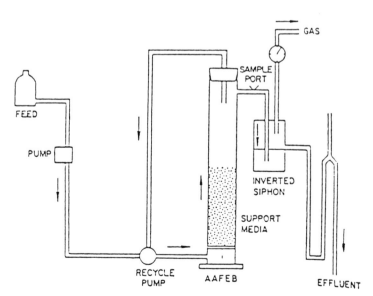

FIGURE 3.14. Schematic of the AAFEB system used in the laboratory study of shock effects.

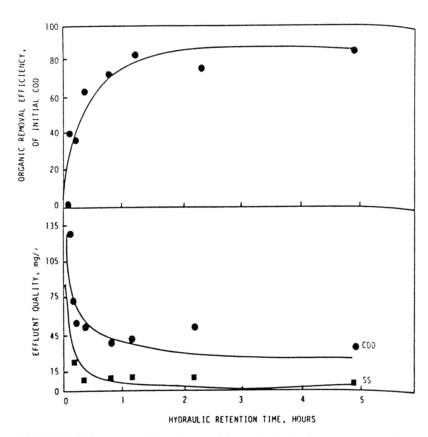

FIGURE 3.15. *Summary of the influence of decreasing hydraulic retention periods on the COD and SS removal capability of the AAFEB treating primary settled domestic wastewater at 20°C.*

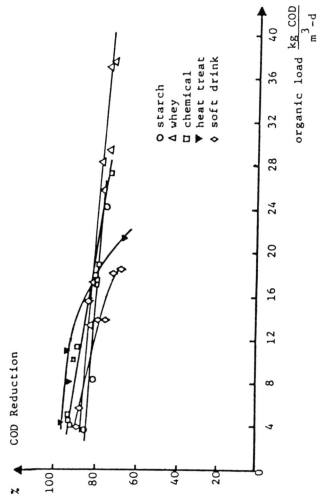

FIGURE 3.16. Anaerobic FBR treatment of high strength wastes.

N-aromatic compounds (indole, quinoline, methylquinoline) which are a major group of compounds in coal gasification wastewater. The results of gas production and COD removal are shown in Figure 3.17 and Figure 3.18. COD reductions after the 75th day in Phase I were accompanied by negligible gas production, suggesting that adsorption was the primary removal mechanism; but the COD reductions in Phases II-V showed a high correlation with the volume of gas produced, and were mainly dependent on biological effects.

As mentioned above, anaerobic fluidized or expanded bed reactors are able to treat at high loadings a wide variety of wastewaters with a wide range of strength, including both municipal sewage and industrial wastewaters. However, some technological problems remain, one of which is that they need a long period for start up because of the small specific growth rates of anaerobic microbes.

3.5 THE ANAEROBIC-AEROBIC FILTER SYSTEMS

Anaerobic treatment processes have a common feature in that they can purify concentrated wastewaters with small energy consumptions and little sludge productions, but are not capable of high degrees of organic compound removal with them alone. On the other hand, aerobic treatment processes are suitable to purify wastewaters to fairly low pollutant levels, though their energy expenditures and sludge productions are large. Taking into consideration the features of anaerobic and aerobic processes, one easily comes to the conclusion that the treatment systems combining anaerobic and aerobic processes are most reasonable. Among such combined systems there is a wide range of use of anaerobic-aerobic filter systems. Moreover, anaerobic-aerobic treatment systems are effective in preventing excessive temperature rises which sometimes are caused by aerobic treatment of concentrated organic wastewaters such as waste molasses, etc.

The other objective of anaerobic-aerobic treatment systems is biological nitrogen removal. In these cases, nitrification, a kind of aerobic process and denitrification, a kind of anaerobic (anoxic) process are combined into a treatment system, with recycling of the effluent from nitrifying filter to denitrifying filter.

The authors et al. [13] treated domestic sewage with a process system consisting of three stages of anaerobic filter (5, 4 and 4 m³) and a single stage of AFBSF followed by a settling basin or a rapid gravity filter as shown in Figure 3.19. The flow rate and the average volumetric BOD loading during the former 200 days and the latter 200 days were 8.4 m³/day, 0.16

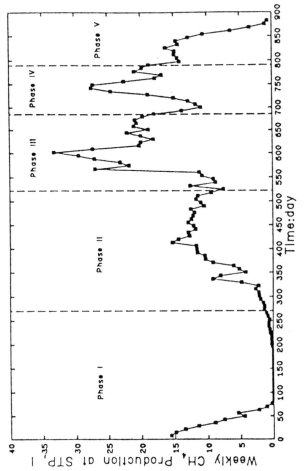

FIGURE 3.17. *Weekly methane production from GAC anaerobic filter treating synthetic polycyclic N-aromatic compound wastewater.*

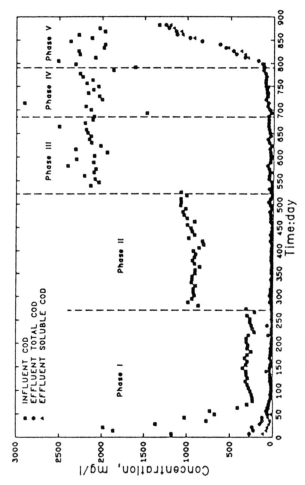

FIGURE 3.18. COD removal in GAC anaerobic filter.

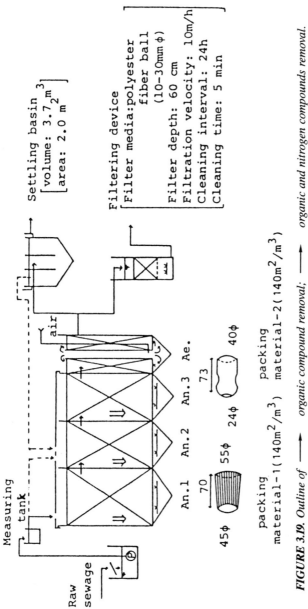

Settling basin
[volume: 3.7 m^3
area: 2.0 m^2]

Filtering device
[Filter media:polyester
  fiber ball
  (10–30mm ϕ)
Filter depth: 60 cm
Filtration velocity: 10m/h
Cleaning interval: 24h
Cleaning time: 5 min]

Measuring tank

Raw sewage

air

An.1 An.2 An.3 Ae.

70   55ϕ 24ϕ 73   40ϕ
45ϕ

packing   packing
material-1(140m^2/m^3) material-2(140m^2/m^3)

FIGURE 3.19. *Outline of ⟶ organic compound removal; ⟶ organic and nitrogen compounds removal.*

171

kg/m³·day, 12.4 m³/day, and 0.23 kg/m³·day, respectively. The change of BOD in effluents from various stages of filter during the experimental period is shown in Figure 3.20. The relationship between BOD loading and removal was as depicted in Figure 3.10. The sludge production in anaerobic filters per unit BOD removed and unit SS removed were 0.32 kg/kg and 0.24 kg/kg, respectively. Next, the pilot plants were operated as biological denitrogen systems. Raw sewage was fed into the second stage of the filter and the supernatant liquid of the settling basin was recycled to the second stage of the filter at the flow rate of 100% of raw sewage in order to operate as two stages of denitrification and one stage of nitrification. The average result obtained is shown in Table 3.8, and mass balances in this system are shown in Table 3.9. More than 90% of the nitrogen removal took place within the first stage of denitrification filter (An.2). About half of the nitrogen removal was caused by denitrification and the remainder was assumed to be caused by microbial uptake and SS separation. When the recycle ratio was increased to 300%, removal percentages of BOD and T-N increased about 10%. Observed sludge productions in anaerobic filters were raised to 0.45 kg/kg-BOD removed and 0.39 kg/kg-SS removed, respectively, which implies that the sludge production by denitrification is about 50% greater than that of methane fermentation.

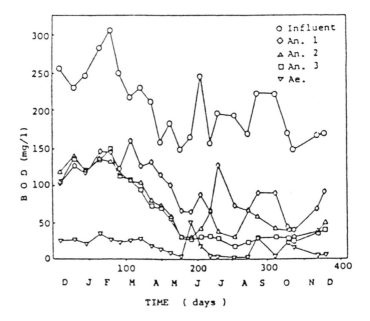

FIGURE 3.20. Change of BOD in experimental period.

TABLE 3.8. Performance Data for Anaerobic-Aerobic Filter System—
Average Water Quality and Removal at 100% Recirculation Ratio—
(Influent Flow 5.3 m^3/d, Recirculation Flow 5.6 m^3/d).

	Influent	An. 1	An. 2	Final Effluent	Removal (%)
pH	6.9–7.5	6.6–7.2	6.5–7.0	6.2–6.6	–
BOD	175(36)	23(10)	21(5)	3.9(0.8)	97.8
COD	95.3(9.4)	21.7(6.5)	16.9(1.8)	9.9(1.9)	89.6
SS	197(62)	12.3(11.1)	7.6(1.6)	3.4(1.9)	98.3
NH_4^+-N	33.1(8.0)	15.3(3.2)	14.0(2.7)	0.7(0.2)	97.9
TK-N	46.9(2.8)	17.2(2.3)	16.5(1.6)	2.2(0.9)	95.3
NO_x-N	0.1(0.2)	0.6(1.4)	0.1(0.2)	14.1(2.2)	–
T-N	47.0(2.9)	17.8(3.6)	16.6(1.8)	16.3(1.5)	65.3
Alk	173(13)	129(8)	134(6)	25.0(5.1)	–

( ) : Standard deviation, water temperature : 27–31°C, DO in aerobic filter : 3.4 (0.7) mg/l.

In Japan, anaerobic-aerobic filter systems are now widespread as small scale domestic sewage treatment facilities, especially as on-site facilities for individual houses. An example of such a facility for 10 persons is depicted in Figure 3.21. Those facilities are designed on the assumption that the flow rate and BOD of normal domestic sewage are 200–250 l/capita·day and 200 mg/l, respectively, and are capable of purifying to 20 mg/l of BOD (the same value as the Japanese standard for a sewage treatment plant's effluent). About 60–80% of influent BOD is removed in anaerobic filters with hydraulic retention time of 1–1.5 day, which contributes to repress sludge production and electric power consumption. The percentage of Nitrogen removal with those facilities is only about 20%, but it can be raised to 60–70%, if the recycling from the aerobic filter to the first stage of the anaerobic filter proceeds at the rate of 2–3 times of average influent flow rate.

Cooper and Wheeldon [40] treated primary effluent of municipal sewage with anaerobic-aerobic (anoxic/oxygenic) fluidized bed reactors as denitrification-nitrification system. Settled sewage was used as a carbon source for denitrification, and high rates of denitrification (5–10 kg-NO_3/ m^3·d) were achieved. The effluent from an anoxic reactor was treated with an oxygenic reactor at high volumetric loadings, resulting in low BOD effluent. Complete nitrification was attained at hydraulic retention time of 1.8 hr.

As an example of industrial wastewater treatment with anaerobic-aerobic fluidized bed reactors, Nutt et al. [41] reported the results of experimental treatment of coke plant wastewater with a pilot scale two stages of fluidized bed reactors (Figure 3.22). Treatment tests were carried out to determine the performances under pseudo-steady-state during which total amounts of

TABLE 3.9. Mass Balance of BOD, Nitrogen and SS in Anaerobic-Aerobic Filter System at 100% Recirculation Ratio.

Influent 5.3 m³/d → (1) → | 10.9 m³/d (2) | (3) An.2 → (4) | (5) An.3 (6) → (7) Ae. (8) → S (9) →

		(1)	(2)	(3)	(4)	(5)	(6)	(7)	(8)	(9)
BOD	(mg/l)	175	88.6	—	23	—	21	—	5.4	3.9
	(g/d)	927.5	965.7	715*	250.7	21.8*	228.9	170*	58.9	20.7
	(g/m³·d)	—	—	178.8	—	5.5	—	56.7	—	—
NH₄⁺-N	(mg/l)	33.1	16.4	—	15.3	—	14.0	—	0.7	0.7
	(g/d)	175.4	179.3	12.5*	166.8	14.2*	152.6	145*	7.6	3.7
	(g/m³·d)	—	—	3.2	—	3.6	—	48.3	—	—
TK-N	(mg/l)	13.8	8.0	—	1.9	—	2.5	—	2.0	1.5
	(g/d)	73.2	87.0	66.3*	20.7	-6.6*	27.3	5.5*	21.8	8.0
	(g/m³·d)	—	—	16.6	—	-1.7	—	1.8	—	—
NOₓ-N	(mg/l)	0.1	7.4	—	0.6	—	0.1	—	14.2	14.1
	(g/d)	0.5	80.6	74.1*	6.5	5.5*	1.0	-153.8*	154.8	74.7
	(g/m³·d)	—	—	18.5	—	1.4	—	-51.3	—	—
T-N	(mg/l)	47.0	31.8	—	17.8	—	16.6	—	16.9	16.3
	(g/d)	249.1	346.9	152.9*	194.0	13.1*	180.9	-3.3*	184.2	86.4
	(g/m³·d)	—	—	—	—	—	—	—	—	—
SS	(mg/l)	197	101.7	—	12.3	—	7.6	—	7.6	3.4
	(g/d)	1044.1	1108.9	974.8*	134.1	51.3*	82.8	0*	82.8	18.0
	(g/m³·d)	—	—	243.7	—	12.8	—	0	—	—

*Removal at each filter.

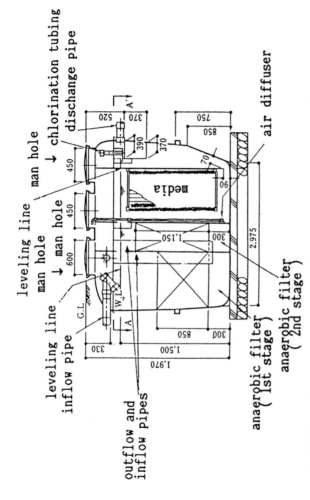

FIGURE 3.21. The cross section of small scale domestic sewage treatment facility for 10 persons.

175

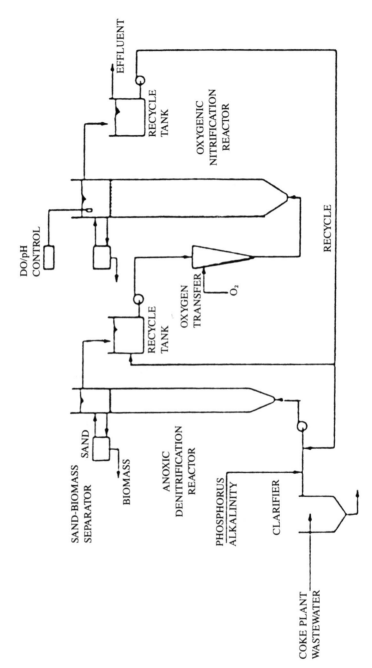

FIGURE 3.22. Two-stage biological fluidized bed process for coke plant wastewater treatment.

biomass in the filters were not exactly constant, and under nonsteady-state. In pseudo-steady-state tests, undiluted coke plant wastewater was treated with extremely short hydraulic retention times for the overall system, ranging from 15.9–44.8 hr, and the resultant average effluent qualities are as listed in Table 3.10. The concentrations of VSS in the fluidized beds (BVS) were very high, ranging 11.9–34.0 g/l and 4.3–13.5 g/l for anoxic and oxygenic reactors, respectively. These high concentrations of BVS seemed to have enabled high efficiencies of treatment. Moreover, SRTs in oxygenic reactor as long as 52–140 days ensured the sufficient accumulation of nitrifying bacteria.

In one approach, Iida et al. [42] achieved an anoxic-oxic treatment system with a single fixed bed submerged filter. That is, they treated municipal sewage, in order to remove organic carbon and nitrogen without addition of auxiliary carbon source, using an intermittently aerated submerged filter with an aeration cycle of 2 hr on and 2 hr off. The nitrogen removal percentages at hydraulic retention times of 7 hr and 11 hr were 55% and

TABLE 3.10. Pseudo-Steady-State Performance of Two-Stage Biological Fluidized Bed System [Median Concentration (mg/l), Based on 9–36 Consecutive Days of Operation].

	Run				
	1	2	3	4	5
HRT (hr)					
Anoxic	9.8	4.8	3.2	1.9	3.0
Oxygenic	35.0	35.0	23.3	14.0	1.40
System	44.8	39.8	26.5	15.9	17.0
COD					
Effluent (mg/l)	270.0	428.0	355.0	348.0	513.0
Removal (%)	91.1	87.6	89.1	89.4	83.4
Phenol					
Effluent (mg/l)	0.08	0.15	0.17	0.14	0.20
Removal (%)	>99.9	>99.9	>99.9	>99.9	>99.9
TKN					
Effluent (mg/l)	9.0	13.3	9.8	14.4	21.5
Removal (%)	95.6	94.6	94.0	92.1	90.7
TCN					
Effluent (mg/l)	4.5	3.8	6.2	6.5	4.5
Removal (%)	50.0	57.8	42.2	18.2	12.5
CNS					
Effluent (mg/l)	1.8	1.8	1.6	2.6	3.0
Removal (%)	99.6	99.6	99.5	99.3	99.2

65%, respectively, but recycling of the settling basin effluent at a recycle ratio of 2.0 elevated nitrogen removal to 80%.

A treatment system which is operated as both attached and suspended growth systems is called activated biofilter (ABF). Viraraghavan et al. [43] reviewed the design and operation of this treatment system. Most information on this process is based on practical facilities, and the lack of fundamental data is pointed out. Potential advantages of ABFs are operational stability and high BOD removal. Kitao also observed such effects as improvements of settling characteristics of sludge, increases of nitrogen removal and operational stability in several ABFs in operation.

3.6 THE PROCESSES USING ROTATING BIOLOGICAL CONTACTORS

On the design, selection of media, operation and performance of rotating biological contactors, reviews by Stover and Kincannon [44], Wu and Smith [45], Wang et al. [46], U.S. EPA [47] and Strom and Chung [48], or a manual by U.S. EPA [49] are helpful.

Phenol-formaldehyde resin wastewater containing 300 mg/l of phenol, 200 mg/l of formaldehyde and 1500 mg/l of COD was treated by Huang et al. [50] with pilot scale rotating biological contactors at the hydraulic retention time of 2.8 hr, resulting in the removals of 99.6%, 93% and 60-90% for each component. Similarly, when influent concentrations of phenol, formaldehyde and COD were 200-400 mg/l, 250-300 mg/l and 1500-2000 mg/l, the removal percentages at the retention time of 4.5 hr were 99.9%, 100% and 90%, respectively. The design parameters proposed by this study were 6.54 g/m²·day, 6.4 g/m²·day and 33.3 g/m²·day of surface loadings of phenol, formaldehyde and COD, respectively, and 35 l/m²·day of hydraulic loading.

In spite of the accepted design criterion of 1.5 g/m²·day, or the maximum empirical value of about 4 g/m²·day as the surface loading of NH_3-N for nitrification of primary or secondary effluent of municipal sewage, Collins et al. [51] achieved a maximum nitrification rate of 11.7 g/m²·day for a pilot scale rotating biological contactor treating a semiconductor manufacturing wastewater containing NH_3-N of 800 mg/l. They concluded that the reason for such a high surface loading is high microbial density of nitrifying bacteria in biofilm grown on the wastewater with very low levels of organic compounds.

Orlands advanced sewage treatment plant near Orlando, Fl operates the largest rotating biological contactor for nitrification, denitrification and BOD removal of sewage from 1981 [52]. The design flow rate was 24 MGD

(90.840 m³/day), and the design influent quality was as follows; BOD 230 mg/l, COD 425 mg/l, SS 250 mg/l, TKN 30 mg/l, TP 10 mg/l and a pH of 7.0. The flow diagram of the treatment system was as follows; Preliminary treatment → Primary settling → Rotating biological contactor for BOD removal and nitrification → Secondary settling with addition of coagulant → Anoxic (entirely submerged) rotating biological contactor for denitrification → Rapid sand filtration → Chlorination → Reaeration → Discharging. Nine stages of air driven Aerosurf contactors were used for BOD removal and nitrification at overall hydraulic loading of 40.7 l/m²·day, hydraulic loading and BOD surface loading of the former four stages for BOD removal being 114.1 l/m²·day and 21.1 g/m²·day, and hydraulic loading and NH_3-N surface loading of the latter five stages being 65.2 l/m²·day and 1.56 g/m²·day, respectively. In the secondary settling basins, phosphorous is removed with the addition of alum and polymer as coagulants. Anoxic denitrifying contactors were entirely submerged types, mechanical driven, and methanol was added as the hydrogen donor. Designed hydraulic loading, NO_3-N surface loading and contact time were 223.9 l/m²·day, 5.4 g/m²·day and 1.7 hr (100 minutes), respectively, and designed effluent T-N was 3 mg/l or less.

Fecal coliform bacteria and Salmonella typhimurium die-off were examined in an experimental rotating biological contactor receiving settled domestic sewage [53]. It was revealed that there was a significant difference in survival of fecal coliform between those attached to disks and those in the bath; the latter died much more slowly. Rotating biological contactors were assumed to be very efficient in removing pathogenic bacteria found in domestic sewage.

Research on methane fermentation with rotating biological contactors are much rarer than those with anaerobic filter or anaerobic fluidized bed reactors. Rotating biological contactors may not be regarded as a process which can utilize the merits of anaerobic reactions most effectively, but some investigations have been carried out for treatment of sewage or industrial wastewaters [54,55].

3.7 REFERENCES

1. Hoshino, Y. and T. Iizuka. 1985. *Ann. Rep. of Research and Development Division of Japan Sew. Wks. Agency,* pp. 22–37.
2. Iwai, S., H. Ohmori and T. Tanaka. 1977. "An Advanced Sewage Treatment Process-Combination of Submerged Biological Filtration and Ultra-Filtration with Pulverized Activated Carbon," *Proc. Int'l. Cong. on Desalination and Water Reuse,* Vol. II, pp. 29–36, Tokyo.
3. Hisakawa, K. et al. 1990. "Estimation of Treatment Function of Three Types Domestic

Wastewater Treatment Processes Developed as Small Gappei Johkasou (small scale wastewater treatment systems for sanitary wastewater), Preprint of 1AWPRC Poster Paper, pp. 763–766, Kyoto.

4. Watanabe, T. et al. 1990. Ibid., pp. 743–746.

5. Iwai, S., T. Kitao, H. Ohmori and K. Inujima. 1977. "Reuse of Drainage from an Office Building," *Proc. of Int'l. Cong. on Desalination and Water Reuse,* Vol. II, pp. 171–175.

6. Iwai, S. and T. Kitao et al. 1973. "Study on Waste Treatment by Submerged Biofilter (3)," *Wat. Purif. and Liquid Waste Treatment,* 14(9):909–914.

7. McHarness, D. D., R. T. Haug and P. L. McCarty. 1975. "Field Studies of Nitrification with Submerged Filters," *Jr. WPCF,* 47(2):291–309.

8. Adachi, T. et al. 1978. "Application of Contact Aeration Process to Secondary and Tertiary Sewage Treatment," *Jr. of Water and Wastes,* 20(5):521–529.

9. Iwai, S., T. Kitao and T. Hirayama. 1976. *Wat. Purif. and Liquid Wastes Treatment,* 17(8):711–719.

10. Gasanov, M. V. et al. 1984. "Microbiological Intensification of Oil Oxidation in Biofilters," *Microbiol.,* 53:128.

11. Costa Reis, L. G. and G. L. Sant' Anna, Jr. 1985. "Aerobic Treatment of Concentrated Wastewater in a Submerged Bed Reactor," *Wat. Res.,* 19:1341.

12. Fuchu, Y. 1983. "Treatment of Organic Waste Water by the Bio-Film Filter," *Jr. of Water and Wastes,* 25(5):477–485.

13. Kitao, T., S. Iwai, H. Ohmori, Y. Yamamoto and M. Fujii. 1986. "Anaerobic Submerged Biofilter for the Treatment of Domestic Wastewater," *Proc. of EWPCA Conf.,* Amsterdam, pp. 736–739.

14. Kitao, T., Y. Kiso and K. Kasai. 1991. "A Study on Removal Performances of Anaerobic-Aerobic Bio-Filter System and Removal Mechanism by Numerical Analysis," *Jour. Japan Sewage Works Association Research Jour.,* 28(324):17–27.

15. Roman, V. and N. Chaklader. 1972. "Upflow Filters for Septic Tank Effluents," *Jr. WPCF,* 44:1552–1559.

16. Kobayashi, H. A. et al. 1983. "Treatment of Low Strength Domestic Wastewater Using the Anaerobic Filter," *Water Res.,* 17:903–909.

17. Yamamoto, Y. et al. 1985. "Anaerobic Submerged Biofilter for the Treatment of Domestic Wastewater," *Jr. Japan Sew. Wks. Assoc.,* 22:12–22.

18. Genung, R. K. et al. 1982. "Energy Conservation and Methane Production in Municipal Wastewater Treatment Using Fixed-Film Anaerobic Bioreactor," *Biotech. Bioeng. Symp.,* (12):365–380.

19. Wheatley, A. D. 1983. "Effluent Treatment by Anaerobic Biofiltration," *Water Pollut. Control,* 82:10–22.

20. Carrondo, M. J. T. et al. 1983. "Anaerobic Filter Treatment of Molasses Fermentation Wastewater," *Wat. Sci. and Tech.,* 15(8/9):117–126.

21. Wheatley, A. D. et al. 1984. "Energy Recovery and Effluent Treatment of Strong Industrial Wastes by Anaerobic Biofiltration," in *Anaerobic Digestion and Carbohydrate Hydrolysis of Wastes,* G. L. Ferrero et al., ed., Elsevier Appl. Sci. Publ., pp. 284–306.

22. Van den Berg, L. et al. 1980. "Anaerobic Waste Treatment Efficiency Comparison between Fixed Film Reactors, Contact Digesters and Fully Mixed, Continuously Fed Digesters," *Proc. of 35th Ind. Waste Conf.,* Purdue Univ., pp. 788–793.

23. Jeris, J. S. et al. 1977. "Biological Fluidized-Bed Treatment for BOD and Nitrogen Removal," *Jr. WPCF,* 49(5):816–831.

24. Kim, H. G. and Y. D. Lee. 1986. "Night Soil Treatment by Biological Fluidized Beds," *Water Sci. Technol. (G.B.)*, 18:199.

25. Remacle, J. and C. Houba. 1983. "The Removal of Heary Metals from Industrial Effluent in a Biological Fluidized Bed," *Environ. Technol. Lett.*, 4:53.

26. Omura, T. et al. 1991. "Biological Oxidation of Ferrous Iron in High Acid Mine Drainage by Fluidized Bed Reactor," *Proc. of 15th Biennial Conf. of the Int'l. Assoc. on Wat. Pollu. Res. and Cont'l.*, Kyoto, pp. 1447–1456.

27. Shieh, W. K. and C. T. Li. 1989. "Performance and Kinetics of Aerated Fluidized Bed Biofilm Reactor," *J. Environmental Engineering* (1):65–79.

28. Jewell, W. J., M. S. Switzenbaum and J. W. Morris. 1981. "Municipal Wastewater Treatment with the Anaerobic Attached Microbial Film Expanded Bed Process," *J. WPCF,* 53(4):482–490.

29. Rockey, J. S. et al. 1982. "The Use of an Anaerobic Expanded Bed Reactor for the Treatment of Domestic Sewage," *Environ. Tech. Letters,* 3:487–496.

30. Lettinga, G. et al. 1983. "Anaerobic Treatment of Raw Domestic Sewage at Ambient Temperatures Using a Granular Bed UASB Reactor," *Biotech. Bioeng.*, 25:1701–1723.

31. Foster, C. F. et al. 1983. "Anaerobic Treatment of Dilute Wastewater Using an Upflow Sludge Blanket Reactor," *Environ. Pollution, Series A,* 31:57–66.

32. Sutton, P. M. and A. Li. 1981. "Anitron System and Oxitron System: High-Rate Anaerobic and Aerobic Biological Treatment Systems for Industry," presented at *36th Ind. Waste Conf.*, Purdue Univ., Lafayette, Indiana, pp. 665–677.

33. Hickey, R. F. and R. W. Owens. 1982. "Methane Generation from High-Strength Industrial Wastes with the Anaerobic Biological Fluidized Bed," presented at *3rd Sympo. on Biotechnol. in Energy Production and Conservation.*

34. Oliva, J. et al. 1990. "Treatment of Wastewater at the El Aguila Brewery (Madrid, Spain). Methanization in Fluidized Bed Reactors," *Wat. Sci. and Technol.,* 22(1/2):483–490.

35. Balslev-Olesen, P. et al. 1990. "Pilot-Scale Experiment on Anaerobic Treatment of Wastewater from a Fish Processing Plant," ibid, 22(1/2):463–474.

36. Sutton, P. M. and A. Li. 1983. "Single Phase and Two Phase Anaerobic Stabilization in Fluidized Bed Reactors," ibid, 15(8/9):333–344.

37. Binot, R. A., T. Bol, H. P. Naveau and E. J. Nyns. 1983. "Biomethanation by Immobilised Fluidised Cells," ibid, 15:103–115, Copenhagen.

38. Jeris, J. S. 1983. "Industrial Wastewater Treatment Using Anaerobic Fluidized Bed Reactors," ibid, 15(8/9):169–176.

39. Wang, Y. et al. 1984. "Anaerobic Activated Carbon Filter for the Degradation of Polycyclic N-Aromatic Compounds," *Jr. WPCF,* 56(12):1247–1253.

40. Cooper, P. F. and D. H. V. Wheeldon. 1982. "Complete Treatment of Sewage in a Two-Stage Fluidized Bed System Part 1," *Water Pollut. Control (G.F.* 81:447.

41. Nutt, S. G. et al. 1983. "Treatment of Coke Plant Wastewater in the Coupled Predenitrification-Nitrification Fluidized Bed Process," *P. 37th Ind. Waste Conf., 1982,* Purdue Univ., Ann Arbor Sci. Pabl., Inc., Ann Arbor, Mich., p. 527.

42. Iida, Y. et al. 1984. "Nitrogen Removal from Municipal Wastewater by a Single Submerged Filter," *J. WPCF,* 36(3):251–285.

43. Viraraghavan, T. et al. 1984. "Activated Biofilm Process for Wastewater Treatment," *Effluent Water Treat. J.,* 24:378.

44. Stover, E. L. and D. F. Kincannon. 1982. "Rotating Biological Contactor Scale-Up and Design," *Water Eng. Manage.*, p. 129, *Reference Handbook,* 4:49.

45. Wu, Y. C. and E. D. Smith. 1982. "Rotating Biological Contactor System Design," *J. Environ. Eng. Div., Proc. Am. Soc. Civ. Eng.*, 108:578.

46. Wang, K. et al. 1984. "Rotating Biological Contactors," *Effluent Water Treat. J. (G.B.)*, 24:93.

47. U.S. EPA. 1985. "Review of Current RBC Performance and Design Procedures," EPA-600/S2-85-033.

48. Strom, P. E. and J. C. Chung. 1985. "The Rotating Biological Contactor for Wastewater Treatment," *Adv. Biotechnological Processes*, 5:193.

49. Brenner, R. C. et al. 1984. "Design Information on Rotating Biological Contactors," EPA-600/2-84-106. U.S. EPA, Munic. Environ. Res. Lab., Cincinnati, Ohio.

50. Huang, C. H. et al. 1985. "Treatment of Phenol-Formaldehyde Resin Wastewaters Using Rotating Biological Contactors," *Proc. 40th Ind. Waste Conf.*, Purdue University, West Lafayette, Ind., p. 729.

51. Collins, A. G. et al. 1988. "Fixed-Film Biological Nitrification of a Strong Industrial Wastes," *J. WPCF*, 60:499.

52. Dallaire, G. 1979. "U.S.'s Largest Rotation Biological Contactor Plant to Slash Energy Use 30%," *ASCE, Civil Eng.*, 49:70–74.

53. Sagy, M. and Y. Kott. 1990. "Efficiency of Rotating Biological Contactor in Removing Pathogenic Bacteria from Domestic Sewage," *Water Res.*, 24(9):1125–1128.

54. Tait, S. J. and A. A. Friedman. 1980. "Anaerobic Rotating Biological Contactor for Carbonaceous Wastewaters," *J. WPCF*, 52(8):2257–2269.

55. Kuroda, M. 1983. "Municipal Wastewater Treatment with Anaerobic-Aerobic Rotating Biological Contactors," *Jr. of Water and Wastes*, 25(10):1029–1036.

HEALTHCARE
DOLLARS AND SENSE

How Classical Chinese Medicine Can Save Your Health—
and Your Company's Bottom Line

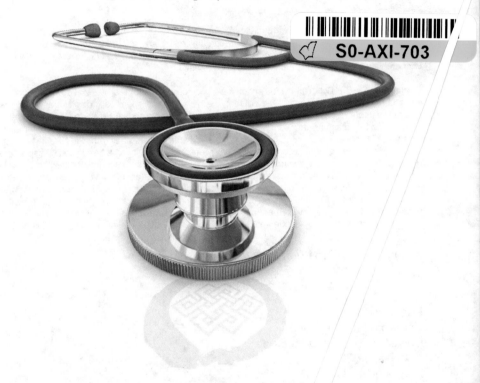

Dr. Tad Sztykowski, ￼.AC
with Bryna René

universal
TAO

HEALTHCARE DOLLARS AND SENSE

How Classical Chinese Medicine can Save Your Health—
and Your Company's Bottom Line

ISBN 978-0-9828384-0-2. (paperback)

Published By:
Universal Tao Publishing
Center for Preventive Medicine
191 Nashua Street
Providence, RI 02904
U.S.A.
401-434-3550
www.thewellnessclinic.net

Back Cover Photo:
Nicole Gesmondi Photographer, LLC
www.nicolegesmondi.com

Acknowledgements

Fifteen years ago, I decided that I wanted to write a book about my experience with Classical Chinese Medicine. I added a file to my file cabinet. I labeled it "book."

For thirteen years, I worked to grow my practice. I saw thousands of patients. I helped a lot of people. But that file was still empty.

Then, I met Bryna René, a writer from Providence. She helped to put the whole project together. My deepest thanks to her for making this long-time dream a reality.

I'd also like to thank Patricia Raskin for her encouragement. Thanks as well to Lisa Tener for giving this project direction (and for introducing me to Bryna).

Thanks to my children—Zosia, Bart, Julia, and Anna—for being so patient and understanding during all the weekends I dedicated to working on this project. And, of course, my deepest love and gratitude to my wife, Kasia, who supports me in everything I do, and without whom I couldn't have done any of it.

Last but not least, I'd like to thank the thousands of patients who encouraged me to write this book, and who helped me to prove, beyond the shadow of a doubt, the efficacy of Classical Chinese Medicine in our modern world.

Table of Contents

Introduction

MY PATIENT CHERYL suffers from Lyme disease. She developed lesions on her brain which caused her to experience short-term memory and cognitive problems, and gave her permanent "brain fog." Seven years of antibiotics weakened her immune system and liver to the point of near non-functionality, and eventually predisposed her to breast cancer. The chemotherapy she received to combat the cancer devastated her immune system even further. She was hospitalized for dangerously spiking blood pressure and shortness of breath, and underwent four surgeries on her knees due to injuries sustained after the Lyme disease compromised her balance.

Cheryl was deemed disabled by the State of Connecticut. She lost her job as a high school teacher, her house (which before her illness had been paid for in full), and her hope for the future. Not including lost work time, her medical bills totaled more than $225,000. But despite all the care she was receiving, she wasn't getting any better.

She came to my office after sixteen years of ineffective treatment by Western doctors, naturopaths, and other practitioners. Within a few visits she went from sleeping thirteen hours a night and waking up exhausted, to sleeping eight hours a night and waking up refreshed. Her brain fog, short-term memory problems, and panic attacks began to dissipate, and her blood pressure stabilized. After three months of treatment, she announced that she was feeling "70% better," and enjoying more days without pain than with it.

Today, Cheryl is more active than ever, and she's beginning to get her life back. She has been able to discontinue all of her prescription medications, including the antibiotics that caused her so many problems. Most days are totally pain-free. Her cardiologist, after numerous tests, gave her a clean bill of health for the first time in more than a decade.

The total cost of Cheryl's treatment by Classical Chinese Medicine? $6,224.

You may be wondering how this is possible. How could a disease that cost a woman her job, her health, and even her home be treated so quickly and successfully through acupuncture and herbal therapy? How could sticking needles in someone succeed where years of potent pharmaceuticals failed?

It's possible because Classical Chinese Medicine gives the body the tools it needs to heal itself. Where pharmaceuticals attempt to correct imbalance and disease in the body by introducing another imbalance in the form of a chemical compound, CCM helps the body achieve internal balance—also known as homeostasis. When Cheryl's body was encouraged to return to a more balanced state, she experienced a cascade of healing reactions.

The concept of balance is truly at the core of Classical Chinese Medicine, and it's this concept which not only encouraged me to leave my practice of Western medicine, but to radically change the way I care for my patients.

If we as a nation want to help people like Cheryl get their lives back after years of chronic illness, and prevent people in the beginning stages of chronic illness from going through the physical, emotional, and financial trauma that Cheryl suffered, we need to look at health care reform in a new light. We need to stop relying on methods and medications that don't work, and change the way we think about health care.

Since 1987, I have worked to further the practice of acupuncture and Classical Chinese Medicine in America. The successes I have seen in my own patients and those of my colleagues have surpassed anything I witnessed while practicing Western medicine.

Although I am a doctor of Classical Chinese Medicine, this is not a book about acupuncture. Rather, it's about how Classical Chinese Medicine can help individuals and businesses reduce health care costs while enabling employees to feel healthier and be more productive. By understanding how chronic diseases, and Western medicine's treatment of them, truly impact the cost and efficacy of our current health care system, we can make proactive care choices which truly benefit our families, our employees, and our nation as a whole.

The data presented in this book, combined with testimonials from my own patients, is eye-opening; some may find it shocking. My intent is to encourage you, the reader, to think about health care in a new way—from a perspective of balance—and to become proactive in caring for yourself and others.

In the five sections of this book, we'll discuss why Western Medicine so often fails to work for the treatment of chronic disease, and why Classical Chinese Medicine is a safe, viable, and cost-effective alternative (or supplement) to standard Western health care not only for individuals, but for small businesses and corporations. We'll explore thirteen common chronic conditions which affect American workers, discover the hidden costs of these illnesses to employers, and learn how, in many cases, the "cures" for these ailments—i.e., pharmaceuticals, surgeries, and other common Western treatments—may not only fail to help patients, but may result in greater illness. In Sections 4 and 5, I'll present ideas to help you create a wellness plan for your company, or enhance the services you already offer. These simple steps can help you and your employees learn to live more healthfully, reduce stress, and help prevent or lessen the impact of chronic disease.

My plan to help individuals and businesses create a "culture of wellness" is a plan for the future. I hope the information contained in this book will open your eyes to the possibilities offered by Classical Chinese Medicine, and enable you to make positive changes in your own life and the lives of your employees, friends, and loved ones.

Yours in Health,
Dr. Tadeusz Sztykowski, MD (Lic. in EU), D.Ac.

A note on CCM: In this book, I refer to the medicine I practice as "Classical Chinese Medicine" or "CCM" rather than the more familiar "Traditional Chinese Medicine" or "TCM" to differentiate it from some of the modalities currently being practiced in this country. My practice draws from the great tradition of the *Huang Di Nei Jing*, the medical text commissioned by the Yellow Emperor, which has been used to great success for more than two millennia. I have combined my knowledge of "the Emperor's medicine" with my experience as an MD and OB/GYN in Poland to create a practice in the United States which I believe truly represents the best of both worlds.

THE GREAT AMERICAN HEALTHCARE SYSTEM

*"What do you mean, mammoths are supposed to be extinct?
I'm here, aren't I?"*

Section 1:
What is Health—and Why Don't Americans Have It?

WHEN IT COMES down to it, Americans are fairly unhealthy people. Two-thirds of our adult population is overweight or obese, and that number rises annually. Heart disease, high blood pressure, high cholesterol, Type II diabetes, and other lifestyle-related conditions are prevalent. Cancer—many instances of which are also related to lifestyle—is one of our nation's biggest killers. To treat these conditions, and the various secondary conditions and symptoms associated with them, we consume 8–10 times more prescription drugs than any other developed nation. The side effects of these medications, and the often unforeseen interactions between multiple medications, have a further negative impact on our collective health.

How sick are we, really? Let's crunch the numbers:
- 67% of American adults are overweight or obese.
- 30.8% of adults over age 20 suffer from high blood pressure (hypertension), and another 18.7% have elevated blood pressure or are "pre-hypertensive" (which means they will likely be hypertensive in the future unless treated).[1]
- 106 million people—nearly half of American adults—have serum cholesterol levels of 200 mg/dL or higher. About 17% of American adults have serum cholesterol levels of 240 mg/dL or higher.[2]
- 10% of our adult population (more than 20 million people) have Type II diabetes, and 41 million more are classified as having "pre-diabetes" or hyperglycemia.[3]
- 1 of every 2 men, and 1 of every 3 women, will develop some form of cancer in their lifetime.[4]

These percentages are staggering, and will only continue to grow until we take some basic steps to help people get healthy.

There are plenty of good reasons for us to take those steps, including the fact that every human being deserves to be as healthy as they can be. From a purely economic perspective, our ill health as a nation, combined with the ineffectiveness of our current health care system, generates billions of dollars in costs for American businesses. Our national health care spending in 2008 exceeded $2.5 trillion, and in early 2009, President Barack Obama determined that health care costs are "the biggest driver of long-term deficits." Obesity alone (exclusive of the deadly and debilitating conditions to which it is considered a precursor) costs our country an estimated $9 billion a year in additional health and disability insurances. Heart attack and stroke, both of which are largely preventable conditions, cost Americans an estimated $448 billion in 2008, according to the Centers for Disease Control and Prevention. Back pain, which many people don't even think of as a disease, racks up a tab of over $100 billion per year.

This information is important to employers for two reasons. First, the billions of dollars spent to treat preventable conditions like obesity, heart attack, and Type II diabetes are coming, in large part, out of the pockets of American businesses. Second, the economic impact of ill health reaches far beyond the cost of treatment into the realms of productivity, employee longevity, and quality of life.

Ill health in the workplace, whatever the cause, creates a "see-saw" effect. As employees get sicker, productivity goes down, while health care and sick leave costs rise. Everyone suffers. On the other hand, healthy employees are more productive, more alert and focused, less likely to take sick leave, and better equipped to deal with stress. When it comes down to it, employers' best interests lie in doing everything in their power to help employees get healthy and stay that way.

If you're thinking that all you'll need to do to achieve this is build a gym on the premises and stock up the cafeteria with apples instead of cookies, I'm sorry to say that it's not always so simple. Although neither of those things will hurt, chances are it will take more than improved diet and exercise habits to make your work force healthy. Even some of the so-called "health nuts" in your company may not be as healthy as they seem; rather, they're just *not sick*.

Section 1:1—What is Health?

To become healthy, first we must understand what health actually is.

Health, as defined by the World Health Organization, is a state of complete physical, mental and social well-being and not merely the absence of disease or infirmity.[5]

If I were to go out onto the street and start asking passersby if they think they are healthy, a lot of them would say "yes," because they have no major symptoms. People tend to ignore minor symptoms in their bodies, which indicate easily-correctable imbalances, until they become the major symptoms which indicate disease. In twenty-five years of dealing with patients from every walk of life, I've learned that people love to minimize, to "soldier on"—as if to admit that they don't feel well would somehow make them less valid as human beings. If one of these "minimizer" types gets a major headache once a week, he will say it's no big deal. If another gets a severe backache once or twice a month, that won't be a big deal either. If a third is too stressed out to sleep three or four nights a week, it's no big deal; he can "learn to live with it."

I ask these people: Would you learn to live with a leaky roof over your head, or a flat tire on your car? Of course, all of us *could* learn to live with these things if we had no other choice, but it would be tremendously inconvenient, and impinge upon our usual quality of life. No person I know would drive on a flat tire for longer than it took them to get to the nearest gas station. Just because people *can* live with the symptoms they are experiencing does not mean they should! The body can take a lot of abuse for a long time, but not forever. Eventually, that little leak in the roof will turn into a cave-in. When things are out of balance for too long, the body starts to break down, and that's when we see evidence of chronic disease.

In order for a person to be truly healthy, there needs to be a balance between his physical, mental, and social health, as expressed in the World Health Organization's definition above. Even a person who spends four hours a day at the gym may be unhealthy and imbalanced if he is under an extreme amount of emotional stress. In time, that imbalance will manifest itself in the form of weakened adrenals, poor sugar metabolism, headache, fatigue, chronic pain, or any number of other conditions triggered by the body's response to stress. In other words, his imbalanced psyche will begin to adversely affect his healthy body, and he will become physically sick, gym or no gym. The reverse is also true: if a normally happy and well-adjusted person sustains a traumatic physical injury, his psyche will likely be affected in a negative way, at least temporarily. Mental health and physical health are inseparable from one another, because both mind and body are part of the whole person.

One of my patients is a personal trainer at a gym in Providence, Rhode Island. On the outside, he is a picture of radiant health: muscular, toned, and tanned. But when he came to my office, our conversation revealed that he was, in fact, anything *but* healthy. The pressures of his job, and the constant need to stay physically "perfect," were beginning to wear on him. His body was constantly achy, and his joints felt inflamed. He took three different medications daily—including an antidepressant, which made him feel groggy and numb. He felt terrible about himself, and terrible in general.

By concentrating so much on his physical health, this patient inadvertently created an imbalance that compromised his mental health. That imbalance, in turn, manifested as physical symptoms. However, once we uncovered the real issue and took steps to address the effects of his emotional stress on his body, the physical symptoms "miraculously" disappeared.

The third aspect of health, social health, is a concept which is infrequently addressed in Western medicine. Social unhealthiness can present in a number of ways, including psychotic, aggressive, and antisocial behaviors. Many times, these behaviors are a direct result of physical or mental ill health.

Perhaps the easiest way to understand this concept is to examine people in your own workplace who suffer from chronic physical ailments. Many—although not all—will have trouble interacting normally with their peers for reasons of depression, social anxiety, low self-esteem, shyness, bitterness, or anger.

All of these symptoms can appear when a person's physical health is poor; in fact, it's relatively common for them to do so. But because we have forgotten the true definition of health in this country, social and emotional conditions are just as likely to be treated as new or separate diseases. This results not in healing, but in greater imbalance and more disease.

We'll explore all of these concepts in greater detail later in this book, but in the end, it all boils down to *awareness*. When we become aware of what health is, when we learn to pay attention to our bodies and the messages they are sending us, when we strive to find real health and not just ways to "live with it," then our whole perception of health as a goal changes. We become more in tune with the inner balance that is the basis of true health.

If we want to help the American work force get its collective health back, we must return to the fundamental definition of health, and start basing all of our diagnoses, treatments, and medical methodology on that definition. We must understand that all aspects of our physical, mental, and social selves are inextricably connected, and that they cannot, by disease or otherwise, be separated into convenient, manageable parts. In the same way that cells viewed through a microscope don't look a thing like the drop of blood or strand of hair from which they were taken, even though they were once part of that whole, we cannot look at one symptom or one bodily system as separate from the rest of the body. We need to see the big picture. When we treat an ill person, we need to treat the *whole* person.

At this point, Western medicine does not do any of these things. And that's why health care in America is failing.

Section 1:2—West vs. East: A Basic Comparison

The science of medicine has exploded over the course of the last century, taking us further than we ever believed we could go. It's given us everything from the polio vaccine to neurosurgery, the laparoscope to the artificial heart. Through this brilliant science and the technology which complements it, Western medicine works miracles for many people. I witnessed many of these miracles myself in my tenure as an MD; I even facilitated a few of them. As a young doctor, it often seemed to me that there was nothing modern medicine couldn't do, and no problem it couldn't solve.

But the more I learned, the more I began to realize that Western medicine is far from perfect. In fact, when it comes to treating chronic or systemic diseases—the kinds of diseases from which Americans currently suffer *en masse*—Western medicine is deeply flawed.

The fundamental problem with Western medicine lies at the core of the practice, in its methods of diagnosis and treatment. Western medicine is totally *reactive*, meaning that it looks for the obvious problem, hones in on that problem, and tries to fix it. This approach works beautifully with acute illnesses or when dealing with traumatic bodily injuries, because in these cases there is a clear and logical progression from symptom to diagnosis to treatment.

For example, in a patient suffering from appendicitis, the appendix becomes inflamed, and produces symptoms like nausea, vomiting, and severe abdominal pain. If the appendix is not removed promptly, it may rupture, and the patient may die. To treat this condition, the Western doctor opens the patient's abdomen, removes the appendix, and closes the incision. After a few weeks and a course of antibiotics to prevent infection, all is well again.

Of course, the above example is a gross oversimplification, but it is nevertheless revealing. Western medicine is at its best when there is an obvious problem and an obvious solution. When it comes to trauma care, there's nothing in the world to match it. Modern medical technology can mend broken vertebrae, repair severed arteries, replace shattered knees and hips—even give someone a whole new face. It's out of these acute, traumatic cases that medical miracles are born.

In my experience, only about 20% of patients who walk into a hospital require care for acute illnesses or injuries. The other 80% suffer from chronic diseases like coronary or respiratory illnesses, systemic illnesses like Type II diabetes or cancer, or inflammatory diseases like osteoarthritis or COPD. These people need a different approach to care, because the solutions to their health problems aren't so clear-cut.

When treating chronic conditions, standard Western practice is to prescribe medications, most of which target the symptoms but not the actual problem. When that fails, Western medicine attempts to improve the affected area—the "broken bone," if you will—with surgery. Thus, even with chronic or systemic illness, Western medicine reacts as if to an acute, linear medical problem. It virtually ignores the fact that every system in the human body is linked together in a vast, interconnected tangle of cells, hormones, and electrical impulses, and that what impacts one bodily system will inevitably have some impact on the body as a whole. For example: while popular statin drugs like Lipitor or Crestor do in fact reduce serum cholesterol, they can also have an intense and destructive impact on the digestive system, the musculoskeletal system, and even the central nervous system. (We'll explore the adverse effects of prescription drugs in more detail in Section 3.)

My point is this: you cannot treat just one area of the body and expect to cure a systemic disease. And yet, that's exactly what Western medicine tries to do.

So, where did we go wrong?

The practice of the Western style of medicine goes back to the Greeks. Through observation and experimentation, Greek physicians were able to draw conclusions about the way certain practices, medicinal compounds, and lifestyle changes worked on the body. They made some startlingly astute observations, and amassed a great deal of information. Evidence of Greek influence is plain even today: many doctors still swear to abide by the Hippocratic Oath, named for the famous Greek physician Hippocrates.

Although they improved little on the pure science of the Greeks whom they eventually conquered, the Romans also made tremendous advances in medicine, particularly in the area of public health. They recognized the connection between poor public hygiene and disease, and took steps to improve sewer systems, drinking water, and the cleanliness of their citizenry.

Roman bathhouses can still be seen throughout Western Europe and the Mediterranean.

Unfortunately, much Greek and Roman knowledge was lost after the fall of the Roman Empire. During the Dark Ages, even the simplest and most ancient medical practices (such as the use of herbal remedies) were all but eradicated by superstition and religious fanaticism. The Renaissance, however, brought a frenzy of rediscovery of classical ideas and ideals, including those related to medicine. The emerging physicians of that era—including Thomas Linacre, founder of the Royal College of Physicians—used the ancient Greek texts as a foundation, and began to add their own research and philosophies. Still, it wasn't until the late 1800s that we began to see methods resembling those we use today.

From the Renaissance to this day, doctors have used cadavers as tools to learn about the human body and the way it operates, in addition to (or even in lieu of) observing their living patients. While this has produced some immensely valuable discoveries, it has also narrowed the focus of Western medicine by fostering the idea that each organ operates like a machine on an assembly line, communicating with the rest of the body only when necessary to perform its assigned task.

I do not say this to belittle the research of those doctors who discovered (and continue to discover) the incredibly complex structures and functions of our organs and systems—but such discoveries need to be integrated with a larger viewpoint in order to maximize their potential use to medicine. A single blade of grass, all by itself under your magnifying glass, will never show you which way the wind is blowing; to see that, you need to look at the whole prairie. In the same way, you will never discover all the ways in which an organ or tissue interacts with the rest of the body by studying its smallest parts, no matter how powerful your microscope. The idea of "one organ, one function" has led Western medicine into a cul-de-sac, and it's having a very hard time finding its way out.

Medicine in Asia evolved quite differently. The practice of Classical Chinese Medicine originated as early as 8,000 years ago. Through observation, tests, and old-fashioned trial and error, the Chinese were able to discern the effects of hundreds of different situations, stressors, herbs, and other treatments on the body—just as did the Greeks, and the doctors of the Renaissance after them.

The core differences between Chinese and Western methodologies stem from the fact that the Chinese made a conscious practice of observing the entire patient, not just the specific system or condition being treated. The body was regarded as far greater than the sum of its parts: not merely a complex and finely-tuned machine, easily repairable if one but knew which cog or gear was causing the malfunction, but rather a universe unto itself, complete with its own internal weather, landscapes, and energetic forces.

The famed Chinese medical text *Huang Di Nei Jing* (which translates as "The Yellow Emperor's Inner Canon") compares the human body to a whole country, governed by rulers and officers in much the same way as any nation: "*The heart is the prince who governs through Shen [the mind], the lungs are his officers and promulgate laws, the liver is a general who works out strategies…*"[6] This perspective gave ancient Chinese doctors a more thorough understanding of the human body *as a single functioning unit* than we have in Western medicine to this day.

It is my belief that Chinese religion and philosophy played a major role in the evolution of their medical practices. The Buddhist and Taoist traditions are more apt to see people not solely as individual, self-contained beings, but as tiny pieces of a greater universal whole. Ancient Chinese doctors looked at medicine in the same way. They knew that nothing in the universe or inside our own bodies exists in a vacuum. Yes, each cell, tissue, or organ is important on its own, and without it the whole person cannot function—but without the whole person to contain it, and other systems to interact with, it has no role to play.

Those ancient doctors also knew that everything in the world, from the whole of Nature to the tiniest cell, needs to be balanced to be healthy. Therefore, the aim of any CCM treatment is to return the body to a state of *homeostasis*, or internal balance. When the body is balanced, it can regulate itself without any external assistance.

One of the foremost modalities associated with Classical Chinese Medicine is acupuncture. While most people have some idea of what acupuncture entails, few actually understand the theory of the practice.

As I stated in the introduction, this is not a book about acupuncture. I will not spend pages detailing the specific acupoints used to treat particular conditions. (If you're interested in learning more about the mechanics of acupuncture, please refer to the Recommended Reading List.) However, in order for you to better understand the information presented in this book, a brief description of the practice is in order.

Acupuncture involves the use of tiny, sterile needles that can, when inserted at certain points on the body, cause the body to react in a certain way. These "acupoints" lie along the fourteen meridians, or energy channels, in the body, which roughly correspond to the major pathways of the nervous and circulatory systems. By stimulating *chi*—which some describe as life force energy, but which literally translates as "air" or "essence of air"—at particular points along the meridians, the body's own attention to a particular problem can be encouraged. The eventual response to this trigger is an improvement of symptoms, followed by a cascade of healing reactions. Healing is complete when homeostasis is restored, and the body is once again able to self-regulate.

When put in these terms, it sounds simple, but the practice of acupuncture is incredibly complex, and relies not just upon the practitioner's knowledge of the meridians and acupoints, but upon careful and meaningful diagnoses. Diagnostic techniques might include visual observation of the patient, meridian and abdominal palpation, Chinese Tongue Diagnosis, pulse diagnosis, and other methods, combined with careful questioning of the patient to uncover all complaints (even those seemingly unrelated to the primary complaint) and ascertain the overall condition of the body. Acupuncture never aims to treat only the obvious symptom—the "broken bone"—as Western medicine does; rather, it looks for the *reason* the bone is broken, and works to ensure that the cause, as well as the symptom, is eradicated.

The diagnostic methods used in Classical Chinese Medicine are geared toward discovering the symptoms of disease—termed "complaints"—but also toward identifying the underlying imbalance or imbalances which caused symptoms to develop in the first place. By searching for patterns in a patient's body, the experienced practitioner can locate the root imbalances which are the underpinnings of disease. Like the iceberg that took down the Titanic, disease lies mostly beneath the surface: 80–90% of possible damages have to occur before symptoms rear their heads above the waterline. CCM strives to find the "icebergs" growing under the surface before they rip a hole in the starboard hull.

Acupuncture is effective because it uses the body's own most efficient resource—*chi*—to relieve symptoms and eliminate disease. It causes no painful side effects, and presents no possibilities for overdose or allergic reaction. Properly performed with sterile needles (as U.S. regulations require), it presents no danger to the patient's health. It can generate immediate and long-lasting symptom relief, and over time can return the body to homeostasis and keep it there.

In short, acupuncture *works*, which is the first and best reason that I practice it.

Other aspects of Classical Chinese Medicine resonate with me as well. Prevention is at the heart of this medicine, and I want to practice medicine in a way that is truly preventive. I don't just want to make sick people well: I want to *keep* people well, and CCM provides the most effective way for me to do that. I feel that I am perpetuating true health among my patients, and that is far more rewarding than anything I experienced while practicing Western medicine.

CCM is deeply rooted in tradition, and I love tradition— when it works. "The proof is in the pudding," as they say: billions of people over thousands of years have used this medicine, and it continues to work for billions around the world today. In fact, even in today's high-tech world, Classical Chinese Medicine treats more people every day than any other medical modality.

Finally, CCM empowers the patient, rather than disempowering him. Western medicine, with its scientific complexity and attention to microscopic detail, tends to be unreachable to the average person. When engaged in Western medical treatment, the patient puts an enormous amount of trust in his doctors, and the system itself, but at the same time he may feel powerless to understand or control what is happening in his own body. If medications and surgeries fail to cure his condition, he begins to believe that he must accept the imbalanced and often painful state of his body as "normal," because there's nothing more to be done for him. He becomes disempowered. On the other hand, CCM requires not only that the patient present himself for treatment, but that he become an active participant in his own recovery. As he learns how to balance his body and take control of his own health, the healing process becomes self-perpetuating. I've seen this happen again and again, in patients with conditions ranging from mild hypertension to Stage 4 cancers.

There are dozens of individual modalities contained within the greater whole of Classical Chinese Medicine, including acupuncture, acupressure, moxibustion, manipulation and massage (including Tui-Na and Gua Sha), herbal therapy, nutritional therapy, and meditation. Acupuncture is the most widely studied of these by Western scientists, and is often the "gateway" for patients who are new to CCM. I use all of the above treatments to some degree in my own practice, but acupuncture is my treatment of choice when working with new patients or people with severe symptoms, because it produces the most rapid and measurable results.

In Section 3 I'll share with you some of the amazing results of acupuncture treatment from both my own clinical practice and national studies, and demonstrate why Classical Chinese Medicine is truly the best medicine for prevention.

Having practiced both Western medicine and Classical Chinese Medicine in my career, I can attest to the merits and pitfalls of each. As we discussed earlier in this section, Western medicine is inarguably superior for the treatment of acute illness or trauma; acupuncture, for all its efficacy, cannot mend a severed artery or realign a shattered spine. On the other hand, none of our modern technologies have made Western medicine adept at treating chronic disease (or at preventing its onset), but Eastern medicine is perfectly suited to dealing with imbalances in the internal landscape which lead to these types of conditions.

If we want to improve our health care system in America, we need to integrate the Eastern modalities and philosophies which have proven effective over the course of millennia. The key is to change our strategy from one of *reactive* treatment focused on a single condition or disease, to one of *proactive* prevention which addresses the body as an integrated whole. Only by doing this can we save our national health, and reduce our out-of-control health care spending.

There is a passage in the *Huang Di Nei Jing* which reads something like this: "To treat an illness is like digging a well when you are already thirsty." The better solution, of course, is to dig the well while one has a good store of water already to hand, in preparation for later want. We need to take steps to protect our well-being while we are still relatively healthy and strong, not when we're tired and achy and full of disease. We need to fix the leaky roof before it starts to rain, so we are not swinging a hammer in a thunderstorm. *We need to become proactive.*

Once we accomplish this, we will be able to provide Americans with the high quality of health care they deserve, without the enormous price tag.

Creating a proactive, efficient, and cost-effective health care system in America is important to everyone—but perhaps most especially to employers, who currently bear the brunt of the cost of our malfunctioning system. We will delve into the reasons why employers should take the initiative in health care in Section 2—but first, let us take a comparative look at treatment versus prevention.

Section 1:3—Treatment vs. Prevention

American medicine in its current state is rather like a national penal system: once you're in, you're probably in for life. This is not because doctors, nurses, and surgeons do not want patients to get well; of course they do. It is because Western treatments for chronic disease don't work.

Doctors are trained to suppress or "manage" symptoms, rather than correct the underlying issues which cause them. Painkillers and analgesics can mask the symptoms of osteoarthritis, but they will not cure the arthritis. Insulin can manage sugar levels in a patient with Type II diabetes, but will not restore his liver and pancreas to normal function. In the same way that it costs taxpayers up to $30,000 per year or more to keep an inmate locked in a state penitentiary, Americans shell out billions, even trillions, of dollars every year to keep their internal imbalances, well, *imbalanced*, without ever realizing what they are doing.

To correct this, we need to start asking the right questions. How much less would it cost this country if we could prevent kids from going to jail in the first place? How much less would it cost if we could prevent, rather than treat, our most common chronic diseases?

The answer is: *billions*. A comprehensive nationwide preventive strategy, coupled with the option of insurance-reimbursed treatment with CCM and other "alternative," non-pharmaceutical modalities, could literally cut our health care spending in half.

So what, you may be wondering, does prevention look like, and how can we implement it?

As an example, let us look at a person you may know—one of your employees, perhaps, or one of your coworkers. She is in her late 40s, maybe early 50s. She is overweight by 50 pounds or more, considered obese by medical definition. She suffers from high blood pressure, high cholesterol, and Type II diabetes. She doesn't sleep well, because she suffers from sleep apnea (caused by her obesity). She has constant and sometimes debilitating lower back pain. She often feels stinging or tingling sensations in her feet and lower legs; her doctor has told her this is due to diabetic neuropathy.

She takes five or more prescription medications per day at a cost of hundreds of dollars per month (a cost from which she may or may not be shielded by her employer-provided insurance). She visits her primary care physician and a host of specialists several times a year (to the tune of thousands or tens of thousands of dollars), and may even have been hospitalized a few times for acute symptoms.

Despite all the treatments she has undergone, and all the pharmaceuticals she takes, she isn't getting any better. In fact, her overall health gets worse every year. And although she still shows up to work on time, and is still good at her job, she takes far more sick days than her coworkers, and often has trouble focusing at work because she's worried about her health.

To top it all off, this person has no real idea how she got to this place. She feels like a prisoner in her own body. She is a slave to her prescriptions, and is limited both physically and mentally by her conditions. And, as far as she can tell, there is nothing she can do about any of it except what she's already doing. She's starting to give up hope that she'll ever be healthy again. Recently, she's been experiencing symptoms of depression, and is considering adding yet another prescription—an SSRI like Prozac—to her daily regimen.

Does this person sound familiar to you? She should. She's one out of five people in this country, part of the collective 20% of American citizens who generate 80% of this country's medical spending.

And every one of her conditions could have been prevented (or at least effectively treated) if the right structures had been in place to support her.

Nobody wants to be sick, but most people have no idea how *not* to be sick. Lack of objective information about healthful lifestyles and prevention, along with the dangerous notion (fostered by pharmaceutical companies) that modern drugs are the answer to all that ails us, have combined to take power away from the patient and put it in the hands of providers.

Now, patients have to turn to doctors, specialists, and pharmacists for everything, because they don't know how to help themselves. Unfortunately, those providers, for myriad reasons, were not able to stop those patients from becoming sick in the first place, and now can only offer an endless cycle of drugs and "disease management."

Before we continue this discussion, I want to make something abundantly clear: I don't blame the good-intentioned, hardworking doctors of this country for any of this. They are doing the best job they can do with the knowledge and training they have. They are just as prone as their patients, if not more so, to manipulation and misinformation: even studies published in major medical journals may be skewed to further the agenda of the drug companies—who, after all, fund 80% of medical research in America, including trials of their own products.[*] Also, the incredibly high overhead associated with operating a medical practice in this country necessitates that patient care be less personalized than it was in the past. Some doctors have to see twenty or more patients before lunch just to pay their bills. In many cases, people spend more time in the drive-thru line at McDonalds than they do with their physician! This time crunch puts a huge strain on doctor-patient communication, and therefore on preventive care.

Now, let's return to our hypothetical patient. Her problems are not new ones—but they aren't lifelong issues, either. Let's assume she visited her doctor about ten or fifteen years ago, when the weight first started to settle around her midsection. At that time, her major complaints were sleeplessness, stress, and an achy lower back. The doctor probably took as much time with her as he could while adhering to his punishing appointment schedule. All of her symptoms were subjective in nature—meaning, there was no way to prove their existence through the kinds of tests usually performed in a doctor's office.

It's likely that, relying on the evidence gathered during this short consultation, the doctor would have prescribed a sleep aid to combat our patient's insomnia, and a pain reliever or muscle

---

[*] While I am not at leisure to discuss the tragic state of American medical "science" in this book, those interested in learning more should pick up Dr. John Abramson's eye-opening book, Overdo$ed America. Please see the Recommended Reading section for bibliographic citation.

relaxant for her nagging lower back pain. Then, he might have advised her to stop eating junk food and lose ten pounds—or, if at that point she still fit (however narrowly) within accepted guidelines for body weight, he might not have mentioned her weight gain at all.

The result of this examination is obvious: the patient received treatment for symptoms she claimed to be exhibiting (insomnia and low back pain), but the imbalances causing those symptoms were never uncovered or even looked for. Maybe the drugs helped for a while, and maybe they didn't, but the information and preventive care our patient needed to halt the progression of her conditions was never offered.

If a preventive philosophy had been applied when her symptoms first became apparent, however, our patient's later ill health might have been avoided, and her quality of life better-preserved. If she had been to see a doctor of Classical Chinese Medicine (or even a Western doctor or naturopath who subscribed to the idea of preventive medicine), our patient's story might have been quite different.

The doctor, able to take the time to complete a thorough interview as well as a physical exam, might have learned that the patient was under severe emotional stress because of a situation at work or at home; that her racing thoughts were keeping her awake; that she was eating too much junk food because she was feeling hungry all the time and sweet, fatty foods were the only thing that satiated that hunger; that she had stopped walking in the morning because she was too tired from lack of sleep; and that her lower back felt achy.

As you can see, things look a lot different when you get the whole story. The underlying cause of every one of her issues wasn't her insomnia, her excess weight, or even her back pain. It was her stress. After hearing this, our preventive-minded doctor might have recommended specific dietary changes based on our patient's individual constitution, such as the reduction or elimination of caffeine and sugar.

He might have asked her to find time to meditate (or just sit quietly) for ten to fifteen minutes a day to settle her racing thoughts, mentioned that soothing music or meditation CDs played before bed can help calm the mind, and explained that even ten minutes of light exercise in the morning could be enough to make her feel more energetic, alleviate her lower back pain, and improve her sleep. The doctor might then have told her that eating junk food late at night can negatively affect sleep patterns, and offered a basic explanation of metabolic function to help her better understand the recommendation.

If the doctor was versed in Classical Chinese Medicine, treatments in those modalities would also have been applied to bolster her adrenal function, combat the effects of chronic stress, and help bring her body back into balance. Lastly, our doctor might have recommended that the patient contact a qualified mental health professional to help her navigate the emotional situation which was the true cause of her physical symptoms.

In this hypothetical situation, not only did the second, preventive-minded doctor uncover the true cause of our patient's complaints—the stressful situation in her life—he was able to give her a sense of control by suggesting specific tasks for her to perform in pursuit of her own well-being. In my experience, this type of "interactive" patient care is by far the most effective because, at the same time as the doctor helps to heal the body, the patient develops positive habits that enable her to maintain her health long after the initial treatment is over.

If our patient had known how to take control of her health fifteen years ago, before her imbalances swelled into major illnesses, she might be a very different person today.

Preventive techniques applied to patients today can reduce the possibility that their health will decline in the future. If they remain healthy, they will not require expensive Western treatments to "manage" their conditions, and will enjoy a better quality of life.

In Latin, the word "doctor" translates literally as "teacher." A *doctoris medicus* is literally then a "teacher of medicine." If American doctors could learn to see themselves as teachers, rather than providers, it would go a long way toward empowering their patients. Instead of just dispensing care, practitioners need to foster the ideas of prevention and personal responsibility in order to help patients achieve what they want most for themselves: good health and long life.

This is also where employers come into the picture. We can be relatively sure that insurance conglomerates and pharmaceutical companies will not encourage a strategy of prevention within the current health care system, because they are in the business of treatment. Employers, on the other hand, are in a unique position, because although they are enmeshed in the current system they are more its victims than its beneficiaries. Therefore, employers can and should strive to create a "culture of wellness" among their employees—and, in doing so, improve their own bottom line and the American health care system as a whole.

Section 2:
Employers and the American Health Care System

SINCE THE 1940s, employers have played a significant role in America's system of health care. Employer-sponsored benefits were a big incentive during the World War II years, when employers were competing fiercely for qualified workers. Today, about 59% of Americans receive their health insurance through their employer. And while employee contributions vary greatly from company to company, in most cases the employer shoulders the bulk of the cost burden.

We are literally the only developed nation with an employer-sponsored health care system. Germany implemented its national health care policy in the 1930s. France, the United Kingdom, Norway, Finland, Canada, Japan, and Taiwan—to name a few—have all instituted single-provider systems, through which the government provides universal health care for all citizens. Other nations, like Switzerland, have opted for programs which mandate health insurance for all citizens, but regulate the premiums which private insurers can charge.

It is significant that, although Western methods of health care are practiced throughout Europe and most of the rest of the world (including China, where attempts have been made to integrate it with Classical Chinese Medicine), no other country's per capita costs are as enormous as America's.

There are several reasons for this. First, doctors in Europe are usually paid a salary, not a commission based on the number of procedures they perform. Because their pay is the same no matter what treatments they recommend, these doctors are far less likely to recommend unnecessary care. This is not to say that American doctors are self-centered beings who care only about their wallets: that would be blatantly untrue. But in situations where doctors believe that a patient may benefit equally from two courses of action—from either angioplasty or aspirin therapy, for example—most see nothing wrong with opting for the more expensive, and therefore more lucrative, treatment. Also, European doctors are generally more open to "alternative" methods of treatment, whether used alone or in an integrative fashion. This may or may not stem from the fact that their personal livelihood is not adversely affected by recommending a non-Western approach to treatment.

Many people will argue that we have better health care than European nations because we can decide who treats us. We can choose our doctors and our hospitals, and therefore maintain some control over the quality of our care. "We have the best health care in the world," these people say, "so it *should* be the most expensive." Having practiced medicine both in Europe as an MD, and here in the U.S. as a Doctor of Acupuncture, I can confidently tell you that the so-called "choice" we enjoy in the United States has not made Americans any healthier.

According to recent, well-publicized studies, we are ranked lower in every major measure of national health—including life expectancy, infant mortality, cancer, heart disease, and obesity rates—than Japan, France, Spain, Italy, Canada, and other developed nations.

Second, although they are certainly expensive, it is not critical care or the lifesaving surgeries and pharmaceuticals used in emergency rooms which drive the cost of health care in America to such boundless heights, but rather the care of chronic conditions—a.k.a. "disease management." Fixing a broken arm, or even performing emergency neurosurgery, is a one-time expense. Managing a chronic disease is a lifelong commitment. And, as you'll see in Section 3, pricey drugs and surgeries are not the most effective ways to treat chronic disease.

Third, Americans are simply less healthy than their European and Asian counterparts. Collectively, we eat more processed, chemically altered foods, and exercise less. We also tend to work longer hours with less vacation time, which results in high stress levels. (We will discuss the effects of stress on health in Section 4.) All of these add up to more disease which needs to be treated.

As if our national ill health isn't bad (or expensive) enough, an incredible 46 million Americans are uninsured at the time of this writing. Even though these people may still have access to our vaunted health care system, statistically they will not receive a quality of care equal to that given to people with insurance.

The truth is, not only are uninsured people more likely to forego routine checkups and tests due to prohibitive costs, they are up to twice as likely to die from both acute bodily trauma and chronic disease as their insured peers. Additionally, many uninsured end up in emergency rooms (some of the most expensive places in the medical world) for minor ailments like colds and flu, because unlike doctors in private practice, emergency rooms offer free care. The average emergency room visit costs about $2,000.

Yes, $2,000 to treat a cold. But it's the best cold care in the world.

Am I suggesting that we should abandon our current health care system in favor of a European "socialist" model? No. While there are advantages to a single-provider or "universal" health care system which are precluded by America's industry-driven approach, no health care system is perfect. Even the widely respected health care systems in Germany and the United Kingdom are flawed, in part because the Western medicine they practice is flawed: after all, reactive medicine is still reactive medicine, even if everyone has access to it. My vision for American health care is of a system in which the focus is no longer on "disease management," miracle pills, and quick-fix surgeries, but rather on *preventive, non-reactive care.*

In other words, I want to create a system that can dig a well before we get thirsty.

The White House Committee on Complementary and Alternative Medicine Policy was established in March 2000.†
This committee made some intelligent suggestions as to how "alternative" modalities (like acupuncture) could be incorporated into American health care policy and practice. While this was a great first step, now it's time to "walk the talk," as they say, and start implementing some of those suggestions.

To begin to move in the direction of a preventive care system, preventive care measures need to be incorporated into every aspect of American life. They need to take place in schools, in government agencies, in hospitals and physicians' offices, in private homes, and in businesses. People in the public view need

† Please see web site for WHCCAMP www.whccamp.hhs.gov.

to take a leading role in talking about, and practicing, preventive care. Practitioners of "alternative" medicine, like acupuncturists, naturopaths, herbalists, chiropractors, and others, need to have a much bigger voice in terms of policy, treatment, and most especially research. Their voices will bring a holistic, "big picture" perspective to balance the microscopic detail of Western medicine. Once these changes begin to take hold, they will trickle down to the average American person, and they *will* have an effect. If people are surrounded by good health, they will become healthy.

This is my hope for the future—my global perspective, if you will. I hope to see this happen in my lifetime.

But it is not enough to talk about change; change has to be created. In my years of practicing medicine, it has become clear to me that the giants of the medical industry—hospitals, pharmaceutical companies, and insurers—are not interested in prevention. Preventive care, after all, will put a big dent in their profit margin, because it doesn't need expensive drugs and surgeries to be successful. In fact, it works to ensure that those types of solutions are seldom, if ever, needed.

On the other hand, employers, upon whose shoulders so much of our national health care costs fall, have much to gain from a shift to preventive care, including lower health care costs and greater employee loyalty, productivity, and presenteesim. Therefore, employers possess the greatest power to create significant change.

Like it or not, at this point our system of health care is completely dependent upon the active participation of employers. If businesses suddenly ceased to be involved in providing health insurance, the public health crisis would be epidemic: without the buffer of employer contributions and the premium discounts afforded to large companies by insurers, many Americans simply would not be able to afford coverage for themselves and their families. They would be forced to go without, further elevating what the media has dubbed "the crisis of the uninsured." Already, America is the only country in the industrialized world where people are routinely forced into bankruptcy by the cost of medical care. In fact, it is estimated that 50% of all bankruptcies and more than 1.5 million foreclosures per year occur as a direct result of medical expenses.[7]

The cost of health care isn't easy for businesses to swallow, either. As of 2006, the average health care expenditure per employee was 6.9% of total compensation for private employers. In the fields of natural resources, construction and maintenance, that number rose to 7.7%; in goods production, 8.4%; and in production, transportation, and materials moving, 9.0%. Municipal and state employers paid the highest benefit rates at 10.6% of total compensation. These numbers show a dramatic increase in benefits costs since 2000, when the average private industry paid about 5.5% of total compensation in health care costs, and municipalities paid about 7.8%.[8]

The rate of inflation in health care seems to have nothing to do with the rate of inflation for the rest of the economy. It is literally growing out of control, far outstripping the ability of the average person to pay for it. The National Coalition on Health Care estimates that by 2016, health care spending will reach $4.1 trillion per year, or 20% of our GDP. In 2008 alone, average premiums for employer-sponsored health plans rose 5%. Small businesses shouldered an increase of 5.5%. Businesses with fewer than 24 employees saw the largest rate hikes, with increases averaging 6.8%.[9] That's a huge chunk of profit going out the window to pay for a service whose quality has not demonstrably improved.

I'm not insinuating that employers should suddenly pull the rug out from under their employee benefit plans. In fact, many are making a commendable effort not to let rising health care costs affect the services they provide for their employees. But if health care inflation rates continue to follow the trends of the last few decades, many companies will be forced to cut their benefits programs just to survive.

Section 2:1—Why Does Health Care Cost So Much?

A lot of media attention is directed at the free care provided by hospitals to illegal aliens and uninsured citizens, but although they make a convenient target, these people are not the reason that health care costs keep jumping. While most hospital emergency rooms don't turn a profit (not, for example, in the way that cancer or cardiac centers do), the real issue is the way the American health care system approaches the treatment of disease for *everyone*.

As we discussed in the previous section, Western medicine is *reactive*; it deals with the problem at hand in the best way it knows how, but makes no real effort to prevent that problem from manifesting. It doesn't make a mountain out of a molehill, as the old saying goes—it ignores the molehill until it becomes a mountain, and then tries to tear the mountain down with a shovel and a few sticks of dynamite, one stone, or one symptom, at a time.

This reactive approach is applied to everyone, not just those who go to emergency rooms seeking free care. Even if no free care were offered, the exorbitant cost of Western medicine would still plague us.

This isn't a new problem. Medicare sent costs soaring when it introduced standardized fees for physicians and hospital visits. While attempts were later made to put the brakes on this, they were too little too late. Then, there's the problem of prescription drugs, our treatment of choice, which are overused by every population group and priced not based on the cost of their manufacture but on what the market will bear. Computer imaging technologies (like MRI machines and CT scanners) and other high-tech equipment can cost millions of dollars, and doctors and hospitals must generate enormous revenues just to pay for them, never mind profit from them. Finally, there are administrative fees to go along with truckloads of paperwork, the expense of ever-more-complicated testing procedures, the high cost of malpractice insurance… The list goes on and on.

The primary goal of the health care industry, like that of any other industry, is to make enough money to cover its expenses, with a tidy bit left over. Decisions made by the industry as a whole, and by its myriad individual corporate entities, are aimed at ensuring that those entities continue to operate in a manner which will produce the most profit for the least expenditure.

Medicine is a service industry: its services are provided in response to the needs of consumers. But medicine is not, at least in my mind, supposed to mirror other service industries in its *approach* to serving those consumers. There should be no room for the "up-sell" in medicine. There should be no attitude of "more care is better care." What is provided should be only what is necessary to restore the patient to health, and keep them healthy. No more and no less.

In other words, the only product the health care industry should be selling is…health!

And yet, the up-sell happens every day. Why use a plain old $300 X-ray when a $1,200 CT scan will do? Why should a patient with arthritis take Advil when they can take a selective COX-2 inhibitor like Celebrex for $150 per month? Why sell lifestyle changes when you can sell Lipitor? And then, the ultimate up-sell: disease management. When there is no cure in sight, the profits are perpetual. If Western medicine suddenly became proactive, the industry would no longer be able to sell its high-tech, expensive treatments, because most people would no longer need them. Quite simply, it is in the best interest of the industry to continue to offer the reactive, ineffective methods characteristic of Western medicine. The mountain that used to be a molehill is the medical industry's gold mine.

Whatever their culpability, the massive cost of American health care is not generated solely by providers. The American consumerist mindset also plays a major role in driving prices skyward. As a nation, we demand the best of everything: the latest technology, the biggest homes, the fastest cars. Economy and reserve fall by the wayside in the face of the newest, best, and brightest. We want it all, and we want it now!

This attitude has, over the course of the last three or four decades, extended to include our health care, and even health itself. Why cut down on salt and saturated fats when you can take a pill to lower your blood pressure? Why bother with stretching and exercise when surgery might cure your low back pain? Why take the long road when there's a quick-fix available?

This "health laziness" is compounded by the efforts of pharmaceutical companies and a few unscrupulous doctors.

The constant and pervasive advertisement of prescription drugs, both on television and in magazines, is a good example of how the medical industry perpetuates the quick-fix mentality. Drug companies know which markets are the most lucrative, and they bombard them mercilessly. Having no real basis for comparison—for example, being unaware of the fact that Keigel exercises can be every bit as effective at dealing with urinary incontinence as Vesicare or Detrol LA—patients flock to their doctors and demand these drugs. "After all," they say to themselves, "if these medications represent the best American medicine has to offer, then I should be taking them!"

As Dr. John Abramson documents in *Overdo$ed America*‡, some patients will actually accuse their physician of depriving them of good medical care if he recommends less aggressive treatment options, or refuses to prescribe the drugs the patients believe they need. In other cases, patients end up with medications they don't need because an unscrupulous doctor receives incentives from the pharmaceutical companies, and because "well, they asked for them."

This brings us back to that other critical issue: the way in which most doctors and specialists are compensated. Like workers on an assembly line, American doctors are paid according to the quantity, not the quality, of the services they provide, and their decisions, sad to say, are sometimes based more on finances than on the true needs of their patients.

Angioplasty procedures, which include balloon angioplasties and stent insertions, are a good example of how this mentality affects patient care. Surgery is a far more lucrative way to prevent a heart attack than an exercise and diet regimen. And although several studies have proved the latter approach to be equally effective—including a landmark study by the COURAGE Trial

‡ Please see recommended reading section for details.

Research Group which demonstrated that percutaneous coronary intervention (PCI) is no more effective than "optimal medical treatment" (read: diet, exercise, and drug therapy) for patients with stable coronary artery disease[10]—cardiologists still perform more than 1.2 million angioplasties every year.

About 50% of those are elective procedures performed on patients with stable coronary artery disease (meaning that their condition was already well-controlled by drugs and/or lifestyle modifications) and can therefore be classified as unnecessary. The cost of each elective procedure: $20,000 or more. In addition, each and every stent costs about $2,000, independent of all other costs of surgery; drug-releasing stents can cost twice as much or more.

For their part, patients, often despite a lack of evidence that surgery will be more effective in treating their condition, continue to demand such procedures, both because their doctors continue to recommend them, and because they genuinely believe that more medical care is better for them. After all, if you can have a state-of-the-art angioplasty procedure at the University hospital, and come out good as new, why *wouldn't* you do it? You can even get a stent with the blood pressure meds built right in! Why own a Hyundai when you could drive a Lamborghini?

The answer is simple: for the vast majority of people, the Hyundai is better. When something goes wrong with your Hyundai, you don't have to fly in an Italian mechanic to fix it. You don't need a performance tune-up every time you take it out of the garage. You might even be able to change the oil yourself.

The human body is remarkably resilient when all its systems are in balance. Unnecessary surgeries and drugs disrupt that balance—sometimes so much so that if you hit one little pothole, one little complication, your whole state-of-the-art Lamborghini suspension system is wrecked forever.

If our modern Western methods of care were actually working, the pitfalls of medical consumerism would be balanced by positive gains. In other words, we would be seeing a concrete return on our investment. But despite an ever-growing arsenal of quick-fixes and miracle pills, people in America are not getting any healthier. In fact, we're getting less healthy with every year that goes by. According to a 2005 report published in the *New England Journal of Medicine*,[11] the current generation of American youth is the first in this country's history to have a shorter life expectancy than their parents. The "more is better" approach, while of great financial benefit to the medical industry, is not helping sick people get well, and it is not keeping healthy people healthy.

And now, we get to the real irony. There is a lot of fear in this country attached to health and health care, because when many people fall ill there is no safety net to catch them. With 1 in 6 people completely uninsured, and another 16 million people under the age of 65 classified as "underinsured," we can expect the number of medical-related bankruptcies and foreclosures to continue to rise over the next decade. As discussed in Section 1, a frighteningly large percentage of Americans will fall ill at some point in their lives; many of them know it, even if they don't know exactly why. They do know that if they get sick they might lose their jobs, their homes, and their quality of life.

This type of constant, lingering fear creates stress, and stress makes people sick. Sick people need treatment, which they may not be able to afford. It is a paradox, and our current *modus operandi* presents not a single solution. Eventually, something has got to give.

The best way to prevent what I see as a potential national health care disaster is not only to address the outrageous cost of health care, but to help Americans become healthier. When people become healthier, they don't need to worry so much about the potential cost of treatments, because they are less likely to need those treatments. Also, people who are educated about, and interested in, their own health are less apt to fall victim to the "more is better" mentality, and more likely to refuse elective surgery and other costly, invasive procedures in favor of a less dramatic (if perhaps not so immediately gratifying) approach.

We can curb the excessive supply of health care in this country in two ways. First, we can reduce demand (which, as discussed earlier in this section, is often manufactured by the industry itself). Second, we can reduce the number of people *subjected to that supply*. The fewer people admitted to hospitals with preventable chronic illnesses, the fewer are likely to receive the kinds of expensive and unnecessary treatments which are driving our costs through the proverbial ceiling.

How do we keep people out of hospitals? We help them get healthy. And preventive medicine, specifically Classical Chinese Medicine, is the best way to do that.

I'm not an economist, but I feel I should address the notion that if the health care industry were to be restrained and deflated, our national economy would collapse—that somehow, as citizens of this nation, we have a responsibility to keep "feeding the beast."

It is easy to understand where this assumption stems from: when an industry as massive as the medical industry flounders, jobs and tax revenues are lost, and the economy becomes depressed. However, unlike other industries, the medical industry hurts our economy as much as it helps, by placing heavy financial burdens on employers and workers alike.

We are seeing this effect right now with the Big Three automakers. Why is Ford in its current tenuous position? Why did General Motors recently declare bankruptcy? In part, due to their exorbitant health care costs! Of course, mismanagement and greedy corporate officers don't help—but when you have to add $1,500 or more to the price of every car just to cover your employee benefit costs, you are in deep trouble. There is no way to stay competitive when your overhead is that high. The fact that those dollars are going to feed the ravenous American medical industry is no consolation to the tens of thousands of auto workers who lost their jobs (and, ironically, their health insurance coverage) over the last few years.

One major difference between the medical industry and other industries is that no corporation within it can be undercut by competitors, because prices have effectively been standardized by Medicare, the high cost of malpractice insurance, and the exorbitant price of medical technology. Even if doctors and hospitals wanted to charge less, they can't afford to. Because of this, health care inflation can only be restrained if consumers become healthier and more educated, and therefore require fewer costly services.

Yes, many people would lose their jobs if hospital revenues fell. Yes, many people would lose their jobs if Pfizer suddenly stopped raking in billions on its cholesterol drugs. But when you consider how even a small increase in overall public health could lighten the burden on employers and taxpayers, and how greatly businesses in other industries could benefit from having healthier and more productive employees, it all starts to equal out.

Lower health care costs would allow U.S. employers to be more competitive in today's global marketplace, because fewer dollars spent on health care could conceivably enable companies to offer higher salaries to skilled employees, invest more money in infrastructure and equipment, and keep important manufacturing jobs here in America.

I believe that when the health care bubble finally bursts, the major effect will not be a catastrophic economic downturn, as proponents of the current system might suggest. In the same way that matter can neither be created nor destroyed, the money bottled up in the health care system will not simply disappear; rather, it will be *redistributed* among other industries, to the benefit of all.

Section 2:2—Prevention Can Save You Money

When I speak of creating a "culture of wellness" within your company, I'm talking, of course, about prevention.

When people make a conscious effort to get healthy—whether through diet, exercise, or therapies like CCM—they are not only helping themselves in the present, they are taking steps to protect their health in the future. It all comes back to the true definition of health: if the body never swings too far out of balance, chronic disease is less likely to gain a foothold.

It seems obvious enough that a healthy person will usually have fewer medical bills than a sick person. But under our current system, if a healthy person and a sick person are part of the same health plan through the same insurer, their costs may end up being comparable, even though the healthy person goes to the doctor only once a year, has no prescriptions, and hasn't been hospitalized since he broke his leg in high school. This is, of course, how insurance companies balance the costs of treatment for sick people. Some companies offer several medical plan options to account for the diverse needs of their employees, but while this approach may offer some benefit, it probably isn't enough to make a real difference in the bottom line.

If a company could actually make prevention part of its operating policy, there could be significant reductions in the cost of health insurance, because the company could address the purchase of health insurance coverage from the perspective of not only the good health of its employees, but their continually *improving* health. (I'll talk about specific ways employers can handle this in Sections 4 and 5.)

Lower health insurance costs are only part of the financial benefit which can conceivably be realized through creating a "culture of wellness" within a company.

There are three ways to improve the profitability of a company: sell more product, cut costs, and increase productivity. When companies sit down to balance their budgets, the first and second items usually get the most attention, because they are the easiest to quantify. When you spend less, your net profits go up. When you sell more, your net profits go up. But how do you create a measurable increase in productivity?

Many business owners and managers have struggled with that question. Often, it seems that no matter how the management team cajoles or threatens, or how many incentives are offered, employees just keep producing at the same level of efficiency. I do not think this is due to any negligence on the part of the employees. At heart, I'm an optimist: I believe that most employees want to do their best, and that they will perform their jobs as well as their bodies and minds allow so long as they feel their efforts are appreciated. But if their health is lacking, productivity will decrease, and no amount of incentivizing or managerial pep-talking or even forced overtime will do anything to increase it. In fact, trying to increase productivity when workers are imbalanced or unhealthy will likely create a negative backlash. Employees will begin to feel they are having trouble keeping up with the tasks management assigns. They will feel alienated and underappreciated. They may begin to show signs of chronic stress, like insomnia or digestive issues; or, conditions from which they were already suffering may worsen because of their stress. As their health declines further, they will start taking more sick days, and company productivity will continue to spiral downward.

A person's health (or lack thereof) impacts everything about his personal and professional life. This is true for everyone, from your janitor to your CEO. People who are in good health—not just good physical health, but good mental and social health as well—tend to be more focused and less susceptible to the negative effects of stress. They take fewer sick days, and have more energy to devote to their work. Quite simply, healthy employees are productive employees, and productive employees are an asset to your company.

The promise of greater productivity is worth investing in. Many companies, both large and small, have already recognized the value of good health to their bottom line. Some have enrolled their employees in the new "pledge plans" offered by major insurers, which require a written commitment to a healthy lifestyle and yearly check-ins with a primary care provider. Others offer rewards to employees who lose weight or increase their level of exercise. There are also a number of statewide initiatives and health challenges, like Shape Up RI[12] here in Rhode Island, in which employers can encourage workers to participate.

"The purple pills will cure your heartburn but may cause headaches. Take a yellow pill for the headaches, side effects include nausea. The white pills are for the nausea; side effects include fever. Green pills for the fever..."

Programs like these are a good start, but we can take the concept of prevention in the workplace a lot further. The modalities classified as "alternative medicine" offer a wealth of possibilities to corporate wellness programs. Of course, these options are probably not going to be presented to employees by hospitals, pharmaceutical companies, or insurers, because it is not in the best interest of these entities to do so. Therefore, it is incumbent upon companies who desire to increase the overall health of their work force to educate employees about the resources available to them, and encourage them to take advantage of these resources in pursuit of health and well-being.

The ultimate goal of any wellness program in the workplace should be to create a sense of *personal responsibility and empowerment*. This is very important. Companies like Toyota (to use the example of automakers again) work to cultivate in their employees a sense of personal responsibility for, and interest in, the corporation through management practices which take an interest in the well-being of the employee. This approach is so radical, and so effective, that it has been dubbed "Toyota-Style Management." The premise is actually quite simple: care about your workers, and your workers will care about you. In a similar way, I believe that a businesses in any industry which endeavors to make employee health a priority, and takes steps to cultivate within its employees a sense of responsibility for their own well-being, will in turn see a marked increase in both loyalty and productivity.

Using a combination of incentives, *objective* information, and alternative care options, you too can begin to create a culture of wellness within your company. Rather than wringing your hands over the monthly payables and waiting for the government to put its foot down, I hope you will use the information in this book to take proactive steps toward creating change. A comprehensive wellness program, properly designed and implemented, can generate incredible returns on a relatively small per-employee investment. This book will help you create such a program.

Section 2:3—The Value of Classical Chinese Medicine

CCM is the ultimate preventive medicine, and this makes it the perfect foundation for any corporate wellness initiative. I strongly believe that employers who include CCM as a fundamental part of their employee wellness initiative will see results far greater than any wellness program has generated to date; and that investment into a CCM-based program will generate returns far beyond their expectations.

There are three measures by which we can quantify the value of CCM to patients and employers: cost (both short-term and long-term), consequences, and results.

Cost

Often, when individuals make the change from conventional Western treatment to CCM modalities (the most common being acupuncture), they are slightly daunted by the up-front cost. Because their treatment may not be covered by their health insurance, their out-of-pocket expenses may be greater than for conventional treatments like prescription drugs. But just because the patient doesn't see the cost of Western care doesn't mean it is not there; it has simply been redistributed through high insurance premiums (often paid for by employers).

In the first year, CCM treatment for chronic disease will generally cost the same or slightly more than the total cost of a person's prescription drugs, but far less than surgery. Over the course of five years, however, the cost of CCM will be only a fraction of that of Western treatments of any type. Why? Because Western treatments for chronic conditions are *perpetual and ongoing*! They "manage" diseases like hypertension, high cholesterol, diabetes, and depression, but never actually make them go away. For example, analyses have shown that annual medical costs for a person with Type II diabetes will very rarely decrease over a five-year period; on the contrary, they are almost guaranteed to go up, even if the person doesn't get any sicker, because of the gross rate of inflation for medical care.

On the other hand, after an initial period of intensive treatment (usually between three months and one year, depending on the severity of the condition), CCM costs no more than the price of a monthly or quarterly "checkup." To continue the example above: I have an 85% success rate in treating patients with Type II diabetes in my practice, because not only do I address their condition using acupuncture and other therapies, I teach them *how to keep themselves well*. Yes, in some cases my treatment costs more in the first year than a typical drug regimen, but after that first year approximately 85% of my patients *no longer have Type II diabetes*! They are more productive, more energetic, and will enjoy a better quality of life for the rest of their lives so long as they maintain the good lifestyle habits they learned in treatment. Once the diabetes has been successfully addressed, I ask these patients to come in for monthly or quarterly checkups, to ensure that their bodies remain in good balance.

It is difficult to make an accurate overall cost comparison between CCM and Western medicine as *systems of medicine*, because to date, few reliable statistics exist as to the success of Chinese Medicine when it is used on people who have not been previously subjected to Western intervention. However, it is possible to compare the costs of both modalities for treatment of specific conditions—as I have done above—and when this is done, CCM invariably emerges as the less expensive of the two methods. In Section 3 we will explore the costs of thirteen common chronic conditions to employers and America as a whole, and compare the price and efficacy of typical Western treatments to that of CCM.

Consequences

Newton's Third Law of Motion states that for every action there is an equal and opposite reaction. The same is true in medicine. The problem is, when it comes to Western methods of treatment, it is not always obvious what that reaction is going to be. And the more complicated the treatment, the less predictable its result.

Let us examine, for example, the 2004 Vioxx recall, a scandal which shook many Western doctors and scientists to the core. Vioxx was a non-steroidal anti-inflammatory drug (NSAID), specifically a COX-2 inhibitor, prescribed to millions of patients with osteoarthritis and other chronic pain conditions. Unlike over-the-counter NSAIDs like naproxen sodium (a.k.a. Aleve), Vioxx was designed to selectively target only COX-2 prostaglandins, and leave the COX-1 prostaglandins which protect the lining of the stomach unaffected. Therefore, it was touted by Merck in numerous advertising campaigns as the optimal solution for patients who experienced stomach irritation when using other NSAIDs.

However, COX-2 also plays a role in the production of prostacyclin, which prevents clotting in the blood. The consequence of using Vioxx, therefore, was an increased chance of blood clots and poor vasodilatation (widening of the blood vessels resulting from relaxation of smooth muscle within the vessel walls). It is estimated that, during the five years in which the drug was on the market, as many as 139,000 people experienced a heart attack or stroke as a consequence of taking Vioxx, and *30–40% of those people died*!

Evidence of this rather significant complication surfaced in numerous clinical trials—including one 8,000-person trial which took place prior to the drug's release—but Merck made no effort to make that evidence known to doctors or patients. In fact, the company manipulated the study results, saying that it wasn't that Vioxx *caused* heart attacks, but rather that naproxen sodium (the drug of choice for many potential Vioxx users) *prevented* them. Armed with these "facts," and other research the company itself had funded, Merck went out of its way to assure everyone that not only was Vioxx safe, it was beneficial. (And, lest anyone seek to give Merck the benefit of the doubt, their claims about Vioxx and naproxen were not innocent mistakes: studies conducted both before and after Vioxx was removed from the market conclusively determined that naproxen does not prevent heart attacks, but may actually increase a patient's risk of cardiovascular events.)

In Western medicine, the potential consequences of using any particular drug, test, or surgery to treat chronic disease need to be weighed and measured against the benefits. This give-and-take is bad enough, especially when you consider that the potential consequences of treating a chronic condition with CCM are virtually nonexistent—but there is a bigger problem. Even the most responsible and educated Western doctors cannot always foresee the consequences of the treatments they use, because they are not provided clear and unbiased information upon which to base their decisions. Pharmaceutical companies fund about 80% of all clinical trials, including trials of their own products,[13] and can now pay the FDA to have their drug evaluation "fast-tracked"—a process which seems disturbingly akin to bribery, and which virtually ensures that the drug in question will go to market whether it is safe or not.

The Vioxx scandal is just the beginning of a long list of cover-ups and campaigns of misinformation. If the results of that 8,000-person study, which showed that patients taking Vioxx were almost twice as likely to have a heart attack as those using naproxen, had been released without Merck's "doctoring," the drug never would have hit the market, and Merck would have lost billions of dollars in sales. Instead, the results were skewed to favor the drug, and a multi-million dollar advertising campaign was launched. Vioxx became a "blockbuster" drug, and a household name. Doctors, subject to the same misinformation as the general public, wrote prescription after prescription, confident that the studies they read were accurate and well-presented—and were forced to shoulder part of the blame when the tower came tumbling down.

As an MD in Poland, I was in the same vulnerable position as any other doctor. And I would be in that position today, had I not realized that you cannot fix a door with a broken hammer, and you cannot fix a patient with medicine that doesn't work. Vioxx was hardly an anomaly among pharmaceuticals: it was just the one that got caught. There are many other drugs (some of which I'll discuss in Section 3) which produce disastrous effects in patients, but are still marketed as safe and beneficial. In other words, it's not a question of *if* other "blockbuster" drugs will be revealed as unsafe or even deadly, but *when*.

The real consequences of Western medicine are evident in the numbers. While heart disease has long been considered the nation's number one killer, claiming just under 700,000 lives per year, Western medicine actually surpasses it. In an eye-opening 2004 study/article entitled "Death by Medicine," researchers Gary Null, Ph.D. et al. determined that "the total number of deaths caused by conventional medicine is an astounding 783,936 per year."[14] That's more than heart disease, cancer, diabetes, or respiratory disease! A large portion of those deaths are attributed to iatrogenesis, a term coined to describe all those pesky unforeseen consequences of medical tests, prescription drugs, and unnecessary surgeries. The rest are due to medical errors, in-hospital infections, and poor or neglectful patient care (unintentional, in almost all cases). The American Iatrogenic Association asserts that the number of *in-hospital* deaths due to "medical error" is 195,000 per year![15]

Even more frightening, some reports estimate that only 5–20% of iatrogenic deaths are ever reported as such. That puts the possible number of in-hospital deaths directly attributable to the practice of Western medicine at more than a million per year, or as many people as live in the state of Rhode Island. Some estimates put the total monetary cost of iatrogenesis, medical error, and medical mortality at $300 billion or more per year.

Many doctors operate under the assumption that in order for something to be effective, risk must be inherent. After all, the most powerful drugs Western medicine has in its arsenal— i.e. chemotherapeutics—also harbor the potential to cause devastating side effects and even mortality. However, while the old adage "nothing ventured, nothing gained" may be an accurate sentiment when it comes to venture capitalism, it is not, and has never been, true in medicine.

Classical Chinese Medicine has a proven track record of safety. The potential consequences of acupuncture are so low as to be virtually non-existent. In nearly all cases, reported "adverse effects" of CCM had to do with improperly-cared-for equipment or unsterilized needles—rare in the U.S., and easily avoidable if a person takes the time to look for a reputable practitioner who uses only single-use, pre-sterilized needles. Even with such incidents accounted for, there were only 76 malpractice claims filed against the nation's 18,000 acupuncturists in the 13-year period between 1990 and 2003,[16] an average of 5.85 suits per year, as compared to an estimated 85,000 medical malpractice lawsuits filed each and every year against the nation's 800,000 doctors, or more than 1.1 million suits over 13 years. So while nearly 10% of doctors can expect to be sued every year, only about 0.0325% of acupuncturists are involved in malpractice suits.

While some of the malpractice discrepancy might be explained by the fact that there's more money to be made by unscrupulous patients and lawyers in suing Western doctors and hospitals, the true reason is that the potential for adverse reaction to acupuncture is negligible at most. In fact, it is one of the safest medicines in the world. Certainly, acupuncture will never cause a heart attack (à la Vioxx), or cripple a patient with a botched spinal fusion surgery, or amputate the wrong foot by mistake.

There are some people who experience what we call a "healing reaction" or "healing crisis" at the start of treatment, exhibiting symptoms like fatigue or mild headache; this is usually a result of the body's natural detoxification process, during which toxins stored in the liver, kidneys, fat cells, and other tissues are released into the bloodstream so that they can be eliminated from the body. A very small percentage of pain patients may also experience increased pain at the site upon initial treatment (again, usually a "healing crisis" in which the body's natural immune reaction causes a temporary increase in inflammation); a reassessment of the chosen acupoints will usually nullify this reaction.

Acupuncture presents virtually no chance of toxicity, allergic reaction, overdose, drug interaction, or any other iatrogenic effect.

Laughably, some doctors prescribe acupuncture to cancer patients based not on their belief that the practice actually works, but because "what's the worst that could happen?" In the end, of course, as numerous studies have documented, patients do in fact experience relief from the torturous side effects of high-dose chemotherapy—but more importantly, such studies conclusively prove that acupuncture is safe and effective even in the most fragile of patients, people whose bodies have been devastated by both cancer itself and by Western medicine's assault on it.

Results

Chronic illness accounts for 75% of all health care spending in the United States, and 162 million people—more than half of all American adults—suffer from one or more chronic conditions. That's a lot of sick people taking a lot of medications and undergoing a lot of treatment, with a lot of potential for iatrogenic complications and unforeseen consequences.

As we learned in Section 1, *Western medicine does not cure chronic disease*. It only makes the sickness manageable. For American employers, this translates into perpetual costs of thousands of dollars per person per year, every year, until the employee retires, dies, or moves on.

Classical Chinese Medicine *can* cure people of their chronic diseases. And while the initial costs of treatment may be equal to the annual costs of prescription drug treatments and doctor visits, they are far less than the costs of elective or unnecessary surgeries, hospitalization, and iatrogenic complications.

I understand that some of the data I have presented may be a little hard to swallow. Most of us grew up hearing that Western medicine is a miracle, that it is practically infallible—and in the areas of acute and emergency care, this may come close to being true. But if we want to get well as a nation, and if American employers want to stop shelling out billions of dollars to pay for health care that not only fails to make their employees well but may actually make them sicker, we need to look at the hard facts instead of the hype and the mystery, and formulate a new plan of action. I believe that CCM should be an integral part of that plan.

Section 3: Common Conditions, Costly "Cures"

BEFORE YOU TAKE steps to create a culture of wellness within your business, it is helpful to have a basic understanding of the health concerns which may be affecting your employees.

An estimated 1 in 3 American workers report productivity losses due to health-related issues. In 2003, 69 million workers reported missing days due to illness, resulting in a total loss of more than 407 million work days. Another 478 million work days were less productive because of workers' inability to concentrate due to their own illnesses, or those of family members. Total costs for lost productivity due to these factors were estimated at $260 billion, and sick pay costs were estimated at $48 billion.[17]

Furthermore, 11.6% of adults are *consistently* limited in their daily activities due to one or more chronic conditions.[18] Some of these people are unable to work at all; others are severely hindered in their daily work capacity, which results in productivity losses for their employers.

Through research and my own clinical experience, I have compiled a list of the thirteen chronic conditions which I believe to be most common among the American work force. I treat these ailments every day in my practice. The underlying imbalances which create or perpetuate these conditions vary widely, and workers in some industries may be more prone to certain of them, but the conditions themselves are extremely common, and occur in people in every job field and of every socioeconomic status. In fact, your employees as a group probably suffer from every one of these conditions, and many may suffer from several.

The thirteen conditions most likely affecting employees are:

1. Obesity
2. Back pain
3. Hypertension (high blood pressure)
4. High cholesterol
5. Asthma

6. Allergies
7. Depression
8. Diabetes
9. Headache
10. Insomnia
11. Repetitive motion injuries
12. Gastrointestinal diseases
13. Inflammatory diseases

You will notice that, although they are some of our biggest killers, I did not mention heart attacks, strokes, or cancer. That's because these events are the result of years of imbalance within the body, and most frequently occur in people who have lived for years with chronic conditions like those listed above (particularly obesity, which increases a person's risk of developing nearly every chronic disease). Therefore, this section focuses on conditions your employees live with every day, at work or at home—the conditions that drain away their energy and productivity, and may eventually lead to the total decline of their health.

This section is crafted as an overview, to help you understand how common conditions affect your work force and your company's bank account. As I stated in Section 1: this is not a book about acupuncture, nor is it a medical textbook. It is not intended to provide the means for anyone to diagnose or treat a chronic condition. If you suspect that one of your employees may be suffering from one of these conditions, please refer them to a qualified practitioner for diagnosis and treatment.

On the following pages, I'll address each condition listed above. I'll explain its pathology, discuss how it reduces functionality and productivity, and briefly outline popular Western techniques for treatment, potential side effects, and estimated costs. Then, I'll describe how the condition can be treated using Classical Chinese Medicine, and cite the probable costs and duration of treatment for my approach. You'll also find testimonials from patients who have been helped by Classical Chinese Medicine, and witness the difference that CCM has made in both their personal and professional lives.

Of course—and I cannot stress this enough—each person is unique; therefore, the how and why of his or her condition will also be unique to some degree. Since CCM diagnostic and treatment techniques allow the practitioner to craft a truly individualized treatment program, there are no cookie-cutter solutions. Therefore, you will not find a "playbook" of acupuncture points, herbal remedies, or dietary recommendations for each condition, but rather a description of how CCM views the condition, and ways in which the condition may be addressed.

In order to discuss in detail how I can help sick people recover their health and productivity, I must first outline the basic procedure by which I diagnose and evaluate each and every patient who comes into my office.

The initial visit is always a diagnostic consultation, in which I examine the patient, evaluate his overall health, and diagnose his specific condition(s). Even if they consider themselves "healthy," most people share at least five major complaints in this first meeting. I use several diagnostic methods in my practice, including simple observation, Chinese Tongue Diagnosis, Palpation Diagnosis (including palpation of the acupuncture meridians and abdomen, and pulse diagnosis), assessment of the musculoskeletal system (including range of motion), and evaluation of any charts, scan images, or other medical materials the patient chooses to share.

Last, and perhaps most importantly, I ask very specific questions and listen carefully to the patient's answers. Often, the root cause of the patient's condition, and the lifestyle patterns which perpetuate it, will reveal themselves during this conversation. Unlike many Western doctors who, for reasons previously discussed, no longer have the time to communicate effectively with their patients, I make sure to conduct this important diagnostic discussion with each new patient.

After I have diagnosed the imbalances at the root of the patient's complaint(s), I begin my initial course of treatment. This "trial period," consisting of 6 to 8 visits over the course of two weeks, determines not only the specific course and duration of the total treatment, but also how well the patient will respond to CCM.

I use only acupuncture during this first stage of treatment because it produces the fastest and most dramatic results. I do not recommend that patients make any changes to their diet and exercise habits at this point, nor do I request that they ask their primary care doctor to reduce or cease their use of prescription drugs. I may do these things later, as we begin to address specific symptoms—but at first, I find it best to treat every case like a scientific study, and keep outside variables to a minimum.

The purpose of this, of course, is to determine the effectiveness of the initial treatment, and the quality of the person's response to acupuncture. With the information garnered during this trial period, I can formulate a plan of action which will be of the highest benefit to the patient, and produce the most substantial results.

While most people respond well and quickly to acupuncture treatment and will see some improvement of major symptoms by the end of this exploratory treatment, there are a few who do not. There are many reasons why this might be so, but the most common is that the patient believes that acupuncture will not, or cannot, benefit them.

These cases tempt some skeptics to assert that CCM is a psychosomatic medicine; that it's "all in their heads." This is most certainly not the case. However, all three areas of health—physical, mental, and social—must be engaged in order for a person to become well, and even the most sophisticated Western treatments will be only minimally effective on those who truly believe they cannot be helped. No matter the "cure," the patient has to want to get better. When such a case arises, I refer the patient to other practitioners in other modalities where they might feel more comfortable.

After the patient has completed the initial course of 6 to 8 visits, I formulate a treatment plan based on his response to acupuncture and the severity of his specific complaints. Physiological, chemical, and genetic factors, as well as lifestyle habits, all play a role in determining the duration of treatment.

Every tissue and organ in the body has a different "timing response," and some heal faster than others. For example, fascia (soft tissue) heals faster than bone, and the cardiovascular system heals faster than either the central or peripheral nervous system. People who have had surgery or who exhibit certain genetic factors (such as a sluggish metabolism) may experience longer response times, and those using prescription drugs of any variety also tend to heal more slowly.

Generally, once the analysis period is complete, I ask patients to allow me to work with their primary care physician to reduce or eliminate their prescriptions as soon as safely possible—especially in cases where I believe the prescriptions are contributing to or causing the patient's complaint(s).

Throughout this section I will reference my "success rate" in treating these thirteen chronic conditions. This is *not* the percentage of patients who experience relief of their symptoms after treatment, but rather the percentage of those patients who have been corrected as completely as possible, who no longer need Western medications to "manage" their disease, and who regain a normal or nearly normal level of functionality.

"Seems a shame to amputate the whole leg just for a bunion."

I do not consider any course of treatment concluded until the patient is returned to a state of homeostasis, or self-regulation, in which the imbalances that caused his complaints are corrected and there will be no recidivism of his condition (unless a return to poor habits triggers a new round of imbalances).

I have been asked by other medical professionals how my success rate can be accurate with regard to the general population. "What about the people who drop out of treatment?" These skeptics say. "They were not successfully treated, nor cured of their conditions." Of course, there are people I treat for a brief period of time, but who never complete the course of treatment. There are also those who are satisfied with only superficial treatment; for example, people who are happy with the initial relief I can provide for their low back pain, but who elect not to continue treatment to address the underlying cause of their pain. These people are not considered when calculating my success rate. This is not some sneaky, underhanded strategy on my part to try to boost my "ratings," but proper research procedure. No major medical study includes dropouts when calculating its results; only participants who complete the study are evaluated. To do otherwise would be akin to calculating a surgeon's success rate based on procedures in which he only performed the first half of the operation, or testing a drug's effectiveness based on the reactions of people who took only half the recommended dose.

I firmly believe that any member of the general population who participates in my full course of treatment will see the same results as the patients upon whose outcomes I have based my calculations. I also believe that the same will hold true for any patient who works with a competent practitioner of Classical Chinese Medicine—because this medicine works.

Section 3:1—Obesity

As I mentioned in Section 1, 67% of American adults over the age of 20 are overweight. About 34% of adults (or half of all those who are overweight) are obese.

The Body Mass Index (BMI) scale is generally accepted as the best (Western) measurement to determine overweight. Those with a BMI of 25 or greater are considered overweight; those with a BMI of 30 or greater are obese.

A 2003 report by the U.S. Department of Health and Human Services (HHS) estimated that in 1994, obesity cost American employers $8 billion in added health insurance, $2.4 billion in additional sick leave, and $1 billion in disability insurance.[19] When those costs are adjusted using the standard rate of medical inflation, those prices become approximately $14.9 billion, $4.4 billion, and $1.8 billion respectively. But even when inflation is considered, those numbers are not an accurate representation true cost of obesity, because they address it *as an independent condition*.

When you consider that obesity dramatically increases a person's risk of heart attack and stroke (price tag: $448 billion in 2008), diabetes ($130 billion), and cancers ($214 billion), it becomes clear that its actual cost is far greater. Other conditions comorbid with obesity include back pain, joint injuries, digestive disorders, non-alcoholic fatty liver disease, asthma, and depression—all of which, as you'll see later in this section, come with their own hefty treatment costs. Also, obese people tend to have a shorter lifespan and a lesser quality of life compared to their non-obese peers.

The cost of obesity to employers has been well-documented. One study, published in the *Journal of Occupational and Environmental Medicine*, concluded that obesity-related absenteeism costs employers $4.3 billion annually.[20] That same study found that *absenteeism costs rise in proportion to a person's degree of overweight*, making moderately to morbidly obese people measurably less productive than their overweight to mildly obese peers.

In a 2003 report, the Department of Health and Human Services concluded that, on a national scale, obesity results in around 239 million days of "restricted activity" and 39 million absentee days per year, as well as an 88% increase in visits to physicians' offices.

A third study conducted for the California Department of Health Services by researcher David Chenowith, Ph.D. found that total lost productivity tied to obesity equaled nearly $3.4 billion per year in California alone.[21] With so many extra costs associated with obesity, it's obvious that employers should want to help overweight employees lose weight and attain better health.

While detox programs, weight loss clinics, and other treatments (like CCM) which help employees lose weight and get in shape may seem expensive initially, when all cost factors are considered they can be a value-added investment. Just as absenteeism and productivity losses increase proportionately to an employee's weight, they have been demonstrated to *decrease* as an employee loses weight.

Dieting

Dieting is practically a national pastime in this country. Spending on weight loss products exceeds $35 billion annually. But despite it all, Americans aren't getting any thinner. This is due in part to Western medicine's approach to the treatment of overweight and obesity, and in part to that all-pervasive "quick-fix" mentality.

I'm not going to delve into the science of weight loss in this book, since there are thousands of books out there on the subject already. However, I *will* say that weight loss, especially for a moderately to morbidly obese person, is never easy. When the body has been brought to such a great state of imbalance, it behaves rather like a pendulum stuck at the end of its arc. The restoration of balance often involves a large push (for example, an emotional/physical push like a major health scare, or a purely physical push like an extended detox program), followed by dramatic swings in either direction which manifest as both emotional and physical symptoms. But eventually, if the person can persevere, those swings begin to calm, and a sense of balance—of true health—is discovered.

It's this "swing" period which frustrates most dieters, and the further out of balance the body is, the more difficult the recovery process. This is where the quick-fix becomes so appealing. The biggest quick-fix of all, bariatric surgery, has become almost commonplace in recent years. While there are benefits, especially for super-obese patients, it's also costly, and comes with a long list of risks (including nutritional deficiencies, impaired metabolic function, and osteoporosis), a 20–40% complication rate, and a 2% death rate within 30 days of surgery. And here's the worst part: even if someone is willing to go through all of that, chances are they'll eventually gain at least a portion of the weight back.

A 2006 study found that, among those with an initial BMI of 50 or greater, 34% regained all or most of the weight within ten years, along with 20% of those with a lower BMI.[22] *As many as 44% of all patients regain enough weight post-surgery to make them susceptible to obesity-related diseases.*

Non-surgical options offered by Western medicine and "nutritional" supplement makers include lipase inhibitors, appetite suppressants, and metabolic boosters. Lipase inhibitors, like Orlistat—marketed as Xenical and the drugstore option Allī, currently under FDA review due to reports of liver injury in users—prevent the body from absorbing fats during the digestive process. While some people do achieve weight loss using these drugs (with Xenical, for example, patients can expect to lose 5–10% of their original body weight), the side effects—which include anal leakage, severe intestinal cramping, and destruction of beneficial gut flora—are far from acceptable.

Other medications work to suppress appetite. The most dangerous of these, Fen-Phen, was banned in 1997. Phenylpropanolamine (present in drugstore-variety appetite suppressants like Dexatrim) was taken off the market in 2001 due to its propensity to cause brain hemorrhage and stroke, especially in people who already suffered from cardiovascular conditions. Now, these chemical cocktails have been replaced by herbal "diet" pills, most of which are high in caffeine. These, along with another favorite, thermogenics, a.k.a "metabolic boosters", can increase heart rate and create feelings of anxiety, and may be dangerous to people with hypertension or cardiac arrhythmia.

While they might produce short-term results, the majority of diet products do nothing for an overweight person except create more imbalances in an already imbalanced body. When that pendulum swings back the other way, the weight will almost invariably come back, unless the person has implemented healthy lifestyle habits along with the "quick fix."

Drugs and surgeries don't cure obesity because they don't address the root imbalance, and health is all about balance. In a great number of cases, overweight and obese people struggle with emotional issues which lead them to seek solace in food. Also, prolonged periods of stress, no matter the cause, can weaken the adrenal glands, causing blood sugar irregularities and increasing hunger. In these ways and so many more, mental and social imbalances combine to create the physical symptom of obesity.

The fact that the underlying causes of obesity can come from imbalances in all three areas of health—physical, mental, and social—means that this condition responds very well to the holistic approach of Classical Chinese Medicine.

Of course, any treatment for obesity must be combined with proper diet and exercise in order to work; this is true even of bariatric surgeries. The difference with CCM is that, unlike Western modalities, it not only encourages the process of weight loss but can ease the discomfort that often comes with it. Acupuncture is extremely effective at moderating those frustrating "swings," and can assist in detoxifying a patient's body in order to facilitate more rapid weight loss. It can also normalize organs and systems which have become imbalanced or diseased due to the stresses placed on them by obesity—in particular, the circulatory system, the liver, and the adrenals.

Numerous studies have proven the efficacy of acupuncture and moxibustion treatment for obesity. A study published in the *International Journal of Obesity* found that acupuncture was significantly better at reducing overall body weight than control of lifestyle alone. It also facilitated more improvement in overall body weight than conventional medications.[23] Adverse effects were described as "minimal." (As we learned in Section

2, side effects from acupuncture do not include anal leakage, hypertensive crisis, or brain hemorrhage—and they never will.)

CCM, properly employed, can also address the mental and social causes of obesity. For example, anxiety and/or depressive tendencies which existed prior to the onset of obesity (often since childhood) need to be treated before the weight can begin to come off. CCM associates emotions with the physical organs and systems; by bringing these systems back into balance, the emotional state may also be balanced.

Western science is beginning to prove that relationships exist between physical symptoms and emotional states; it calls these "functional somatic syndromes." One UCLA study concluded that, "The emerging neurobiological models of allostasis/allostatic load and of the emotional motor system show striking similarities with concepts used by Traditional Chinese Medicine (TCM) to understand the functional somatic disorders and their underlying pathogenesis."[24]

When it comes to treating obesity, it's all about finding the underlying cause. That cause can be habitual, genetic, physical, chemical, emotional, social, or any combination of the above.

One thing I've observed in my practice is that obesity is in many cases a *toxic* problem. When the body is unable to eliminate the myriad toxins we ingest, inhale, and absorb on a daily basis, those toxins are stored in our fat cells; in this way, the body attempts to protect the organs and systems from damage by these substances. The more chemicals ingested, the more fat cells stored. Prolonged exposure to certain environmental and food-borne toxins can lead to weight gain even in naturally thin people. When a person is obese, their toxic load can be enormous.

When an obese person begins to lose weight and the stored fat is burned as energy, these toxins are released into the bloodstream, leaving the person sick, weak, and tired—and less likely to succeed at his weight loss program. Acupuncture can help the body eliminate these stored toxins more efficiently, which not only helps the person feel better in the short term but helps him shed pounds faster over time. Reducing body toxicity can also curb sugar and salt cravings, two of the biggest causes of diet failure.

Since toxicity is such a major issue, the first course of treatment for any obese patient—more important even than weight loss—is detoxification. Acupuncture can aid in normalizing and restoring the proper function of organs and systems. After the detox period—the duration of which is determined in the initial 6 to 8 visit analysis—nutritional and lifestyle counseling, supplemented by acupuncture to stimulate metabolic function and speed weight loss, can commence. In cases where prescription medications are implicated in the person's weight gain, I address the conditions which necessitate the use of the drug(s) before any weight loss program is considered; often, once the underlying condition is rectified and the drug is no longer necessary, the weight will come off on its own.

Most of my obese patients come into my office not simply because they want to lose weight, but because they are suffering from another chronic condition which is not responding to conventional Western medicines. Conditions common to my obese patients are high stress levels, low adrenal function, insomnia, dysglycemia (disturbed blood sugar regulation, which can be present even in non-diabetics), high blood pressure, and high cholesterol. It's only when I explain how obesity may be contributing to these conditions that the issue of weight loss comes to the forefront.

After the initial 6 to 8 visits, I can typically help an obese person lose 25 to 30 pounds over a period of 30 to 50 visits, at a cost of $2200 to $4000. This includes acupuncture and/or moxibustion, detoxification, herbal supplements, and nutritional counseling. Treatments in other CCM modalities, such as massage, may also be beneficial. People who are overweight but otherwise in fair health may be helped in as few as 12 to 20 visits, at a cost of $1800 or less. Major weight loss of 100 pounds or more will take more time and effort; however, after the first six months the patient may be well enough to continue the process on his own, with only monthly check-ups, because he will be in better control of his eating habits and lifestyle.

Recidivism is high in obesity, just as with addiction. In fact, physical addiction to sugar is common in my obese patients, and can, over time, impact the liver and digestive system as profoundly as alcoholism. I recommend that weight loss patients visit my office at least four times a year even after their entire weight loss process is complete, to ensure that they don't return to their old destructive habits. The cost of these quarterly "preventive maintenance" visits is less than $400 a year—less than some people spend on fast food and soda in a month.

In conclusion, I'd like to reiterate that, in the majority of cases, obesity is a disease caused by poor habits. Those habits are often present in people whose BMI is still in the "normal" range; it could be said that these people are predisposed to obesity. Once imbalances in the body reach a certain point, weight gain is exacerbated, and weight loss made more difficult. Therefore, CCM is valuable not only to those who are already obese, but to those who are on the road to obesity. It is in these cases—where burgeoning problems can be "nipped in the bud," as they say— that we see the true preventive power of CCM, and its ability to preserve and enhance health as well as restore it.

Section 3:2—Back Pain

The spine, especially the lumbar spine, is one of the most vulnerable areas of the human body. Back pain is a condition from which an estimated 80% of Americans will suffer at some point in their adult lives, and which costs this country an estimated $100 billion per year. It's also the leading cause for workers' compensation claims, resulting in a total of 333,000 lost-time claims per year at a price of more than $1.4 billion.

As treatment costs for back pain have risen in the last few years, so have premiums for workers' compensation insurance. Nationwide statistics show a cost increase of 50% between 2000 and 2004. Some areas, including California, saw even steeper rate increases. While there has been some effort to install programs in workplaces which address the preventable causes of back strain (and therefore minimize the incidence of back-related workers' compensation claims), they have been only mildly successful.

Traumatic injury (caused by impact, as in a car accident or a fall), heavy lifting, prolonged sitting, poor posture, repeated twisting or reaching with improper alignment—all of these can lead to back strain in the form of muscular strains or tears. A single major strain or injury can also cause discs to herniate or rupture. Causes of chronic back pain include degenerative conditions like spinal stenosis, spondylosis (degenerative osteoarthritis of the spine), and spondylolisthesis (anterior displacement of vertebrae)—but also dehydration, organ dysfunction (particularly of the kidneys), chronic stress, poor sleep, diabetes, obesity, and non-arthritic inflammatory conditions like fibromyalgia. Many doctors, viewing back pain as a musculoskeletal problem, fail to identify the latter conditions as possible culprits; for this reason, among others, it is estimated that 85% of low back pain patients cannot be given a precise diagnosis.

Many people live with back pain for years before taking action—but when a person does decide to treat his condition, the first appointment he usually makes is with his primary care doctor.

Depending on the degree of pain, our theoretical patient's doctor may recommend application of heat combined with pain-relief creams and over-the-counter pain relievers like ibuprofen (Advil) or acetaminophen (Tylenol). Depending on the actual cause of the pain, these remedies may or may not be helpful. If the pain is severe, or if it persists for a period of more than a few weeks, the doctor might prescribe a more powerful pain reliever, a muscle relaxant, an oral corticosteroid, or some combination of the above. (Some script-happy doctors might skip directly to step two, regardless of the severity of pain.) These prescriptions can cost upward of $100 per 30 pills, and may or may not actually help the condition.

Depending on which brands of medication are prescribed, and how the physician directs the patient to take that medication, total costs can reach $500 or more per month. That's a lot of money to spend on drugs that may or may not help the patient's condition. Despite the fact that insurance acts as a buffer (in fact, many people have no idea what their prescriptions really cost, only what their co-payment is) someone somewhere has to pay for all those drugs. As an employer, it's probably you. Insurance companies are very good at spreading costs around.

As with all prescription medications, the drugs commonly used to relieve back pain come with risks. With narcotic pain relievers like Vicodin, Percocet, and OxyContin, there's a very real possibility of addiction, even over a short term of use. Common side effects include digestive issues like nausea, vomiting, and constipation, as well as dry mouth, headache, and excessive sweating. These drugs may also cause severe adverse reactions including respiratory depression, apnea, hypotension (low blood pressure) or circulatory depression, and respiratory arrest, especially when taken in combination with certain skeletal muscle relaxants or alcohol. Also, they impair cognitive response and motor skills, rendering the person unable to function normally at work or behind the wheel.

Muscle relaxants like Flexeril work by modifying signals from the brain that cause muscle spasm or contraction. They should not be taken with other sedatives (like narcotic pain relievers), because of the risks mentioned above, but unfortunately this advice is often ignored by patients. Common side effects of muscle relaxants include nausea, dizziness, heartburn, and drowsiness, all of which may hinder normal daily function. Other potential side effects include tremors, anxiety, depression, hypertensive crisis, jaundice (indicating liver dysfunction), respiratory distress, fainting, and seizures.

In other types of muscle relaxants, like Skelaxin, the method of operation is related to the drugs' sedative properties. Reported reactions include hemolytic anemia, leucopenia (low white blood cell count), and jaundice. Digestive disturbances and dizziness are common, and nearly everyone who takes them will experience a "stoned," sluggish feeling.

Corticosteroids (i.e. Prednisone) are usually prescribed to "kick-start" the recovery process when a person suffers a back injury. They work by suppressing the immune system, which reduces inflammation at the site. The backlash is that the patient becomes slower to heal and more likely to contract infection—which seems more than slightly counterproductive. Corticosteroids can slow or even stop the growth and function of the adrenal glands in children and teenagers. They also increase blood pressure, make it difficult to regulate insulin levels in diabetic patients, and can increase the likelihood of bone disease not only in older women (considered the most likely candidates for osteoporosis), but in men and women of all age groups. The likelihood that a person will experience these side effects increases with the length of time they take the drug. Finally, corticosteroids cause weight gain, sometimes *major weight gain* of 20 pounds or more. Aside from the obvious fact that increasing the burden on the spine is a poor way to treat back pain, major weight gain can make a person more susceptible to the myriad health problems that accompany obesity.

In the end, of course, medications of any type don't correct back pain; they only mask it. Sometimes this works: in cases of acute low back strain, for example, narcotics like Vicodin can dull the pain and allow the muscles to relax while the body heals itself. But when it comes to chronic back pain—the kind of back pain from which the majority of people suffer—prescriptions are rarely as helpful as they are purported to be. Even if our theoretical patient experienced no major side effects from his prescriptions,

TESTIMONIAL

Back Pain

PATIENT: James M., Providence, RI
AGE (at time of treatment): 40

CONDITION(S): Back pain, Scoliosis, degenerative discs.

CONDITIONS TREATED WITH WESTERN MEDICINE:
Disc problems.

PRESCRIPTION AND NON-PRESCRIPTION DRUGS USED BEFORE CCM TREATMENT:
Flexoril.

SIDE EFFECTS FROM MEDICATIONS:
Sluggishness, constant fatigue and "hung over" feeling.

TOTAL ESTIMATED COST OF WESTERN MEDICAL TREATMENT
OVER THE COURSE OF THE CONDITION(S):
$10,260

HOW DID YOUR CONDITIONS AFFECT YOUR WORK LIFE?
It was very hard to get out of bed in the morning. I had to move slowly
and carefully.

PLEASE DESCRIBE YOUR EXPERIENCE WITH DR. SZTYKOWSKI AND CCM:
I took Flexoril for at least two years before starting acupuncture. I don't like
taking drugs of any type, and the muscle relaxants made me feel exhausted
and hung over. And despite the drugs and a number of visits to specialists,
I wasn't experiencing real relief from my back pain. I contacted Dr. Tad after
consulting a number of holistic practitioners. Dr. Tad was the only one who
didn't tell me over the phone that he could "cure" my condition: he insisted
that I come into his office for an in-person diagnosis. I appreciated this
personalized approach.

Since starting my treatment program with Dr. Tad, my back pain has mostly
gone away. Some days after a hard workout or physical activity, I know I need
to see him for a session, but the daily aches and pains I once experienced are a
thing of the past.

WHICH PRESCRIPTION MEDICATION(S) ARE YOU STILL TAKING TODAY?
None—not even aspirin.

TOTAL COST OF TREATMENT WITH CCM:
$6,260

they would still leave him feeling groggy and unwell. He would be far less productive at work, and his quality of life would be compromised.

When his back pain doesn't go away after three to six months, our patient may decide to see a specialist at a cost of $250 or more per visit. After several visits, the specialist may suggest physical therapy, which can cost $100-$200 or more per session. Or, as is becoming increasingly common, the specialist may recommend surgery. That's where things get complicated—and very, very expensive.

I believe that surgery should always be a last resort, reserved for the treatment of bodily trauma and life-threatening acute conditions. In my experience, elective surgery is almost never a good idea, especially for back pain. Surgeries—and the preliminary tests, anesthesia, drugs, and therapies which accompany them—are expensive and risky, and the results are anything but stellar.

The 650,000 back surgeries performed last year in the United States incurred a total cost of over $20 billion,[25] or more than $30,000 per surgery. That total does not include the cost of the drugs prescribed to relieve the pain of surgery, nor the antibiotics used to prevent infection at the incision site. Rehabilitation and physical therapy may total tens of thousands of dollars. Finally, there is the cost to the person's employer in terms of lost work hours, reduced presenteeism during work days both before and after surgery, and (especially for smaller companies) possible insurance rate hikes as a result of the surgery.

The amazing thing is that, in many cases, back surgeries do little or nothing to improve people's lives. More than 200,000 people undergo lumbar spine disc surgeries every year, but a study published in *International Orthopedics* found that *53% will see no symptom relief*! The success rate for spinal fusion surgeries is even worse. In *Overtreated,*ʼ her groundbreaking book on medical economics, Shannon Brownlee writes that "the best that can be said for spinal fusion is that one in four patients are helped by surgery."

---

* Please see recommended reading section.

Therefore, it's not surprising that the 2006 Spinal Patient Outcomes Research Trial (SPORT) found no discernible difference between back pain patients who underwent surgery and those who opted for more "conservative" care.

A two-year study by Consumer's Medical Resource (CMR) found that Fortune 500 companies spend approximately $500 million a year on more than 13,000 avoidable back surgeries for their workers. These surgeries resulted in total indirect costs (including lost productivity and sick pay) of $1.5 billion.[26]

Even when surgeries seem successful initially, many patients experience a relapse within three years. This is likely because neither surgery, nor drugs, nor physical therapy were able to address the underlying cause of their back pain. At that point, many patients opt to go through the entire process all over again, and pay the same exorbitant costs a second time, in the hope of a better result—or, they're forced to resign themselves to lifelong use of the only other treatment Western medicine can offer: painkillers.

Those are the success stories.

In tens of thousands of cases per year, patients who were able to function at a nearly normal level while managing their back pain with non-surgical methods end up permanently disabled after elective back surgery. For them, surgery was more than just a bad investment. Its failure will cost them their quality of life, hinder their professional aspirations, and make every day a misery of chronic pain.

Also, it will cost hundreds of thousands, even millions, of dollars in additional medications and treatments over the course of their lifetime.

This phenomenon is so common that there's even a name for it: Failed Back Surgery Syndrome, or FBSS. It's seen in 10% or more of people who undergo back surgery. One study conducted in the Czech Republic found that FBSS can occur in 5–50% of back surgery patients, depending on the nature of the surgery.[27]

Another issue of concern is the relative unimportance of precise diagnosis in Western medicine when it comes to back pain.

While some causes, particularly musculoskeletal conditions like spinal stenosis, are easily recognizable, most of the time Western doctors simply don't know what's causing a person's back pain. They may fail to recognize an otherwise non-symptomatic organ dysfunction as cause for the patient's discomfort. They may look at X-rays and CT scans of the patient's spine and see virtually no abnormalities—and yet, the pain is still there.

Herniated discs are cited as a primary cause of back pain, but cases exist where a patient may have one or more herniated discs and virtually no discomfort; additionally, surgeries may be performed to correct or replace herniated discs with no subsequent reduction in pain for the patient.

The issue in these cases is not merely the lack of a definitive diagnosis; back pain is a complicated issue, and often there is more than one cause. No, it is the fact that many Western doctors treat the pain *in spite of that lack*. It's a game of chance, a hit-or-miss strategy—the medical equivalent of dropping your transmission before you know what's wrong with your car. If it turns out the problem is your muffler, you've just spent a lot of time and money for nothing. Your car will probably never run as smoothly again, and if your mechanic forgot to tighten a bolt somewhere, you'll end up back in the shop before you know it.

Western medicine's reckless approach to the treatment of back pain, in which potent drugs and dangerous surgeries are tossed around willy-nilly, leaves the door wide open for medical error and the myriad facets of iatrogenesis we learned about in Section 2.

I feel I should also mention that there are dangers inherent to the Western approach to treating back pain which extend beyond drugs and surgeries. The very methods used to determine the cause of a person's back pain could be detrimental to his health, and add thousands of dollars to the cost of treatment. I'm talking, of course, about CT scans.

In the West, we've invented hundreds of machines and tests and protocols to assist us in "fixing" our patients. These inventions were supposed to make health care more successful, more efficient, and less costly than it was before—but the

Back Pain, Bone Disease, Depression

PATIENT: Christine L., RI

AGE (at time of treatment): 56

CONDITION(S): Back pain, degenerative bone disease, depression, high blood pressure.

CONDITIONS TREATED WITH WESTERN MEDICINE:
All of the above.

PRESCRIPTION AND NON-PRESCRIPTION DRUGS USED BEFORE CCM TREATMENT:
Morphine, Vicodin, cortisone injections (3 series), phenerol, Avapro (irbesartan), Lasix (furosemide), Budeprion XL (bupropion hydrochloride), Mobic (NSAID pain reliever).

SIDE EFFECTS FROM MEDICATIONS:
It was discovered that I am allergic to cortisone injections. Morphine made me feel like a walking zombie, and I couldn't it take frequently because I had to drive.

TOTAL ESTIMATED COST OF WESTERN MEDICAL TREATMENT OVER THE COURSE OF THE CONDITION(S):
$450,000+

HOW DID YOUR CONDITIONS AFFECT YOUR WORK LIFE?
I took numerous short days, and was ready to go on permanent disability. I was out of work for weeks or months with each of my surgeries. When I was able to work, I had to find different ways to do things: for example, I couldn't stand at the copy machine, so I had to wheel my chair over. I also had to find parking spots closer to the building, and I tried not to get up from my desk if I could help it.

PLEASE DESCRIBE YOUR EXPERIENCE WITH DR. SZTYKOWSKI AND CCM:
Prior to starting treatment with Dr. Tad, I was ready to go on full disability. I was working, but could only do short days, and my back pain was so severe that I had to find alternate ways to do almost every task. I'd already had two knee replacements, left foot reconstruction, bilateral carpal tunnel and hand surgeries, and had nodules removed from my vocal cords. I was on morphine for two years and Vicodin prior to that, and I underwent three series of cortisone injections until it was discovered I was allergic to them.

(continued next page)

When I came in to start my acupuncture treatment, I walked with a cane. My Western doctors wanted to operate on both wrists again for the carpal tunnel syndrome, and to perform spinal fusion. But I felt my body had already undergone too many surgeries, and I was sure that if I went forward with the spinal fusion I would never work again. I would be forced to go on disability, and would lose my home.

Now, I manage my conditions through acupuncture and herbal therapies recommended by Dr. Tad. I can concentrate at work, and I feel I'm better able to fulfill my duties as a result. I didn't need any of the surgeries my Western doctors were recommending.

My recovery has been a struggle at times, but my results with CCM have been remarkable, and I think that my decision to try acupuncture was one of the best I've ever made. I have finally found a way to live without pain!

WHICH PRESCRIPTION MEDICATION(S) ARE YOU STILL TAKING TODAY?
Occasional arthritis medication.

TOTAL COST OF TREATMENT WITH CCM:
$12,344

majority have done precisely the opposite. The Computerized Tomography (CT) scan is fast becoming one of our most popular diagnostic tools, but not only does each scan cost upward of $1,000, one single-organ CT scan can deliver up to 100 times the radiation of an X-ray.[28] Researchers have discovered that patients who undergo frequent scans have an increased chance of developing radiation-related cancer (which, incidentally, is likely to be treated with more radiation). Also, the contrast dye used with CT scans is iodine-derived, and can put a lot of strain on the kidneys, liver, and thyroid. Reactions in people who are already experiencing imbalance in these organs can be severe.

There are times when a CT scan does offer a better chance for an accurate diagnosis in a patient with chronic back pain, but these instances are not as common as one might think. In the majority of cases, an old-fashioned X-ray is sufficient to reveal spinal abnormalities. Still, these expensive scans are used when a simple X-ray would do, either because the patient insists upon it (believing that an old-fashioned X-ray constitutes a poorer quality of care), because the doctor mistakenly believes that an accurate diagnosis is more likely with a CT scan than with another test, or because CT scans are an easy way for the provider to turn a profit. Whatever the reason, the end result is the same: thousands of patients exposed to unnecessary radiation every month, for arguably minimal diagnostic gain.

Magnetic Resonance Imaging scans (MRIs) offer a better chance for precise diagnosis in back pain cases because they can reveal problems in fascia, or soft tissue, as well as bone. They also present a perfect opportunity for misdiagnosis because, as discussed earlier, not all herniated discs or spinal abnormalities are the cause of their owner's back pain. Also, it's very easy to misread an MRI, and to see abnormalities where in fact there are none, especially for new doctors.

In fact, MRIs are at the root of thousands of unnecessary back surgeries.

Whenever possible, I like to have my back pain patients bring in their previous MRI results for me to review. Because I know what to look for—and because I'm looking through the lens of

Classical Chinese Medicine—I am often able to use MRIs to assist in my own diagnostic and treatment processes, or (more often) to verify what my CCM diagnostic methods have already told me.

In a few very complex cases, I have asked a patient to schedule an MRI (with a recommendation from his primary care physician) before proceeding with acupuncture treatment, in order that I might construct the treatment program in the most effective way possible.

This is not to say that I encourage everyone with back pain to have an MRI scan. In fact, I usually encourage the opposite, for three reasons. First, each scan can cost $2,000 or more, so if it is not absolutely necessary to obtain an accurate diagnosis an MRI is a waste of resources. Secondly, the chance of misdiagnosis resulting in unnecessary surgery is higher among patients who undergo MRI.

Finally, the gadolinium-based contrast dyes used to enhance imaging can trigger a condition known as nephrogenic systemic fibrosis, or NSF, in patients with kidney and liver disorders;[29] as both kidney and liver problems can cause or contribute to back pain, a slight possibility exists for this reaction to occur in patients with undiagnosed kidney or liver dysfunction.

Like my colleagues who practice Western medicine, I see a lot of back pain patients. Sometimes the cause of pain is obvious: an injury suffered while shoveling snow, or while lifting a heavy box at work. Other times, a "differential diagnosis" is required, meaning that the correct diagnosis is arrived at through an educated process of elimination. Often, the back pain is revealed to be a symptom of another chronic condition (like insomnia or inflammatory disease), and will correct itself once the primary condition has been addressed.

Classical Chinese Medicine often points to dysfunction of the kidneys and insufficient blood flow (stagnated chi) as a cause of low back pain. As in many other chronic conditions, the degree of pain is closely related to the degree of inflammation in the body, which can be either systemic or local and may affect bones, joints, and fascia.

Because the ways in which CCM diagnoses and treats back pain (and other chronic pain conditions) are very different from Western methods, it is difficult to make a true comparison between the two as far as success rates. Western medicine attacks the symptom (pain), while CCM looks for the root imbalance, which may on the surface seem to have nothing to do with the pain. Therefore to say that CCM can "cure" back pain may be accurate—but was the back pain cured because a kidney imbalance was addressed, or the person's sleep habits improved? In such cases, the cure is not of the back pain itself, but of the condition to which back pain was merely an accompaniment. As you can see, it all becomes a bit hard to quantify—like comparing apples to oranges, as the old saying goes.

However, many well-executed Western medical studies have proved the efficacy of acupuncture in treating chronic back pain by examining statistical results only in the areas of pain and patient functionality, and not in improvement of conditions in which back pain is a symptom. These results have been unwaveringly positive.

A 2006 study published in *Medical Acupuncture* tested the effect of scalp acupuncture on chronic back and neck pain. Relief of 40–100% of initial pain occurred within 10 to 30 minutes of the start of treatment, and lasted from 1 to 5 weeks after one visit *with no follow-up*.[30] A 2005 U.K. study concluded that, at a 24-month follow-up, acupuncture was "significantly better at relieving low back pain than usual care," and that "GP referral to a service providing traditional acupuncture care offers a cost-effective intervention for reducing low back pain over a 2-year period."[31]

One German study found that, after 6 months of treatment, 45.5% of acupuncture patients surveyed reported significant improvement in functional ability, and that the mean number of days with pain decreased by half or more. Employed participants also reported a 30% decrease in lost work hours.[32] Another interesting Japanese study found that acupuncture treatment in steel factory workers who suffered from low back pain significantly decreased standardized medical expenses and hospital visits.[33]

Also of note: the acupuncture group in this study exhibited a marked decrease in "mood disturbance" scores, suggesting that their pain was no longer exacting a psychological toll.

In my practice, I will not address back pain (nor any other complaint, for that matter) until I determine exactly what's causing it; to do so would be to ignore the possibility of greater disease or imbalance within the patient, and potentially put his or her health at risk. Also, I don't like wasting my patients' time—or my own—with trial-and-error treatment methods. A thorough diagnosis helps me to keep my costs down, and my treatment prices low.

For the purposes of comparison, let's examine a case in which a patient in otherwise satisfactory health is exhibiting back pain as her primary symptom. The pain is caused by spinal stenosis, which is fairly common in older people. Once I make my diagnosis, I will treat this person using acupuncture, possibly combined with massage and herbal supplementation. I will give her a customized series of stretches to perform at home, and perhaps suggest that she sign up for yoga or Tai Chi classes to improve her flexibility, circulation, and blood oxygenation. It is likely that I will also suggest adjustments to her sleeping and eating habits which will help to reduce inflammation in her body and prevent the back pain from recurring.

Once treatment begins, a patient like the person described above may start to feel relief in as little as one week. After a month or two, she will see an improvement in functionality at work and at home, and at 6 to 8 months she will likely be completely corrected. In patients who are on several prescriptions, or who have had back surgery, treatment may take longer—however, even in cases where I believe that medications are interfering with my treatment, I will never advise a patient to discontinue his medications without the consent of his primary care provider. The one exception to this is pain medication, which a person can safely stop using once he is no longer in pain.

At the present time, my personal success rate in patients who complete my course of treatment for back pain is about 90–95%—far better than either drugs or surgeries.

PATIENT: Jerry J., Blue Jay, CA

AGE (at time of treatment): 61

CONDITION(S): Hypertension

CONDITIONS TREATED WITH WESTERN MEDICINE:

Hypertension

PRESCRIPTION AND NON-PRESCRIPTION DRUGS USED BEFORE CCM TREATMENT:

Kalan (calcium channel blocker), Lisinopril (ACE inhibitor), Benicar HTC (olmesartan medoxomil with diuretic).

SIDE EFFECTS FROM MEDICATIONS:

I was hospitalized at the UCLA Medical Center for an entire night after taking just one of the calcium channel blockers. My reaction was so severe that I started choking. I thought I was having a heart attack.

TOTAL ESTIMATED COST OF WESTERN MEDICAL TREATMENT

OVER THE COURSE OF THE CONDITION(S):

$350,000+

HOW DID YOUR CONDITIONS AFFECT YOUR WORK LIFE?

I travel extensively for my job. When my condition was at its worst, I was unable to travel, and my work was severely compromised.

PLEASE DESCRIBE YOUR EXPERIENCE WITH DR. SZTYKOWSKI AND CCM:

I work in a high-pressure environment in locations around the world. Although I've never struggled with depression or anxiety, I do have a lot of stress. I have struggled with hypertension since 1992. The medications I was taking weren't managing my blood pressure, but I refused to take more aggressive medications because I didn't want to deal with the side effects. I was also experiencing back pain and lower abdominal pain, both related to my high stress levels.

At one point my condition became so bad that I was trapped in Los Angeles for three months. I practically lived in doctors' offices. My blood pressure was so high that it brought on vertigo, and I couldn't even open my eyes without feeling as though I was flying through space. I underwent MRIs and CT scans, but the doctors only recommended more medication.

(continued next page)

Prior to beginning CCM treatment, my blood pressure was 180/105, even with the medication. Today, with only half the medication originally prescribed by my physician, my blood pressure averages about 145/85. One day, I had a blood pressure of 165/100, and I came into Dr. Tad's office for a treatment. By the time I went to bed that night, I was measuring 126/67!

Acupuncture not only helps me control my hypertension, it also gives me an incredible sense of well-being, and helps me feel balanced in my busy life. Although the pressures of my job haven't decreased, I feel that I can absorb more without being affected. My back pain and stomach issues have completely disappeared, and I feel a sense of clarity. Some days, I even feel that my vision has improved.

My treatment with Dr. Tad has cost only a small fraction of the Western treatments I received, and it's been far more effective at managing my condition. I don't have time or money to waste on treatments that don't work. I choose acupuncture because it does work.

WHICH PRESCRIPTION MEDICATION(S) ARE YOU STILL TAKING TODAY?
Lisinopril, Benicar. I take only about half of the dose originally prescribed by my physician.

TOTAL COST OF TREATMENT WITH CCM:
$5,962

The average duration of treatment for degenerative conditions like spinal stenosis or spondylosis is usually 30 to 50 visits, at a cost of $3,000 to $5,000. That's a big step down from the tens of thousands of dollars the same patient (or his insurance plan) would pay for surgery and drugs. Simple back strains may be helped in as few as 6 to 12 visits, at a cost of $1,000 or less.

Section 3:3—High Blood Pressure

High blood pressure, or hypertension, affects nearly one third of the adult population. Nearly 90% of Americans aged 55 and over will develop hypertension.

Blood pressure is a measurement of the flow of blood multiplied by the resistance that blood meets while passing through the blood vessels. When that resistance is increased, the result is hypertension. According to Western studies, the specific cause of hypertension is unknown in a staggering 90% of cases. In these cases, the condition is termed "idiopathic hypertension," meaning of unknown or non-specific origin.

The other 10% of cases are classified as "secondary hypertension," in which elevated blood pressure occurs as a direct result of kidney disease, thyroid or adrenal diseases, alcohol abuse, hormonal irregularities (usually caused by birth control pills or hormone replacement therapy), or pregnancy. Elevated blood pressure, left untreated over time, can result in damage to the heart, blood vessels, kidneys, and brain.

The American Heart Association estimated the total direct and indirect cost of hypertension at $59.7 billion in 2005.[34] That number is frightening enough on its own, but when you consider that hypertension is a factor in 69% of heart attacks and 77% of strokes in the United States,[35] and that the annual cost to treat heart attacks and strokes is estimated at $448 billion, the true economic impact of hypertension becomes apparent.

What does this mean to employers?

According to the American Hospital Association's 2007 Trendwatch report, hypertension accounts for 2.8 million missed work days annually in California alone. The average cost of reduced presenteeism is $247 per employee with hypertension per year.[36] In a separate study, the overall economic burden of hypertension, including absenteeism and reduced presenteeism, was estimated to be $392 *per year per eligible employee.*[37] Since hypertension can be aggravated by stress, employees with hypertension may be more affected by workplace stressors, take more sick days, and experience greater productivity loss than their non-hypertenisve counterparts. In workplaces such as warehouses, manufacturing facilities, and retail/service industries where some physical exertion is necessary, employees with hypertension are more likely to be limited in the duties they can perform.

Traditionally, hypertension has been considered a disease of the elderly. While it's true that the majority of people with hypertension are aged 50 and over, its prevalence in young people is increasing steadily, along with obesity rates. Therefore, you may not only be seeing hypertension-related productivity losses in your older employees, but throughout your entire company.

It's also important to note that, no matter what their age, employees with hypertension are statistically more likely to have other conditions which impair their ability to function in the workplace. About 75% of the hypertensive population is overweight or obese. About two-thirds of those with hypertension also have high cholesterol (hyperlipidemia), the presence of which can further decrease cardiac function and increase sick time and medical expenses.

Obesity, poor diet, lack of exercise, prolonged stress, high salt intake, age, prescription and non-prescription drug use, and numerous other factors are generally acknowledged by Western doctors to be contributors to hypertension, and a whole battery of studies has shown that regulation of these factors can have a positive effect on blood pressure. When lifestyle changes fail to work (or when the patient does not actively participate in his own care by making those lifestyle changes), medications and eventually cardiac surgeries are prescribed.

When it comes to treating hypertension, drugs are the option of choice for Western practitioners. Several varieties of prescription drugs are used in treating high blood pressure, including diuretics (water pills), beta blockers, channel calcium blockers, Angiotensin-Converting Enzyme (ACE) inhibitors, Angiotensin Receptor Blockers (ARBs), adrenergic agonists, and direct vasodilators. Often, two or more drugs are used in combination.

The success of drugs in managing high blood pressure has been well-documented—but *management* is all they can do. Pharmaceuticals do not cure high blood pressure, because they do not address the root cause of the problem. Also, there is the question of whether the benefit of these medications to the patient offsets the risks and monetary costs associated with their use.

For example: "water pills," although they sound innocuous, contain hydrochlorothiazide, which is also used to treat edema. These drugs block the re-absorption of fluid and salt in the kidneys, increasing urination. This seems logical when treating patients with edema, a condition in which the body retains excess fluid to the detriment of organs and tissues. It makes less sense when treating hypertension. In fact, the exact mechanism of this drug's effect on hypertension is unknown. What *is* known is that hydrochlorothiazide can upset sugar levels in the body, causing electrolyte imbalance, hyperuricemia (abnormally high levels of uric acid in the blood), and hyperglycemia (high blood sugar). Hyperglycemia can cause a person to experience excessive hunger, especially for foods and beverages that are high in sugars or carbohydrates. When these foods are consumed in excess, the hyperglycemic imbalance perpetuates itself, and the slightly overweight person with high blood pressure can easily become a moderately to severely overweight person with worse high blood pressure.

Other possible side effects include dizziness, lethargy, muscle pains or weakness, gastrointestinal discomfort, and headache,[38] all attributable to dehydration and potassium loss due to frequent urination. Whatever their cause, none of these side effects encourage the patient to become more active and healthy. Therefore, although they do lower systolic and diastolic blood pressure, the side effects of water pills can actually make it harder for people to lose weight and therefore correct their high blood pressure.

Beta blockers work by reducing the effects of adrenaline (epinephrine) on the circulatory system, therefore slowing the heartbeat and reducing blood pressure. Because less blood is pumped with less force through the body, poor circulation in the extremities, fatigue, weakness, and shortness of breath are common side effects. Calcium channel blockers work by dilating the arteries, therefore reducing the load on the heart. ACE inhibitors block the chemical angiotensin II, which is manufactured in the body and can cause blood vessels to constrict. Clonidine, a popular adrenergic agonist, works by stimulating certain receptors in the brain and decreasing the output of messages from the central nervous system to the body. As a result, the heart rate slows and blood vessels are relaxed—but digestive function can also be impaired, lethargy and headache may be experienced, and severe rebound high blood pressure may occur if the drug is stopped without a "step-down" period.

In short, none of these drugs do anything to address the actual cause of a person's hypertension. They do not heal; rather, they create greater imbalances in the body, and therefore more sickness. The cost of these ineffective treatments can range from under $20 (for generics) to hundreds of dollars per patient per month. Collective costs add up to billions annually.

If a person's high blood pressure continues despite "management" with drugs, or if damage to the blood vessels occurs as a result of a prolonged hypertensive state, surgery becomes an option. Usually this takes the form of balloon angioplasty or stent insertion which, as discussed in Section 2, can cost upward of $20,000 per operation, and may offer only small returns. This is not to say that angioplasty is not beneficial in some cases, even life-saving—but for many patients, its impact on quality of life and longevity is indifferent at best.

While it is considered a disease unto itself by Western medicine (and is treated as such), hypertension is really a symptom of one or more underlying imbalances. Classical Chinese Medicine is able to identify and correct these imbalances, which are often (but not always) related to dysfunction of the kidneys, liver, and/or adrenal glands, excessive stress, dehydration, poor metabolism, obesity, and/or sedentary lifestyle. Addressing the imbalances which cause hypertension will result in normalization of blood pressure without drugs or surgery.

About 50% of the people I treat for hypertension are on at least two types of medications. Many take three to five types of medication per day to treat both hypertension and other conditions (such as high cholesterol). These people are said to have "controlled" blood pressure. But if they were to stop taking their medications, their high blood pressure problem would still exist, because they are not cured by the drugs they are taking.

The other 50% of people I see have irregular or borderline-high blood pressure, and are not taking medication for hypertension, although they may take medications for other conditions. These people tend to respond more quickly to treatment, because their bodies are not confused by their medications.

If a person is in relatively good health, takes two or fewer prescription medications daily, and maintains a fairly active lifestyle and good diet, I can usually address his high blood pressure and correct the underlying problem in 20 to 40 visits, at a cost of $1,750 to $3,500.

My success rate for those who complete my program for hypertension is 92–95%, with success indicated by the complete normalization of systolic and diastolic blood pressure, and no further need for pharmaceutical regulation. Response to treatment may be seen in as little as two weeks, and the person may be able to discontinue his blood pressure medication (with the consent of his primary care provider) in as little as one month.

There are several techniques employers can introduce in the workplace to help alleviate high blood pressure in their employees—including meditation, which has been shown to be quite helpful in managing moderate hypertension. Noise reduction in the workplace has also demonstrated positive benefits. These and other stress reduction methods will be discussed in greater detail in Section 4.

Section 3:4—High Cholesterol

High cholesterol, one of the lipid disorders classified as *hyperlipidemia*, is a common condition among older Americans, but is lately becoming more prevalent among people of all ages. Young adults in their twenties, even children, are now being diagnosed with elevated cholesterol levels, a trend which is connected to rising obesity rates. According to the American Heart Association, nearly half of Americans over age 20 have serum cholesterol levels of 200 mg/dL or higher—but because it is not a symptomatic condition, many people don't know they have it until their doctors tell them.

Once they begin treatment for high cholesterol, however, even young people may be classified as "risky" patients by insurers because of the correlation between high cholesterol and coronary diseases, particularly atherosclerosis. The more risk factors an employee has, the higher his insurance costs. This applies not only to health insurance but to life insurance as well.

According to the U.S. Department of Health and Human Services' Agency for Healthcare Research and Quality (AHRQ), spending for treatment of hyperlipidemia increased by more than 350% between 1996 and 2003, from $4 billion to more than $18 billion. Over 90% of the cost increase is due to the widespread use of one particular class of drugs: statins. This family of medications includes such "blockbuster" drugs as Lipitor, Crestor, and Zocor. In fact, lipid regulators are now the most prescribed drugs in America!

The explosion in prescription numbers can be attributed to a change in guidelines for the treatment of high cholesterol, enacted by the National Cholesterol Education Program (NCEP) in May 2001. The adjusted guidelines increased the number of potential candidates for statin treatment from just over 13 million people to more than 36 million people, blowing the market wide open for the pharmaceutical companies and their new crop of statin drugs. With the new guidelines to hand, and a timely advertising blitz directed at both doctors and patients, Big Pharma was able to triple sales of cholesterol drugs in record time. Global revenues from Lipitor were reported at more than $12 billion in 2008.

Despite advertising claims that they can reduce a person's chances of heart attack and stroke, statin drugs do not in fact reduce the risk of coronary disease in patients, and make no difference in mortality rates, except in patients who have already experienced a cardiac event, according to the recent and enormous ALLHAT (Antihypertensive and Lipid-Lowering Treatment to Prevent Heart Attack Trial) study.[39]

So why are millions of people taking them?

Research has, in fact, shown that people with lower serum cholesterol levels are less likely to suffer a heart attack or stroke. Excessive cholesterol in the blood can lead to atherosclerosis, or buildup of cholesterol deposits inside the arteries, which in turn can lead to heart attack or stroke. If you can reduce serum cholesterol, therefore, it seems logical that the risk of atherosclerosis will also be reduced.

Unfortunately, this logic is flawed, because high cholesterol isn't the only risk factor for atherosclerosis. As many as 55 million Americans meet the diagnostic criteria for metabolic syndrome, a condition which includes symptoms like high triglycerides, low HDL cholesterol levels, elevated glucose levels, and high C-reactive protein levels, and which is believed to dramatically increase a person's chances of developing heart disease. Like cholesterol levels, high triglyceride levels are linked to poor diet and a sedentary lifestyle, and are significantly higher in obese people and diabetics. And while triglyceride levels may still be dangerously high in people who lower their LDL cholesterol levels with drugs, those who lower their cholesterol levels through CCM treatment and lifestyle modification usually lower their triglyceride levels simultaneously.

Furthermore, many people develop atherosclerosis without ever having had high cholesterol; these people often live a sedentary lifestyle, suffer from chronic stress and a poor diet, and have other lifestyle-related or genetic risk factors. Conversely, individuals with untreated high cholesterol but no other risk factors may never develop coronary disease.

The cost of treating high cholesterol with statin drugs varies depending on the type of statin prescribed and whether the drug is generic or brand-name. The more expensive drugs can cost upward of $150 for a 30-day supply, or more than $1,800 a year.

People who take statins can generally expect to use them for the rest of their lives, and the cost of use is compounded over time. One U.K. study found that in the year 2000, the average per-patient cost of statin use per life-year gained was £27,828. When statins were used in younger men, aged 35–44, the cost per life year gained jumped to £69,373, and could reach £94,645 for each life year gained when prescribed to younger women.[40] Adjusted for standard US medical inflation rates, and at the June 2010 pounds-to-dollars-conversion rate of 1:1.46, these costs equal USD $151,403 per life year gained for younger men, and $206,559 per life year gained for women. That's a lot of money to spend on drugs whose benefits are questionable at best, and which will never cure the condition they were prescribed to treat.

While there is no conclusive evidence that statins reduce the incidence of heart attacks and stroke, there is plenty of evidence to demonstrate their harmful effects.

Although there is cholesterol in the food we eat, it is not of a type which can be immediately used by the body. As a matter of fact, only about 5% of the cholesterol we eat has a direct impact on our serum cholesterol. The cholesterol which shows up in our blood is manufactured by the liver, and is necessary to many bodily processes. Statins work by inhibiting HMG-CoA reductase, a rate-controlling enzyme necessary to cholesterol synthesis. Limiting this enzyme stimulates LDL receptors, and thereby reduces the amount of free-floating cholesterol in the bloodstream.

There are six statin drugs currently on the market: lovastatin, simvastatin, atorvastatin, fulvastatin, pravastatin, and rosuvastatin. Each has a slightly different mechanism of action within the liver, but all share similar side effects, including constipation, nausea, vomiting, flushing, insomnia and other sleep disturbances, drowsiness, weakness, dizziness, paresthesia (numbness in extremities), and myalgia (general muscle pain). These reactions generally stop when the drug is discontinued.

More severe side effects can include myositis (muscle inflammation) and myopathy (a condition in which muscle fibers cease to function, resulting in muscle weakness), either of which may lead to rhabdomyolysis (breakdown of skeletal muscle), and potentially renal failure and death. Serious muscular side effects are attributed to the fact that statins raise blood levels of creatine kinase (CPK), a muscle enzyme produced in the liver; also, statins deplete the body of Co-Enzyme Q-10, which is vital to the "breathing" mechanism of cells and therefore to cardiac and muscular function.

Because statins work primarily in the liver, serious liver problems including enzyme derangement have occurred. The risk of liver damage goes up with every year of use, but many doctors have virtually stopped testing CPK levels and liver function in the general patient population, choosing to monitor only those patients at high risk for major side effects, such as those with existing liver disease or fibromyalgia. This selectivity allows the negative impact of statins on a moderately healthy liver to go unnoticed—and untreated—for extended periods of time, and results in a greater need for expensive and invasive treatments.

Other side effects which have been documented but not well-studied include pain and damage to joints, tendons, fascia, and nerves. There is no hard evidence (yet) which links statin drugs to the onset of neuropathy—a condition affecting the peripheral nervous system which can cause numb or tingling extremities, pain, and mild to severe loss of function—but I have seen tremendous evidence in my own practice to support this theory. While not all people on statins experience neuropathy, fully 100% of my neuropathy patients are using statin drugs when they first come to my office. Once the drugs are discontinued (with the approval of the primary care physician), about 50% of symptoms are alleviated immediately. Also, the remaining symptoms will respond more quickly to acupuncture treatment than they did while the patient was still taking statins. Since Western medicine labels nearly 80% of neuropathy cases as idiopathic (of unidentified origin), it seems reasonable to assume that at least some of them might be caused by the prolific use of statin drugs.

Based on my research and my own clinical experience, I feel justified in saying that the fewer people who take statins, the better. I believe that their risks, once fully understood, supersede even their perceived benefits, and far outweigh their actual benefits. Not only do statins do nothing to reduce the risk of death from heart attack or stroke, their side effects generate millions, perhaps even billions of dollars in additional treatment costs, and most definitely increase the annual iatrogenic mortality rate.

The saddest part is that with simple corrective measures like a proper diet and moderate exercise, almost no one would need to take statins in the first place. But today, as the guidelines for statin use become even more relaxed, prescriptions are flying out of doctors' offices, and pharmaceutical companies are looking for new markets for their statins—including the pediatric market.

Although there are several other types of drugs used to treat high cholesterol, statins are by far the most popular, which is why I've confined the discussion thus far to them. In my research, I discovered that most "patient information" sites barely even mention other methods of treatment (nor, interestingly, is much information given about specific diet and exercise methods which might reduce the need for statins).

Despite their relatively limited use, there are a number of non-statin drugs on the market. Some, like Zetia (Ezetimibe) work by blocking the absorption of cholesterol in the intestine; these drugs may cause serious side effects such as liver dysfunction, severe muscle pain, personality changes, irritability, sexual side effects, and memory loss. Worse, Zetia is often prescribed as a complementary treatment to statins, which can hasten the progression of liver problems and compound serious muscular side effects like rhabdomyolysis. Another class of drugs, called "bile acid sequestrants," work by binding with bile acids in the intestines and causing the acids to be excreted from the body. In addition to causing serious intestinal complaints, these drugs can increase triglyceride levels—and therefore actually increase a person's chance of a heart attack or stroke, since high triglycerides

are an even bigger risk factor for coronary disease than
high cholesterol.

A third class of drugs, known as fibrates, enable the liver
to absorb more fatty acids, thereby reducing the production of
triglycerides and lowering LDL cholesterol. Like other cholesterol
drugs, fibrates can increase the risk of rhabdomyolysis and renal
failure; when prescribed as a complement to statins, they can
increase the risk of rhabdomyolysis more than twelvefold.

Many of the people I treat for high cholesterol and other lipid
disorders do not initially come to me seeking help for these
conditions. Rather, they visit for more symptomatic conditions,
like back pain, and we decide in the consultation to treat the
high cholesterol (or not to treat it) simultaneously with the
other complaints.

Because high cholesterol is not a pain disorder, and produces
no discernible symptoms, it is hard to measure the effectiveness
of my treatment through the subjective information a patient
can provide. Testing is necessary to determine serum cholesterol
levels at different points throughout the treatment. In people
who are taking statins, these test results will obviously be
affected by the drugs. Therefore, I encourage patients to allow
me to work with their primary care doctors to discontinue all
cholesterol medications as soon as safely possible, both to allow
the effectiveness of the acupuncture treatment to be accurately
measured and to alleviate any harmful side effects of
the medications.

I can usually normalize a patient's cholesterol levels in
15 to 20 visits, at a cost of $1,300 to $1,800, if the problem is
primarily related to diet and lifestyle. In addition to treating the
hyperlipidemia itself, I work with my patients to improve their
eating and exercise habits. I may also prescribe certain herbal
formulas—including Shan Zha (Hawthorne), which is now being
used by some Western doctors because of its efficacy in reducing
LDL cholesterol and preventing the formation of arterial plaque.
Often, patients can replace their statin drugs with Shan Zha
(under the supervision of their doctor) within a few weeks of
starting my treatment.

In my clinical experience, about 80–90% of hyperlipidemia patients will also have some type of liver imbalance. Whether this is caused by long-term statin use or by another condition in the body depends on the patient; in most cases, there are multiple contributing factors. In people with severe liver dysfunction or other complications, the treatment period may be as long as 50 visits.

Once treatment is complete, 80–85% of my patients enjoy a reduced risk of heart attack and a healthier serum cholesterol level, and have no need to continue taking cholesterol-lowering drugs. In short, acupuncture treatment can deliver everything the statin drugs claim to provide, with no risk, no side effects, and minimal expense.

"I've got it, boss! Contrived Ruinous Aliment Phobia.
Fear of made-up diseases!"
"CRAP? That's brilliant! Now, there's only one question: How do we sell it?"

Section 3:5—Asthma

Asthma is a condition in which inflammation and constriction of the airways and excessive mucous production lead to shortness of breath, wheezing, coughing, and major respiratory distress. Attacks are generally triggered by one or more environmental factors (including but not limited to cigarette smoke, pollen, mold, perfume, or air pollutants), by the body's response to emotional stress, or by physical exertion. Daily symptoms can range from mild to severe. In persons with severe asthma, attacks can result in hospitalization and even death.

Survey data collected by the CDC indicates that in 2006, about 7% of the total adult population of the United States suffered from asthma.[41] In 2004, asthma resulted in an national average of 927 sick days per 1000 employees; these numbers were far higher (more than 1000 sick days per 1000 employees per year) in the Northeast, Pennsylvania, the Pacific Northwest, and Alaska. In fact, over 8 million work days are lost annually to asthma in New York alone![42]

A recent study found that people with severe or hard-to-treat asthma experienced a greater level of "impairment" at work (28% vs. 14%), and in daily activities (41% vs. 21%), than their mildly to moderately asthmatic counterparts.[43] This demonstrates that productivity and quality of life decrease as the severity of asthma increases—but if even mildly asthmatic employees are reporting 14% impairment at work due to their condition, it's clear that asthma of any severity can have a profound impact on a company's bottom line.

The cost of asthma treatment is also relative to the severity of the asthma. National costs for asthma treatment were estimated to be $57 billion in 1999, with direct treatment costs accounting for about $51 billion (88%) of that amount. Hospitalization costs were estimated at that time to be $3,102 per patient for an average stay of 3.8 days, while outpatient treatment costs were about $234 per visit.[44] When these numbers are adjusted for inflation, we see that today the same services might cost more than $4,811 and $360, respectively.

Obesity can increase both treatment costs and the severity of an asthmatic's condition. Obese people are statistically more likely to develop asthma in their lifetime than their non-obese peers. Nearly 75% of people who visit emergency rooms for asthma attacks are obese. In many of these people, obesity predates their asthma. Excess weight puts pressure on the lungs and diaphragm, making it more difficult for the lungs to fully expand with inhalation. Also, obesity causes changes in hormone levels, particularly leptin, which can contribute to systemic inflammation and create or worsen asthma symptoms.

The cause of asthma is not known to Western science. There is no way for Western medicine to prevent its onset, or to cure it once it occurs. Thus it has become one of a long list of conditions for which a regimen of "disease management" is prescribed, a course of action which generates billions in revenue for pharmaceutical companies but does little or nothing to improve the health of most asthmatics.

The most common treatments for asthma are inhaled corticosteroids and short-acting bronchodilators. Corticosteroids, as we explored in Section 3:2, can have a variety of negative effects on the body including sugar imbalance (which can lead to weight gain and thereby worsen the severity of the asthma), elevated blood pressure, reduced immune function, and osteopenia (a precursor to osteoporosis). These effects are considered to be less severe with asthma treatments because the dosage administered through an inhaler is far lower than the oral dose prescribed for back strain and other conditions. Still, risks are present, and while oral corticosteroids are usually only administered for a short period of time, inhalers may be used for years, even decades, and the risks of corticosteroid use are cumulative. Also, severe asthma attacks may be treated with oral or intravenous corticosteroids, thereby increasing the likelihood that the patient will experience adverse effects.

Short-acting bronchodilators, including the popular albuterol, work by relaxing the smooth muscles of the airways, reducing constriction and allowing more oxygen to enter the lungs. These are the fast-acting inhalers asthmatics use when they feel an attack coming on. While they do work in the moment to avert severe asthma symptoms, these inhalers come with a long list of potential side effects: some of the most common include anxiety or nervousness, insomnia, headache, muscle pain, diarrhea, dry mouth, dizziness, and tachycardia (rapid heartbeat). These are classified as "less serious" side effects because they aren't life threatening, but to me they indicate imbalance in the body, which, as we learned in Sections 1 and 2, is a precursor to disease.

From a purely practical standpoint, such side effects will most certainly reduce a person's productivity at work and in daily activities. The "more serious" side effects associated with albuterol are reactive bronchospasm, rapid and/or irregular heartbeat, tremor, and severely elevated blood pressure resulting in headache, blurred vision, cardiac arrhythmia, and seizure. Albuterol is also known to have negative interactions with a number of other pharmaceuticals including diuretics, beta blockers, and other types of bronchodilators. Overdose can be fatal, and is a major concern with children and young people who may not be as responsible with their medications as older asthmatics.

Some courses of treatment attempt to "cure" asthma by desensitizing the person to the identified trigger. While this has some preventive effect—and is certainly better for the patient than prescription drugs—it is not usually enough to alleviate the asthma completely. This method is most effective in people with allergic asthma whose symptoms are worsened by specific environmental triggers like dust, mold, and pollen.

In the same vein, while it won't cure the asthma, removing allergic triggers from the home and work environments can have a profound effect on overall well-being, response to treatment, and recovery time after attacks. We'll explore ways in which employers can reduce the presence of asthma and allergy triggers in the work environment in Section 4.

TESTIMONIAL

Asthma/Allergies

PATIENT: Tressa H., Swansea, MA

AGE (at time of treatment): 35

CONDITION(S): Severe asthma/upper respiratory allergies.

CONDITIONS TREATED WITH WESTERN MEDICINE:
Asthma, upper respiratory allergies.

PRESCRIPTION AND NON-PRESCRIPTION DRUGS USED BEFORE CCM TREATMENT:
Advair Discus 500/50 (salmuterol), albuterol, Claritin D (OTC allergy
medication), Mucinex. Also regular use of antibiotics, corticosteroids.

SIDE EFFECTS FROM MEDICATIONS:
Mood swings, jitters, rapid heartbeat. Damage to bones and eyes from steroid
use. Compromised immune system.

TOTAL ESTIMATED COST OF WESTERN MEDICAL TREATMENT
OVER THE COURSE OF THE CONDITION(S):
$500,000+

HOW DID YOUR CONDITIONS AFFECT YOUR WORK LIFE?
I took many sick days. I was also admitted to the hospital several times to treat
my asthma. Even when I was able to go to work, caring for my children when I
got home was difficult, and became nearly impossible when I had a flare-up.

PLEASE DESCRIBE YOUR EXPERIENCE WITH DR. SZTYKOWSKI AND CCM:
I was diagnosed with asthma before my first birthday. One year before writing
this testimonial, I was sicker than I had ever been. I was unable to enjoy my
children due to my lack of energy and focus. The medications I had to take
every day caused drastic mood swings and jittery spells. My eyes and bones were
damaged as a result of long-term steroid use. Even the slightest cold would often
develop into a sinus infection, and would require a dose of antibiotics to treat.
I received constant medical attention, but my health continued to decline.

As a last resort, I decided to try acupuncture. After my very first treatment, my
energy level was through the roof! After the first two weeks, I was sleeping better,
and my mood swings were resolved. My allergies disappeared.

(continued next page)

Two months into my treatment, I was able to completely discontinue my medications—with the exception of Albuterol, which I continued to use occasionally. I probably could have come off my medications sooner, but I was so conditioned to reach for my inhalers that I continued to use them even though I didn't physically feel the need for them. In a way, the process became a mental and emotional detoxification as well as a physical one.

During the course of my treatment, Dr. Tad prescribed Chinese herbs and dietary modifications. Once I implemented these changes, I felt lighter, clearheaded, and more energetic.

One year ago, I felt worse than I'd ever felt in my life. Today, I am totally allergy-free, and only rarely have asthmatic flares. I am off all my medications (although I keep my Albuterol around just in case), and I feel better than I ever imagined possible. I am living, BREATHING proof that Classical Chinese Medicine really does work!

WHICH PRESCRIPTION MEDICATION(S) ARE YOU STILL TAKING TODAY?
Albuterol inhaler, very rarely.

TOTAL COST OF TREATMENT WITH CCM:
$6,145

In CCM, we find six to seven distinctive patterns in the body which cause asthma. Some people will exhibit only one; others will have two or three. The main difference between the way CCM treats asthma and the way Western medicine treats it is that CCM does not see asthma as an independent ailment. Rather, it acknowledges asthma as a symptom of a deeper imbalance in which all organs and systems which support immune response are involved.

A 1995 review found that acupuncture reduced the need for asthma medication in 91% of study cases.[45] Another study found that when acupuncture was incorporated into a standard Western treatment regimen, patients with chronic obstructive asthma experienced statistically significant improvement in several areas —including on the St. George's Respiratory Questionnaire (SGRQ), where scores showed an 18.5-fold improvement in the acupuncture group.[46]

Acupuncture is also effective in short-term intervention. One study conducted in Taiwan demonstrated how acupuncture can equal the effects of short-acting bronchodilators by creating immediate improvement (over 20% as compared to baseline) in forced expiratory volume (FEV).[47] In two out of three patients, pulmonary function returned to baseline within four hours after acupuncture.

Another study traced the exact mechanism of acupuncture in improving symptoms of bronchial asthma. The anatomical locations of dorsal root ganglia (C7-T5) correspond to the location of the Back-shu acupoints along the midspinal line commonly stimulated in the treatment of asthma. When these points are stimulated through acupuncture or moxibustion, they help to regulate the synthesis and release of neurotrophic factors, which are found at high levels in patients with bronchial asthma. [48] By inhibiting the uptake of neurotrophic factors, therefore, acupuncture can reduce neurogenic inflammation in the airways and relieve symptoms of asthma.

I have an 80–85% success rate when treating asthma patients. By success, I mean that at the conclusion of treatment, the patient is completely asthma-free—or, in cases where the lung tissue has been permanently damaged by prolonged inflammation, that the asthma is controlled as completely as possible. After treatment, my asthma patients enjoy increased lung capacity, and are able to stop using prescription drugs (with their primary doctor's consent). Often, I'll tell them to hold on to their fast-acting inhalers "just in case," but most never need them again.

My treatment for adult asthma, with no other symptoms presenting, takes about 50–60 visits. The length of treatment depends on which patterns the asthma follows, the severity of the person's response to allergic triggers, and the length of time the person has been exhibiting symptoms. The total cost of my asthma treatment is about $4,000 to $5,500—about the same as a single emergency hospitalization.

The use of oral or inhaled steroids makes asthma harder to treat, since steroids, as we learned in Section 3:2, suppress the immune response, and therefore limit the ability of the body to heal itself. If the asthma is comorbid with obesity, inflammatory disease, or other systemic conditions, the duration of treatment may be longer, perhaps 70–100 visits. In children and young adults, who tend to respond more quickly to acupuncture, the treatment plan may span only 25–30 visits.

Section 3:6—Allergies

Many types of allergies affect American workers, including food and chemical allergies, but for the purpose of brevity this section will concentrate on respiratory allergies (also known as "indoor-outdoor" allergies), since these are among the most common types of allergies, and therefore the most likely to impact your bottom line.

According to the Asthma and Allergy Foundation of America, about 50 million Americans—one out of every six people—suffer from allergies. About 40 million of those claim an indoor-outdoor allergy such as hay fever as their primary allergy; other major triggers include pollen, dust, and pet dander. Most people with allergies are prone to more than one trigger; for example, people with severe hay fever are more likely to be allergic to certain foods, latex, penicillin, or another non-respiratory trigger.

The AAFA reported that, in 2000, allergies resulted in 17 million visits to physicians' offices and cost a total of $7 billion annually. Over 4 million sick days, plus substantially reduced presenteeism in employees who did come to work, equaled total productivity losses of over $700 million. When we adjust that number for inflation, we see that lost productivity could equal more than $1 billion in 2010.

EHS Today reported in 2003 that, according to a study presented at the annual meeting of the American Academy of Allergy, Asthma and Immunology, unmanaged allergic rhinitis (nasal allergies) cost employers a total of $597 per employee per year. Productivity losses due to allergy symptoms were estimated at 2.3 hours per employee per symptomatic day, for a total of 113 lost hours per year.[49] Another study, which polled more than 5,000 employees, found that 34% of employees missed an average of 1 to 5 days of work per year as a direct result of allergic rhinitis, and that 82% reported a 26% decrease in productivity and effectiveness at work on days when they were affected by allergy symptoms.[50]

Treatments for allergies include over-the-counter antihistamines (i.e. Benadryl), decongestants (Sudafed), corticosteroid nasal sprays (Nasonex, Veramyst), and "allergy shots" (immunotherapy), as well as several other options. Drug therapies, in my opinion, do nothing for patients with allergies except cover up their symptoms. With extended use, they may even be detrimental to a person's health.

Antihistamines work by attaching themselves to histamine receptors, thereby preventing histamine from causing a reaction in the body. The most common side effect is drowsiness, which can impact productivity and impair judgment; this is a particular concern for employees who work with machinery or operate automobiles while on the job. Antihistamines should also be avoided for individuals with hypertension or glaucoma, since they can elevate blood pressure—a warning which is seldom observed by those purchasing them in the "cold and flu" aisle of the local drugstore. Combination medications which include a pain reliever can, with extended use, cause inflammation of the liver, and bleeding in the stomach or intestines.

Decongestants work by narrowing the blood vessels, thereby reducing blood supply to swollen membranes. This effect can be localized, as with nasal sprays, or more generalized, as with oral tablets. Because of their mechanism, these medications can cause rapid increases in blood pressure, irregular heartbeat, severe headache, even convulsions or seizures. When blood flow is reduced, the ability of the body to heal itself is impaired, because tissues and organs are cut off from part of their oxygen supply. In CCM, we would say that they impair the flow of Chi, the "essence of Air." This reduction in blood supply to the brain also causes what some have affectionately dubbed "Sudafed head" —the woozy, foggy feeling some people experience after taking a decongestant.

This altered state has a drastic effect on productivity, making a person more prone to mistakes and accidents. Finally, like antihistamines, decongestants can only offer the "quick-fix," a short reprieve from symptoms. What's worse, many people, believing that allergies are incurable and unavoidable, use decongestants as the sole form of treatment for their allergy symptoms, and therefore not only fail to address their underlying condition but create bigger imbalances in their bodies, which lead to new and more serious complaints.

Corticosteroid nasal sprays work by suppressing substances in the body which trigger inflammatory response. These products are similar to those used to treat asthma, except that they are delivered through the nostrils rather than through the mouth. Because these are delivered directly to the nasal membranes, many believe that they do not affect the body in the same way as oral steroids.

I disagree: in order to be effective, the inhaled corticosteroids must be absorbed into the bloodstream through the thin nasal membrane. In the same way that cocaine, when inhaled, eventually reaches the bloodstream—and the brain, where it creates the "high"—the corticosteroids in nasal spray travel beyond the nasal passages to the rest of the body. Therefore, they have the potential to cause the same side effects as their orally-administered counterparts. Also, people have a tendency to overuse nasal sprays, making the likelihood of adverse effects greater than controlled studies would indicate. The acknowledged side effects of nasal sprays are limited to nosebleed, headaches, and nasal dryness, with a possible warning that long-term use may lower resistance to infection and contribute to glaucoma. (See Section 3:2 for a full list of possible side effects of corticosteroids).

Immunotherapy, in which small doses of allergy triggers are introduced to the body via shots in the upper arm, can be helpful, and certainly constitutes a more holistic approach than medication. But unless an effort is made to address the underlying imbalances which cause the body to react so forcefully to allergenic triggers, the effects of this treatment may be short-lived.

The costs of drugs and immunotherapy can vary greatly. Over-the-counter medications like antihistamines and decongestants can cost less than $10 for a week's supply, while nasal corticosteroids generally cost about $100 per unit (about a 1-month supply). Immunotherapy costs can vary depending on the frequency of visits, and are generally less than prescription drug costs after the first year, but I do not consider it a cost-effective solution because it requires a commitment of 3–5 years for treatment, with visits scheduled every 4 to 6 weeks. Using acupuncture, I can usually help a person become free from upper respiratory allergies in less than 30 visits, or about 3 months. Acute sinusitis can often be alleviated in 3 to 5 visits, or about one week.

Although Western science claims that there is no cure for respiratory allergies, I have had great success in treating them with Classical Chinese Medicine. Like asthma, allergies occur as a result of the body's reaction(or, more accurately, its over-reaction) to external triggers, and can be alleviated by correcting the internal imbalances which cause these acute responses.

A randomized, sham-controlled trial published in the *American Journal of Chinese Medicine* in 2002 found that participants who received acupuncture treatment experienced significant improvement in sneezing, nasal itch, and nasal discharge, as well as eye irritation and watering. The authors also concluded that "…the effect is not associated with any side effects and usage of relief medications other than acupuncture given."[51]

As mentioned above, I can usually eliminate a person's upper respiratory allergy symptoms and correct the underlying immune imbalance in 15 to 30 visits, at a cost of $1,350 to $2,500. Some people may be completely corrected in as few as 12 visits.

Also, since most adults spend nearly a third of their week in the workplace, it behooves employers to address factors in the work environment which may contribute to asthma and allergies among their employees. Preventive measures at work can help to reduce the overall severity of symptoms for most employees. For some, it may even reduce the need for treatment (depending on individual triggers).

For more on allergy triggers and how to reduce them in your workplace, see Section 4.

Section 3:7—Depression

Antidepressants are among the most prescribed drugs in America today, second only to cholesterol-regulating drugs and codeine combinations. Nearly 165 million prescriptions were filled in 2008. That means that more people take antidepressants than take prescription medication for high blood pressure, asthma, allergies, or contraception.

Outwardly, it would appear that we have a real issue with depression in this country. But when you look deeper, you discover that Americans are probably no more depressed than they ever were. It is the *definition* of depression that has changed, not our experience with it as a society.

Depression is a real illness, and people who suffer from it need to be treated in the most effective way possible. In some cases, psychotherapy combined with pharmaceuticals may provide the best solution. However, I feel confident in saying that the majority of people currently taking antidepressants don't actually need them, because they are not, in fact, depressed.

Merriam-Webster's Medical Dictionary defines depression as: *a mood disorder marked especially by sadness, inactivity, difficulty with thinking and concentration, a significant increase or decrease in appetite and time spent sleeping, feelings of dejection and hopelessness, and sometimes suicidal thoughts or an attempt to commit suicide.*[52]

Under the umbrella of this definition, several different types of depression are included: major depressive disorder (MDD), dysthymia, bipolar disorder, postpartum depression, seasonal affective disorder (SAD), and psychotic depression. Of these, major depressive disorder and dysthymia are the most common.

When I was in medical school, dysthymia was not classified as depression *per se*, but rather as "low mood." Meaning, the dysthymic person isn't happy and bubbly, but neither is he overly sad, suicidal, antisocial, lethargic, or apathetic, as is so often the case in people with major depressive disorder. Also, while MDD tends to be episodic, dysthymia is usually a more chronic, if fluctuating, low mood.

Of course, every effort should be made to improve quality of life for dysthymic patients, but to classify all people with low mood as "depressed" is not only seriously inaccurate, it's irresponsible—and potentially dangerous when you consider the side effects associated with antidepressants (which we'll explore later in this section.)

Pharmaceutical companies are great at massaging the facts in order to sell more products, and nowhere is this fact-twisting more evident than in the case of antidepressants. Today, dysthymia and depression have become interchangeable terms. Listen to advertisements on television describing the symptoms of depression: the narrator could be talking about anyone who's had a rough week. These commercials ask, "Are you irritable? Do you have low energy? Do you have trouble sleeping? Do you feel sad?" Answering "yes" to these questions may indicate major depressive disorder, true—but there are many other reasons for people to feel this way, not the least of which are pre-existing physical conditions such as obesity, sugar addiction, atherosclerosis, or diabetes. Sedentary lifestyles also tend to perpetuate dysthymia, in part because blood flow to the brain may be lessened or restricted due to lack of cardiovascular stimulation, and because the chemicals which regulate mood are not as efficiently distributed. The truth is, just because people *feel* depressed doesn't mean they *are* depressed.

Unfortunately, most people aren't aware of this distinction when they listen to the drug company's ads—and so, having no knowledge to the contrary, they start to think of themselves as depressed. They then go to their doctors and ask for antidepressants. Most of the time, they get them.

A large part of the problem is that the diagnosis of depression is entirely subjective. If a patient say he is experiencing symptoms congruent with the latest definition of depression, he must be depressed. In a way, patients are allowed to self-diagnose. And with a large percentage of prescriptions coming from general and family practitioners (23% and 21% respectively) as opposed to psychiatrists (29%),[53] it's safe to say that a lot of people on antidepressants have never tried any other form of treatment.

While any responsible family doctor will suggest counseling as an adjunct to medication, most will not require proof that the patient is following through with a program.

In the case of depression, our national quick-fix obsession once again shows its dark side. The issues people face in their lives are often painful, and very difficult to confront. It's much easier to take a pill than to address the situations which are causing their dysthymic state, and to make the hard but necessary changes which would enable them to move forward in their lives. As Dr. Ronald Dworkin, author of the book *Artificial Happiness: the Dark Side of the New Happy Class*, told CNN in a July 2007 interview: "Doctors are now medicating unhappiness." [54],[55]

Finally, a portion of antidepressant prescription sales can be attributed to the propensity of pharmaceutical companies to "highlight" certain conditions, or create new ones, in order to sell more of a particular drug. Let's look, for instance, at Premenstrual Dysphoric Disorder (PMDD), a condition created to classify women who have more-dramatic-than-average mood swings during ovulation, sometimes accompanied by physical symptoms. The most common treatment? Antidepressants! In 2000, Eli Lilly released Serafem (fluoxetine hydrochloride), which uses the same active ingredient as the company's blockbuster, Prozac. Coincidentally, almost no one had ever heard of PMDD before the Serafem marketing campaign was launched.

But there's a problem: the majority of women exhibiting PMDD as a primary symptom *aren't depressed*. It's likely that they are not even dysthymic. They are simply experiencing greater-than-normal shifts in the levels of estrogen and progesterone in their bodies. And although excessive hormonal shifts *can* cause a temporary dip in serotonin production (usually lasting a few days), that's hardly a reason to prescribe a full-time regimen of SSRIs! In most cases, reducing sugar and caffeine, getting adequate sleep, and exercising regularly will alleviate all or most of the symptoms of PMDD. If that's all it takes, the side effects associated with antidepressants hardly seem worth the benefits— yet SSRIs remain the most popular treatment for PMDD.

The CDC has observed that "the substantial increase in prescriptions for antidepressants also suggest widespread "off-label" (other than FDA-approved uses) use for subsyndromal mental health conditions and a variety of physical disorders." According to a 2007 report compiled by their offices, antidepressant use more than tripled between the periods of 1988–1994 and 1999–2002; by 2002, more than 8% of the population were using these drugs. Use was more than twice as high in women (10.6%) compared to men (5.2%).[56]

You may have realized by now that I don't agree with the way depression is diagnosed. I believe that much of the estimated cost of depression stems, at least in part, from other conditions which produce depression-like symptoms. However, statistics about the cost of depression in the workplace are still enlightening, even if they reflect misdiagnosed depressive-type symptoms in addition to actual depression.

The National Institutes for Mental Health (NIMH) estimated that in 2000 the total annual cost of depression in the United States was $83 billion. Adjusting for inflation, that total could conceivably reach $120 billion per year or more by January 2011.

A 1996 study estimated the annual health- and work-related cost per depressed worker to be $6,000; of that, $4,200 was paid by the employer. In 2011, these costs could be as high as $9,900 and $6,680 per worker respectively. Work-related costs were assessed by another survey study in 2003 which concluded that, in the prior year, depression cost employers about $44 billion in lost productivity, and that average productivity losses were 5.6 hours per week per employee (as compared to an average of 1.5 hours per week for those without depression).

The researchers asserted that the majority of that cost was "invisible," resulting from lost productive time while at work, and not from sick time.[57] A third study estimated that the direct cost of depression, including hospital visits and pharmaceuticals, was equal to only 31% of the actual cost of the condition; the remaining 69% was indirect cost, including lost productivity and sick time.

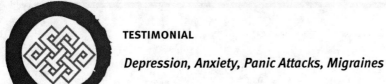

TESTIMONIAL

Depression, Anxiety, Panic Attacks, Migraines

PATIENT: Gail P., Foster, RI

AGE (at time of treatment): 56

CONDITION(S): Depression, anxiety, panic attacks, migraines, obsessive/compulsive disorder.

CONDITIONS TREATED WITH WESTERN MEDICINE:
All of the above.

PRESCRIPTION AND NON-PRESCRIPTION DRUGS USED BEFORE CCM TREATMENT:
Xanax, Wellbutrin, Paxil, Sudafed, aspirin, Tylenol.

SIDE EFFECTS FROM MEDICATIONS:
Weight gain, grogginess. Xanax rendered me non-functional.

TOTAL ESTIMATED COST OF WESTERN MEDICAL TREATMENT
OVER THE COURSE OF THE CONDITION(S):
$180,000+ in 5 years, $600,000+ in the last 15 years.

HOW DID YOUR CONDITIONS AFFECT YOUR WORK LIFE?
I was forced to change jobs, although I did remain with my employer.

PLEASE DESCRIBE YOUR EXPERIENCE WITH DR. SZTYKOWSKI AND CCM:
I have suffered from depression, anxiety, and panic attacks for more than 20 years. When I first came to see Dr. Tad, my panic disorder was so extreme that it was disabling. I couldn't be in an elevator, never mind get on a plane to take a vacation. I was so afraid of being away that I was trapped inside a 20-mile radius of my home. At times, my migraines were so severe that the sound of a ticking clock or the click of my car's directional brought on excruciating pain. I had such terrible sinus pain and inflammation that I couldn't even wear my glasses, because they put too much pressure on the bridge of my nose.

My prescriptions cost hundreds of dollars per month, but sometimes they hurt more than they helped. Xanax made me non-functional. I would go to work for three hours only to have to be picked up and driven home so I could crash for ten hours. Paxil helped for a little while, but it didn't control the panic attacks. After a while, my body grew accustomed to the medications, and my doctor increased my dosages again and again.

(continued next page)

I made frequent trips to the emergency room—sometimes two or more times per week—thinking I was suffering from heart attacks, paralysis, and numerous other issues. I even spent a week in a psychiatric hospital, but all the doctors there did was to prescribe more drugs.

I was completely weighed down by my illnesses. My quality of life was minimal at best. Then I learned about acupuncture through a close friend, and decided it couldn't hurt to give it a try. After all, nothing else was working for me, and I didn't want to weigh three hundred pounds and be depressed all the time. I wanted to be a functional, happy person.

At the time of this writing, I have been seeing Dr. Tad for about nine months. I feel happy and stable. The depression has lifted completely, and my panic disorder is no longer controlling my life and my schedule.

I haven't had a single migraine since I began my treatment, and my sinus pain and pressure is completely gone. My outlook on life and day-to-day functionality has improved so drastically that people actually ask me what drugs I'm taking!

For the first time in two decades, I feel like my body belongs to me, and that I have control over how I feel. I've traveled to more places in the last year than I have in my entire life; I even drove to Canada by myself! I'm off all my medications, and my head is no longer fogged up by narcotics. It brings me to tears when I think about the person I was a year ago, and compare her to the woman I am today. Acupuncture has changed my life, and I recommend it to everyone I meet who's suffering from a chronic condition.

WHICH PRESCRIPTION MEDICATION(S) ARE YOU STILL TAKING TODAY?
None.

TOTAL COST OF TREATMENT WITH CCM:
$4,973

According to the National Health and Nutrition Examination Survey (2005–2006), 80% of people with depression reported "some level of functional impairment" due to their condition, and 27% reported serious difficulties at work or at home. Interestingly, less than 30% of these people reported that they had contacted a mental health professional in the last 12 months.[58]

Now that we've touched on the astounding overuse of antidepressants in this country, let's look at what they actually do.

There are several types of antidepressant medications, the most common being MAOIs, TCAs, and SSRIs. The question—particularly with the "blockbuster" SSRIs like Prozac, Effexor, and Zoloft that make up the bulk of new antidepressant prescriptions—is whether the benefits they offer are worth the risks.

MAOIs

Monoamine Oxidase Inhibitors, or MAOIs, were first introduced in the 1950s, and were some of the first antidepressants on the market. They are prescribed far less frequently today than they were a few decades ago, in part because of their tendency to cause serious reactions when combined with other drugs. However, they are still prescribed for people for whom other treatments like SSRIs have proven unsuccessful—and, interestingly, some doctors are now using them to aid in smoking cessation. Brand names include Nardil and Emsam (a transdermal antidepressant patch).

MAOIs work by inhibiting the absorption of monoamine neurotransmitters like serotonin, norepinephrine, and dopamine, thereby increasing the body's available store of these chemicals and improving mood. However, they also interfere with other neurotransmitters like dietary amines. This interference can result in a number of serious issues. Because they render the body unable to rid itself of excess norepinephrine, MAOIs can cause hypertensive crisis when foods containing tyramine (like aged wine and cheese, chocolate, processed meats, and fermented foods) are consumed; if untreated, this can lead to stroke, cardiac arrhythmia, and intracerebral hemorrhage. Hyperserotonemia (extremely elevated serotonin levels) can occur after consuming foods like turkey, which contain tryptophan, or when MAOIs are combined with other serotonergic drugs; this may result in seizure, hallucination, and coma.

Less catastrophic side effects include insomnia, weight gain, sexual side effects, rash, nausea, vomiting, and dizziness, to name a few. Other adverse effects, like vivid dreams, hallucinations, convulsions, and psychosis, may present themselves when patients discontinue the drug suddenly, and can be quite severe. These withdrawal symptoms may prompt people to stay on MAOIs longer than they otherwise might, since getting off them can be even scarier than staying on them.

As I see it, the main problem with MAOIs is this: rather than rectifying the specific chemical imbalance which Western doctors believe causes depression—the inability of the brain to use serotonin and norepinephrine normally, an explanation with which I agree only in some cases—these drugs create new chemical imbalances which affect not only the brain, but the whole body. So in the end, even those people who might have in *theory* benefitted from chemical regulation of their monoamine transmitter levels now end up with a whole new set of imbalances—imbalances which could potentially turn an innocent chocolate overdose into a brain bleed.

TCAs

Tricyclic antidepressants are named not for their method of operation, but for their molecular structure, which contains three distinct rings of atoms. (These are not to be confused with tetracyclic antidepressants, which have four rings in their atomic structure.) Brand names include Norpramin (desipramine), Vivactil (protriptyline), and Surmontil (trimipramine). Like MAOIs, TCAs work by inhibiting the reuptake of serotonin and norepinephrine, however they do not seem to cause the same complications with tyramine and tryptophan in the digestive tract.

Side effects associated with tricyclics include drowsiness, dry mouth, blurred vision, decreased gastrointestinal secretions (which can lead to poor digestion and constipation), nausea and vomiting, difficulty with memory, hyperthermia (increased body temperature), excessive sweating, and dizziness. Increased appetite with weight gain and sexual side effects are common. More severe side effects include cognitive difficulties, muscle twitches, breakdown of muscle tissue (rhabdomyolysis), hypotension, and irregular or rapid heartbeat.

TCAs have a severe rate of morbidity and mortality in overdose, due to their neurological and cardiovascular toxicity. They also have a high rate of serious interaction with other popular medications, including oral contraceptives, sleep aids, aspirin, antihypertensives, diabetes medications, and antibiotics, to name a few. The iatrogenic possibilities are endless.

Of major concern for patients and employers are the cognitive effects associated with TCAs, especially memory loss. These effects can have a marked impact on productivity, and can become dangerous in situations where the employee is operating equipment or an automobile. While some studies claim that cognitive impairment is temporary and usually rectifies itself once the patient adjusts to the drug, many doctors have stopped prescribing TCAs as a precaution, especially in older patients who are more susceptible to cognitive decline.

SSRIs

Selective serotonin reuptake inhibitors, or SSRIs, are the giants of the antidepressant drug world. Prozac, Zoloft, Celexa, Paxil, and Luvox all fall into this category.

These drugs work by preventing the reuptake of serotonin into presynaptic nerve cells, leaving more of it available to bind with postsynaptic receptors. A good way to understand this process is to think of the bellows you use to start a fire in your fireplace, with serotonin replacing the air. When physical or emotional triggers prompt the bellows to close, serotonin is released, fueling the flame of "happiness" and inducing a feeling of warmth and contentment. SSRIs prevent a "backwash" of serotonin from being sucked back into the bellows when it opens to deliver another blast, thereby giving the brain more time to utilize the available store of serotonin.

It's important to note that the theory which explains depression as a lack of stimulation at postsynaptic receptors is just that: a theory. And on the basis of that educated guess, millions of people around the world are now having the synapses between their nerve cells flooded with excess serotonin, so that their "faulty" receptor cells can recognize the serotonin signal for a second, third, or fourth time. Yes, this can alleviate depressive symptoms, but it can also trigger a whole cascade of reactions in the body, because emotions aren't the only functions controlled by serotonin.

SSRIs are more selective in their activity than MAOIs, and therefore do not pose the same threat of hypertensive crisis from over-retention of norepinephrine, but they can still have profound effects on the digestive system. Rapid weight gain (or, rarely, weight loss) due to changes in blood sugar levels, impairment of liver and kidney function, nausea, and severe decrease in libido are all listed as "common" side effects of these drugs. Other potential side effects include dizziness, photosensitivity, vivid dreams or hallucinations, drowsiness, akathisia (a feeling of restlessness often coupled with anxiety), and tremors.

Nearly every patient who takes an SSRI will experience one or more of these side effects when he begins treatment. Symptoms are expected to subside once the body adjusts to the drug. However, although the person may no longer notice these symptoms after a few weeks or months on the drug, the imbalances which caused these symptoms to present in the first place still exist, and continue to wreak havoc in the body as long as the use of the drug is continued.

Weight gain is a side effect which has been reported with every SSRI on the market, but obesity is a major contributor to depression. A 2006 study published in the *Archives of General Psychiatry* found that obesity is associated with a 25% increase in diagnoses of general depression, bipolar disorder, and panic disorder (agoraphobia).[59]

Therefore, it stands to reason that the use of SSRIs, which cause weight gain in a significant percentage of people, might in some cases be counterproductive. While they may "manage" or cloak depressive symptoms, over the long term SSRIs may actually compound the problem, especially in people for whom the root cause of depressive symptoms is a physical imbalance such as—to use the most ironic example—obesity.

Studies have also found a correlation between the use of SSRIs and increased risk of adult onset diabetes (Type II), presumably attributable to two common side effects: weight gain and impaired kidney function.

One of the most disconcerting potential long-term effects of SSRI use is one which isn't listed on any prescription bottle. It's the possibility of a physiological backlash, and the creation of the very imbalance doctors seek to correct when they prescribe these drugs.

Here's how such an event might occur. The constant presence of excess serotonin in the synaptic gap (the space between transmitter and receptor nerve cells) can prompt presynaptic cells to "reassess" their serotonin output, creating a serotonin deficiency. This persists until the sensitivity of the autoreceptors on presynaptic cells becomes dulled—when, we might say, the patient's body begins to adjust to the drug. Also, post-synaptic receptors may be put out of commission when serotonin levels are too high, altering the ratio of transmitters to receptors. When these changes occur, a person can literally become dependent on the SSRI, because now he or she actually has the chemical imbalance the SSRI was prescribed to treat. This is particularly frightening when you consider the number of non-depressed people (i.e., women with "PMDD") who take these drugs every day. Our vaunted "cure" for depression may be creating a nation of truly depressed people!

Even people who truly suffer from major depressive disorder may be harmed by the long-term use of SSRIs. One study stated that, "according to conservative estimates, 10–20% of people with major depressive disorders…fail to respond to conventional antidepressant therapies."[60] In these cases, it seems logical that depressive symptoms must be caused by something other than abnormal receptor cells, and therefore the theory upon which treatment by SSRIs is based is not applicable. It also seems logical to suppose that for those 10–20% of MDD sufferers, long-term use of SSRIs may create lasting damage to the nervous system where no similar damage existed previously.

The study referenced above also concluded that treatment-resistant depression (TRD) places a significant economic burden on employers—$14,490 per employee per year as opposed to $6,665 per year for depressed employees who were not "TRD-likely"—due to increased use of medical services and greater productivity losses among TRD patients. In my experience, many TRD patients experience depressive symptoms because of an imbalance not in their brains, but in their bodies. Because Western medicine chooses to separate the brain from the body in its methods of treatment and diagnosis, it is hindered in its search for a "cure" for people with TRD.

Where antidepressants are a "Band-Aid," an attempt to cover the problem by altering the body's internal chemistry, Classical Chinese Medicine attempts to address the underlying causes of depressive symptoms, and correct them. A 2008 Chinese study found that "the therapeutic effect of acupuncture on depressive neurosis possibly is better than or similar to that of Prozac, but with less side-effects."[61] The effective rate in the acupuncture group was 86.4%, versus 72.7% in the Prozac group.

When I meet a patient who claims to be depressed, I look at a number of factors to determine whether he is actually suffering from major depressive disorder, or whether he is simply exhibiting symptoms congruent with the current definition of depression.

Sometimes, depression is of psychological origin, stemming from a single incident or a series of traumatic events in the person's life. These events can be recent, or decades old. Especially in cases where abuse is present, the emotional imbalance caused by trauma will often lead to physical symptoms which can range from mild to severe. Until the mind is healed, the body will remain imbalanced; therefore, I recommend these patients to a qualified mental health clinician with whom they can work while undergoing CCM treatment.

Conversely, I find in many of my patients a physical imbalance which has caused their emotional health to decline, resulting in depressive or dysthymic symptoms. In fact, I believe that about 90% of dysthymia is directly related to physiological imbalance. People who are physically unhealthy are less equipped—both emotionally and physically—to deal with external stressors. They are often sedentary, and have poor adrenal function. They are prone to chronic viral infections and hypothyroidism, and they may have circulatory and hormonal problems. In other words, their bodies, and therefore their minds, are chronically distressed.

These are the majority of the "depressed" people I see in my practice. They are also the ones who see the biggest relief from acupuncture, because once their physical imbalances begin to heal, their emotional states come naturally into balance.

In my practice, I have about an 80% success rate in the treatment of depressive symptoms. At the end of my course of treatment, 80% of my patients are able to function normally without medication. However, there are some depressive patients who do not respond to CCM; these people, I recommend to other practitioners in other modalities. I can usually determine the quality of a person's response in the initial 6- to 8-visit trial period.

The average course of treatment to alleviate the symptoms of depression and correct all underlying problems is 40 to 60 visits. After the first 30 visits, many people can stop taking their antidepressants (with the consent of their primary care doctor). The total cost for my treatment plan ranges from $3,500 to $5,000. People working through deep-seated emotional crises may require a longer period of treatment in conjunction with counseling or psychotherapy.

Acupuncture is not a perfect solution for everyone. Patients who suffer from major depressive disorder should continue to see a mental health professional while undergoing acupuncture or any other treatment, and should never stop taking their medications without the support and supervision of their doctor. Psychotic disorders, like schizophrenia and borderline personality disorder, which can be far more complex and dangerous than simple depression, are not in my purview as a Doctor of Acupuncture to diagnose or treat. While acupuncture may be an effective addition to treatment for these conditions, additional psychiatric therapy and possibly drugs may be necessary to help patients regain a normal quality of life.

That said, I can attest to the fact that acupuncture is far safer than Western treatments for MDD and dysthymia, and far more effective in cases where the underlying cause of depressive symptoms is not emotional or neurological.

Regardless of whether you agree with my opinion of the way depression is diagnosed and treated in this country, the connection between poor physical health and depression is evident. Of course it's not your job as an employer to diagnose or treat your employees—especially when those employees are dealing with serious illnesses like major depressive disorder—but the more you understand about the detrimental side effects of antidepressants, and how physical imbalances are misconstrued as mental illness in today's medical culture, the more you'll begin to realize the potential benefits of a well-structured wellness program. The fact is, people who are in balanced physical health are less likely to be depressed or dysthymic, and more likely to be happy, focused, and productive.

Section 3:8—Diabetes

Diabetes mellitus is defined as *a condition in which the pancreas no longer produces enough insulin or when cells stop responding to the insulin that is produced, so that glucose in the blood cannot be absorbed into the cells of the body.*[62] The inability of the pancreas to produce insulin is indicative of Type I diabetes, where the "resistance" to insulin by the cells of the body is characteristic of Type II diabetes.

Onset of Type I diabetes usually occurs before the age of twenty, the result of an immune dysfunction which causes the body's own immune cells to attack the insulin-producing cells of the pancreas and damage the insulin production mechanism. Secondary diabetes, which is similar to Type I diabetes, occurs when a non-immune factor, such as surgery, cystic fibrosis, or another degenerative condition, causes damage to the pancreas, and renders it incapable of producing enough insulin to adequately regulate blood sugar levels.

About 3 million people in the United States have Type I diabetes. These people must maintain strict control over their diet, exercise regimen, stress levels, and medication use in order to prevent devastating hyper- or hypoglycemic episodes, and preserve the healthy function of their organs. Most sufferers require daily insulin injections and frequent blood sugar testing.

Type II diabetes (also called "adult onset diabetes"), accounts for 90–95% of all diabetes cases in the United States, according to the American Diabetes Association. Unlike Type I diabetes, Type II diabetes is preventable, and curable in most cases, because it is primarily a lifestyle-related disorder. Approximately 90% of all those with Type II diabetes are obese. Other predisposing factors include lack of exercise, smoking, excessive alcohol consumption, hypertension, high triglyceride levels, and a high-fat/high-sugar diet. Age is also considered a risk factor, as diagnoses of Type II diabetes increase dramatically in patients over 45; however, the increase in obesity over the last few decades has contributed to a rise in diagnoses amongst younger people and even children.

For the purposes of this discussion, I will concentrate on the costs, issues, and secondary health concerns inherent to Type II diabetes. Type I diabetes, while both debilitating and costly, can be neither prevented nor cured by any treatment or modality, and while patients will certainly benefit from wellness initiatives, their condition can never truly be reversed. On the other hand, people with Type II diabetes can, through conscientious changes to their habits and a carefully-planned CCM treatment regimen, be fully and completely cured of their condition.

The American Diabetes Association estimates that the annual cost of diabetes in the United States exceeded $174 billion in 2007. Of that, $58 billion was attributable to absenteeism, disability, reduced productivity, and early mortality. Total costs have increased by $42 billion since 2002—more than $8 billion every year. Annual per capita health care costs for diabetics are an astounding $11,744 per year ($6,649 of which is directly related to their condition). One in five American health care dollars is spent caring for someone with diabetes.[63]

In addition to the approximately 20 million people in the U.S. with Type II diabetes, 6 million more are believed to be undiagnosed, and another 57 million are believed to have "pre-diabetes"—that is, they have fasting glucose levels which are higher than normal but not (yet) high enough to classify them as diabetic. What percentage of these people will eventually develop Type II diabetes is not known, however the future looks grim. The CDC estimates that one of every three persons born in the year 2000 will develop Type II diabetes in their lifetime. If we do not take steps to reverse this trend, the monetary and human cost of Type II diabetes as we see it now—when only about 8% of the population suffers from this disease—could be just a drop in the bucket compared to the costs we'll see in the future.

Diabetes is of particular concern from a clinical standpoint because of the numerous secondary conditions which can result from it. Diabetic neuropathies, which occur in 60–70% of all people with diabetes, are nerve disorders causing pain, tingling, numbness, and loss of feeling.[64] Symptoms are most commonly felt in the extremities, but damage can also occur in the heart, lungs, digestive tract, sex organs, and elsewhere in the body.

Consistently high blood sugar levels damage pericyte cells, which form the walls of smaller blood vessels. Diabetes are therefore particularly susceptible to bleeding disorders like diabetic retinopathy, a condition in which blood from the capillaries leaks onto the retina and interferes with vision. The blood-brain barrier may also affected by sustained hyperglycemic conditions, causing loss of cognitive function, poor focus, and memory issues. Because of this, diabetics are at increased risk for vascular dementia and cerebral hemorrhage, and may be more likely to develop Alzheimer's disease.[65]

Disruption of and damage to the blood-brain barrier becomes more likely the longer a person suffers from diabetes: one study found that the relation between diabetes and impaired cognition is not age-related, but instead related to the onset of the disease. Therefore, even people in their 30s and 40s with Type II diabetes may experience discernible cognitive impairment, particularly in the areas of executive function and speed.[66]

It's important to remember that Type II diabetes is a condition primarily related to lifestyle, and that it can be combated in the home office as well as in the doctor's office. Pre-diabetics especially can benefit from health education and wellness programs. If these people can take control of their lifestyle before their condition blossoms into full-blown diabetes, they can preserve their health and quality of life—and save a lot of money in the bargain.

While Type I diabetes is nearly always treated with insulin, many Type II diabetes patients do not require insulin shots. To fill this market void, pharmaceutical companies have created a slew of drugs targeted toward insulin-resistant patients.

One of the newest, Januvia (sitagliptin) works by blocking the enzyme GPP-4, which helps regulate the balance of insulin secreted from the pancreas and glucose secreted from the liver. By confusing GPP-4 receptors, the drug slows the release of glucose from the liver and increases the insulin output of the pancreas. Side effects including kidney problems and severe allergic reactions have been reported. Also, the drug has never been studied with insulin—which seems strange, because although it's not being targeted toward insulin-dependent patients, it is being marketed to diabetics.

Another relatively new drug, Byetta (exenatide injection) works by slowing the rate at which glucose enters the bloodstream, thereby regulating the amount of insulin produced by the pancreas. Side effects can be unpleasant, and include nausea, vomiting, diarrhea, and abdominal pain. Hypoglycemia can result when Byetta is used in combination with other drugs like sulfonylureas. Pancreatic inflammation has also been reported.

Sulfonylureas (Glucotrol, Amaryl, Micronase) and meglitinides (Prandin, Starlix) prompt the pancreas to make more insulin by stimulating beta cells. Sulfonylureas have been in use since the 1950s, and are among the most commonly prescribed drugs for Type II diabetes, but they are far from benign. Because they disrupt hormone and blood sugar levels, they can wreak havoc on the liver, kidneys, and thyroid, especially when taken improperly. Glucotrol in particular is known to increase the risk of death from cardiovascular disease, and comes with a long list of possible drug interactions.

Thiazolidinediones work by increasing insulin efficacy in muscle and fat. The first drug in this class, Rezulin, was taken off the market because of its propensity to cause severe liver problems. The remaining thiazolidinedione drugs, Avandia and ACTOS, have not yet been proven to share these risks, but neither have the risks been disproven. However, it clearly states on Avandia's boxed warning label that the drug can "cause or exacerbate congestive heart failure in some patients."[67] (The label goes on to note that, should said heart failure occur, a reduction in the dose of Avandia "must be considered.")

$12.5 billion was spent in 2007 on drugs to treat diabetes—up from only $6.2 billion in 2001. A good portion of this cost is attributable to the new (and very expensive) drugs which hit the market in the intervening years. Avandia costs a staggering $225 a month, and more than 7 million prescriptions have been written since the drug's release. Januvia's pre-release price was

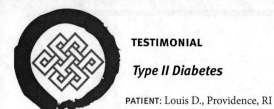

Type II Diabetes

PATIENT: Louis D., Providence, RI

AGE (at time of treatment): 67

CONDITION(S): Obesity, Type II diabetes, high blood pressure, blood clots (two of which moved into the lungs), diverticulitis, melanoma, basal cell carcinoma, tinnitus.

CONDITIONS TREATED WITH WESTERN MEDICINE:
All of the above except tinnitus.

PRESCRIPTION AND NON-PRESCRIPTION DRUGS USED BEFORE CCM TREATMENT:
Januvia 50mg (Type II diabetes medication), Lisinopril 40mg (ACE Inhibitor), Fluoxetine 20mg (SSRI antidepressant), Lipitor 10mg (cholesterol medication), Actos 45mg (diabetes medication to reduce insulin resistance), Tricor 145 (fenofibrate, to lower triglycerides), Glyburide 5mg (Type II diabetes medication), Hydrochlorothiazide 25mg (diuretic), Lovaza 1gm (to lower triglycerides), Warfarin 5mg (anticoagulant), aspirin.

SIDE EFFECTS FROM MEDICATIONS:
Hypoglycemia, kidney problems, anxiety, depression, changes in sexual ability, fatigue, weight gain.

TOTAL ESTIMATED COST OF WESTERN MEDICAL TREATMENT
OVER THE COURSE OF THE CONDITION(S):
$165,000, nearly $60,000 of which was spent on prescriptions.

HOW DID YOUR CONDITIONS AFFECT YOUR WORK LIFE?
I was forced to take temporary disability leave, and filed a claim for missed work time in the amount of $12,700. My focus was severely impaired, and I took a lot of sick days.

PLEASE DESCRIBE YOUR EXPERIENCE WITH DR. SZTYKOWSKI AND CCM:
After assessing my condition, Dr. Tad explained his treatment plan, which included working with my prescribers to reduce my medications and alleviate the side effects I was experiencing. He substituted some of the medications with Chinese herbal medicines.

(continued next page)

He told me that regaining my health would be a challenge because of my numerous conditions, and that I would have to become an active participant in my own recovery, especially in the areas of diet and exercise.

I have been seeing Dr. Tad for six months at the time of this writing, and I am definitely on my way to recovery. My breathing has improved, I have more energy to run my business, and I've lost 38 pounds so far. I have discontinued (through gradual reduction) more than half of my prescription medications, and look forward to stopping the others in the near future.

Dr. Tad and his staff are professional, knowledgeable, and caring. I have recommended Dr. Tad and Classical Chinese Medicine to all my friends and family.

WHICH PRESCRIPTION MEDICATION(S) ARE YOU STILL TAKING TODAY?
Warfarin, Atenolol, Chlorthalidone, Glyburide, Lovaza.

TOTAL COST OF TREATMENT WITH CCM (TO DATE):
$8,700

set by manufacturer Merck & Co. at $4.86 per tablet (about $146 a month). There are a number of older generics on the market which offer similar treatment for less money, but prescription numbers for these less recognizable drugs are falling; whether this is due to patient demand for advertised brand-name drugs or physician bias or both is unknown, but this predilection for "designer" drugs certainly isn't doing anything to rein in the cost of diabetic care.

Go to the manufacturer's web site for any of the drugs I've just listed, and you'll see the words, "manage your diabetes." The truth may be couched in pretty marketing-speak, but even Big Pharma has to admit that not one of the pharmaceutical options available to Type II diabetics offers even the remotest possibility of a cure. And yet, Americans are paying billions to obtain these drugs, because they are presented as the best solution to the diabetic problem.

Only hard work, weight loss, exercise, and lifestyle change can make a real difference for a person with Type II diabetes. CCM can help to correct the imbalances which have manifested as a result of poor habits, and help prevent those imbalances from ever recurring. But first and foremost, the patient has to take responsibility for his own health—and that responsibility starts with education.

When most people sit down to a meal of greasy, fried food and a sugary dessert, they probably don't understand the implications of that choice—not because they don't care about their health, but because they don't understand the processes at work in their own bodies. I firmly believe that once people learn *why* their habits are destructive, and discover the consequences those habits might have, they take a step closer to change. This is where, as a doctor, I also become a teacher.

Studies conducted to date on the efficacy of acupuncture for the treatment of Type II diabetes support the successes I've seen in my own practice. One Chinese study used rats to determine the effect of acupuncture on plasma glucose levels; researchers concluded that the electro-acupuncture method "reduced plasma glucose concentration in an insulin-dependent manner."[68] Another study, conducted at a Chinese teaching hospital with a group of 100 patients with Type II diabetes, tested depression scale, anxiety scale, and fasting blood glucose levels in patients before and after treatment with acupuncture; the data showed that acupuncture can improve blood glucose levels while simultaneously improving a patient's mental state.[69]

A third study, published in the *Journal of Alternative and Complementary Medicine*, found that auricular acupressure (which uses the same acupoints as acupuncture) was able to raise antioxidative properties in the blood for patients with high-risk diabetes mellitus. Free radical formation is high in people with diabetes, and can contribute to organ and tissue damage. Researchers found that serum concentrations of Superoxide Dismutase (one of the most potent antioxidant compounds) significantly increased in subjects who received auricular acupressure.[70]

Acupuncture has also been determined to be highly effective in the treatment of peripheral diabetic neuropathy. A 1998 study found that 77% of patients treated with acupuncture saw significant improvement in primary/secondary symptoms. At follow-up, it was found that 67% of participants were able to eliminate or significantly reduce their use of pain medications. Only 24% of patients required further acupuncture, and 21% reported that their neuropathy was completely cured after the initial course of treatment.[71]

Of course, research must be conducted in a controlled environment, and subjects are not asked to change their daily lifestyles during the course of a study. However, I believe that this also limits measurement of the true efficacy of CCM for Type II diabetes. Treatment of diabetes with Classical Chinese Medicine rarely consists of acupuncture alone: nutritional counseling, herbal supplementation, even meditation or Qigong may all be part of the treatment program. If acupuncture treatment is received by a person who continues to practice his old destructive habits, the treatment will not be fully effective; rather, it will act as a barrier, providing stabilization of the condition while staving off further damage to body by the diabetic condition. However, any positive gains from treatment will eventually be undone by the continued perpetuation of the underlying imbalances by the patient himself.

In my practice, I ask my patients to make significant changes to their habits after the initial 6- to 8-visit analysis period. This certainly contributes to the success rate I see in my practice, but does not account for it entirely. When a person changes his or her lifestyle, it's possible that Type II diabetes might resolve itself on its own, especially if the person is young and has not lived long with his condition. However, in most patients, prolonged diabetes will have impaired their kidney and liver function, so that even radical lifestyle changes will fail to restore homeostasis if they are not supplemented with acupuncture or another treatment.

Once a person has committed to changing his lifestyle I can begin to use CCM to correct the imbalances in his body. The issues I see in my patients with Type II diabetes are manifold, and may include adrenal dysfunction, kidney and/ or liver dysfunction, hyperlipidemia, circulatory issues, obesity, neuropathy, and numerous other problems comorbid with or resulting from diabetes.

The treatment duration for Type II diabetes can be relatively short—about 30 to 50 visits at a cost of $2,600 to $4,300—provided the disease has not yet contributed to secondary conditions like those listed above. In more advanced or complex cases, treatment will take longer, but even in complex cases my methods have demonstrated an 80–90% success rate, with no hyperglycemia and no need for insulin-regulating drugs at the conclusion of the treatment period.

During treatment, I work with the patient and his primary care doctor to develop healthful eating and exercise habits, and to safely decrease and eventually eliminate prescription medications. Unlike typical Western dietary guidelines, which claim that there is one "right" way of eating for everyone, Classical Chinese Medicine's approach to diet is entirely individual, and uses foods to counter specific imbalances in the body. I usually recommend that my Type II diabetes patients use Chinese herbal formulas to support the function of the liver, kidneys, and pancreas, and to help detoxify the body. I also help them develop an exercise regimen which includes 20–30 minutes per day of moderate cardiovascular activity such as swimming, biking, walking, or jogging.

Once the treatment is concluded, the patient's state of renewed health can persist indefinitely, so long as he or she continues to maintain the good lifestyle habits developed during the treatment program. I recommend that Type II diabetes patients return for follow-up visits at least a few times per year after treatment is finished. To my former diabetics, I become a partner in accountability; a "wellness coach," if you will, helping to keep them on track as they work to achieve their personal goals for health.

In the case of Type I diabetes, even Classical Chinese Medicine can offer no cure. It can, however, offer relief to the patient by helping to regulate blood sugar levels, balance immune function, and maintain full functionality of the organs most commonly affected by the condition (such as the pancreas, liver, kidneys, and heart). Usually CCM treatment for Type I diabetics is ongoing; it may take 60 to 70 visits to see the full benefit of treatment, and consistent visits thereafter to maintain an optimal state of balance in the body.

Section 3:9—Headache

There are several major types of headache disorders, including tension-type headache, migraine, cluster headache, and chronic daily headache syndrome. Other causes of recurring headache include injury to the head, neck, or back; infection; medication or substance overuse or withdrawal; and glaucoma. Headache disorders are chronic conditions, and can persist for months or even years at a time.

According to the World Health Organization, tension-type headache alone affects 80% of all female adults and 66% of males at some point in their lives, and one adult in twenty experiences a headache every day or nearly every day.[72]

Although their debilitating effects have not been as well-publicized as those associated with back pain or sleep disorders, headache disorders place substantial emotional, financial, and social burdens on those who suffer from them. Almost all migraine sufferers report that their professional and social lives are inhibited by their condition, and that their work capacity is noticeably reduced; 60% of tension headache sufferers report similar impairments. For some people, fear of the next headache can be nearly as paralyzing as the headache itself, and can prevent them from participating in work or social activities even when headache is not present.

Overall medical costs are substantially higher than average among people who suffer from headache disorders, due in part to a high rate of comorbidity. For example, headache patients are three times more likely to be diagnosed with depression than their healthy peers. Other comorbid conditions include hypertension, asthma, and sleep disorders.

Migraine

Migraine is possibly the most debilitating of the headache disorders, because of the degree of pain it causes. A severe migraine attack can leave a person literally unable to function. If left untreated, pain can linger for 24 hours or more.

The National Headache Association estimates that about 28 million Americans suffer from migraine. The condition is more than twice as common in women as in men. In 40–50% of female sufferers, migraine accompanies the menses; in 25%, attacks occur four times or more per month. The use of oral contraceptives can trigger headaches in some women. Other triggers include weather changes, sensitivity to certain foods or chemicals, lack of sleep, missed meals, and emotional stress. Obesity is believed to increase the frequency and severity of migraine attacks. Genetics also appears to play a role, as four out of five migraine sufferers report having a close family member with the condition.

The cause of this type of headache is still, to a large extent, a mystery to Western science, although the mechanism is loosely understood. When migraine occurs, inflammatory and pain-producing chemicals are released around the nerves and blood vessels of the brain, triggering the headache. About 60% of migraines occur on only one side of the head; others are more centralized, and may initially be mistaken for severe sinus headache. Accompanying symptoms include sensitivity to light and sound, nausea, and vomiting.

The impact of this condition in the workplace is evident in the estimated annual costs: more than $28 billion in the United States alone. About half of this is direct cost (medications, doctor and hospital visits, etc.) while the other half is indirect cost (absenteeism and reduced productivity). Lost work days alone cost more than $8 billion. A 1999 study found that in the United States, "migraineurs" required a total of 112 million days of bed rest annually—an average of 3.8 days per person for men and 5.6 days for women. Costs were highest for patients between the ages of 30 and 49. This study also concluded that "…the economic burden of migraine predominantly falls on patients and their employers in the form of bedridden days and lost productivity."[73]

Another study, conducted by United BioSource Corporation's Center for Health Economics and Policy found that the indirect cost of migraine to three major U.S. employers over a three-month period was $404,660 per 1,000 migraine sufferers, and that sufferers worked an average of 8.9 days over the three-month period while experiencing symptoms. Direct costs added an additional $200,410 per 1,000 sufferers.[74]

According to a NIMH-sponsored 2008 study, migraineurs are three times more likely than non-sufferers to exhibit some form of mental disorder such as depression, and also three times more likely to have comorbid pain conditions. Researchers also found that migraineurs experienced role disability on more than 25% of the previous 30 days.[75]

Western treatment for migraine usually involves one or more types of prescription and/or over-the-counter medication.

The first type of drugs are prophylactic—meaning, they stop or slow the progression of the headache, therefore lessening its severity. Aspirin, ibuprofen, acetaminophen, and naproxen can all reduce inflammation and migraine pain. All have been associated with stomach and gastrointestinal issues; acetaminophen (Tylenol), in particular, is implicated in liver damage and even liver failure in high doses or if taken without food.

Ergotamines, which are derived from the ergot fungus (and which are precursors to LSD), are vasoconstrictors, and act on neurotransmitters to reduce pain signals. They also act on dopamine and noradrenaline receptors, and can have significant side effects if overused, including thrombosis and gangrene (due to their powerful vasoconstrictive properties).

Tryptans are serotonin receptor agonists which constrict blood vessels to relieve swelling in the brain; like serotonergic drugs used for depression, these can include flushing, tingling in the extremities, dizziness, weakness, and tightness in the chest, and should not be used by people with hypertension or other predisposing factors for heart disease. Imitrex (sumatriptan), a popular drug in the tryptan family, is a hepatoxin (liver toxin), and has been demonstrated not only to increase the risk of heart attacks and strokes but to actually cause them. The list of drugs which "should not be used with Imitrex" is long enough to warrant its own book.

Tricyclic antidepressants or SSRIs may be prescribed for migraine because of their effect on serotonin levels. However, the imbalances these drugs can create, especially in non-depressed people (see Section 3:7 for more on the risks of antidepressants), in my mind outweigh any positive impact. Anticonvulsants have also been used, as they appear to decrease the frequency of migraine in about one of every four patients—but their side effects, which include weight gain, nausea, fatigue, and dizziness, were enough to discourage 14% of trial participants from their use.[76] Other possible prescriptions include beta-blockers and calcium-channel blockers for their effects on the vascular system.

Botox has been cited as a "miracle cure" for migraine due to its effect upon nerve sensitivity. While it's true that Botox injections may reduce the need for drugs, and may improve a person's functionality at work, the possible implications of injecting a nerve-deadening toxin into the face, head, and neck must be considered. Also, the concentrations of Botox used in migraine treatment (25 to 100 units) are far greater than those used in cosmetic applications, and have not been tested for long-term safety. Botox has not been approved by the FDA as a treatment for migraine at the time of this writing.

Tension-Type Headache
Muscular tension in the neck and shoulders is widely believed to be the cause of tension-type headache (TTH). Usually, these headaches are episodic, as with migraine, and pain can persist anywhere from a few hours to a few days. Rarely, chronic tension-type headaches can persist relentlessly for weeks or months; these cases fall into the category of chronic daily headache syndrome.

TTH is less severe than migraine, generally manifesting as mild to moderate pain with a feeling of "tightness" in the head which sometimes spreads into the neck. The symptoms of TTH are also less recognizable than those of other types of headache, and are often ignored or viewed as an insignificant complaint.

Research on TTH and its economic burden is relatively scant, however there are a few studies which document how this type of headache affects employees. One, which examined health care workers in a Brazilian hospital over a period of 6 months, found that when employees quantified the effect of their tension-type headache on daily activities, 35% reported a mild impact, 14% reported a moderate impact, and 2% reported a severe impact. A negative effect on work efficiency (mild, 28%; moderate, 11%; severe, 4%) was also reported.[77]

The most common treatments for tension-type headache include over-the-counter medications like aspirin, ibuprofen (Advil), naproxen (Aleve), and acetaminophen (Tylenol), all of which have been associated with gastrointestinal disturbances. Overuse of aspirin in can cause bruising and bleeding disorders. Also, patients should be warned—but often are not—that ongoing use of pain relievers for tension headache, including OTC medications and opiate analgesics like OxyContin, may actually contribute to the development of chronic daily headache syndrome, especially when taken more than two times per week..

Cluster Headache

Cluster headaches are short in duration but extremely painful, and recur frequently in "clusters" over periods of 6 to 12 weeks. Often, these headaches recur at the same time each year, and each episode lasts the same number of weeks. Pain is usually one-sided, and may be accompanied by a stuffy or runny nose and drooping eyelid on the affected side.

Cluster headache affects nearly six times as many men as women. The disorder is relatively uncommon, affecting only about 0.1% of the population. In a very few cases, cluster headache can become chronic.

Tryptans, lidocaine, intranasal capsaicin, and oxygen therapy are all suggested treatments for cluster headaches. Prophylactic treatments include the calcium channel blocker Verapamil, prednisone (oral steroid), melatonin, and ergotamines. Tryptans and ergotamines were discussed in the section on migraine, and we'll explore the adverse effects of lidocaine and prednisone in Section 3:11 (repetitive motion injuries). Suffice to say that none of these drugs address the root cause of cluster headaches, and therefore can make no progress toward actually curing them.

Chronic Daily Headache Syndrome

Although only about 4% of the population experiences chronic daily headaches, this condition is debilitating enough to warrant consideration by employers.

In some patients, headache is a primary symptom; in others, it occurs as a direct result of another, often systemic, condition. People who experience chronic daily headaches often suffer from depression and/or anxiety disorders, have irregular sleeping and eating habits, and have a lower immune threshold than non-sufferers.

More than with any other type of headache, there is usually a definable underlying cause to chronic daily headache, and therefore the best way to treat the headache is to ferret out the cause. Western medicine recognizes this: the American Academy of Family Physicians (AAFP), in its diagnostic guidelines, states that: "A pathologic underlying cause should be considered in patients with recent-onset daily headache, a change from a previous headache pattern, or associated neurologic or systemic symptoms."[78]

The AAFP estimates that at least one-third of chronic headache patients overuse their medication. Considering that the most common medications used to treat chronic headaches are beta-blockers, antidepressants, and antiepileptic drugs, this presents a major health concern. Pain relievers—over-the-counter, prescription, or both—also become a part of a chronic headache patient's daily routine, and although side effects vary widely, all contribute to imbalance in the body.

Acupuncture has proven effective for the treatment of headaches in several clinical trials. One study conducted at the University of North Carolina's School of Medicine found that acupuncture, when used in conjunction with medical management, significantly reduced daily pain and increased quality of life for patients with chronic daily headache. After 6 weeks, the acupuncture group was 3.7 times more likely to report less suffering. Researchers also noted that "headache-specialty medical management alone was not associated with improved clinical outcomes among [the] study population."[79]

Another study, conducted in Italy, pitted acupuncture against transcutaneous electrical nerve stimulation (TENS) and infrared laser therapy for transformed (chronic) migraine in a randomized study over ten sessions. Acupuncture not only significantly reduced the number of headaches per month, but also proved the most effective of the three methods over time.[80] A German study compared the outcomes of patients treated only with acupuncture to those treated with metoprolol (brand names: Lopressor, Toprol-XL). Not only did a larger percentage of patients respond to acupuncture (61% vs. 49% experiencing a 50% or greater reduction in migraine attacks), but both patients and physicians reported fewer side effects with acupuncture.[81]

In the practice of Classical Chinese Medicine, headache is not considered a disease, but rather a symptom. Even when the headache is the primary complaint, there is usually a deeper imbalance causing it. When I treat patients for headache, I strive not only to relieve their pain, but to seek out and correct the underlying cause of that pain. Migraine, for example, may be triggered by stress, food sensitivities, hormonal fluctuations, or one of several other causes. Until I identify that cause, I use acupuncture to provide safe, effective pain relief.

The typical duration of treatment for migraine or other headache disorders is 20 to 50 visits, at a cost of $1,500 to $4,500. In cases where chronic daily headache syndrome is caused by medication overuse, the duration of treatment may be longer, as I must work with the patient and his doctor to address the pharmaceutical issues before I can treat the cause of the headache. Also, comorbid conditions such as depression or insomnia may increase treatment time.

I consider my treatment for chronic headache successful only when the patient can stop taking daily medications such as tryptans or antidepressants (with the consent of his or her primary care doctor), and is headache-free at least 95% of the time (not including, of course, occasional headache due to colds, flu, sinusitis, or other unrelated conditions). I advise my headache patients to keep prophylactic medications (such as ibuprofen or other OTC analgesics) on hand, but most find that when my treatment is complete they no longer need any medications.

Headache patients, perhaps even more so than those who suffer from other chronic ailments, have a sense of powerlessness when it comes to their condition. Western medicine does nothing to remedy this, as its only answer to failure is more and different medication. CCM, because of its focus on maintaining balance, is more empowering. In order to prevent the recurrence of migraine or other headache syndrome once treatment is complete, the patient must understand what caused him to suffer the condition in the first place, and work to correct any habits which may have contributed to it. I find that a thorough explanation of the underlying imbalance and its cause, coupled with a detailed outline for a preventive routine, gives the patient a sense of control over his own health, and the motivation to keep his body in balance.

Section 3:10—Insomnia

The average adult needs 7 to 9 hours of sleep to maintain an optimal physical and emotional state. The National Sleep Foundation's 2008 poll discovered that the average American adult sleeps only 6 hours 40 minutes per night, and many sleep significantly less than that. From my own observation, most people believe that sleep deprivation is a normal part of being a parent, growing older, etc., and that it is therefore something to be taken in stride, and simply "dealt with." What these people don't realize is that their chronic lack of restorative sleep is contributing to or even causing their overall ill health.

Poor sleep habits cause more than just irritability, dark under-eye circles, and inconvenient napping during staff meetings. Sleep is essential to the rejuvenation of the body. During healthy sleep, cells regenerate, organs and systems "reset," and an optimal physical state is achieved. When sleep time is decreased by even one hour, the body can no longer revitalize itself efficiently, and it becomes harder to maintain homeostasis.

Lack of sleep can perpetuate sugar imbalances as well as kidney, liver, and adrenal dysfunction. The resulting hormonal shifts—an increase of ghrelin and a decrease in appetite-suppressing leptin—can trigger excessive hunger, especially for sweet and/or fatty foods. An analysis by researchers at the University of Warwick Medical School (U.K.) found that chronic sleep deprivation nearly doubled a person's chances of becoming obese.[82] Lack of sleep also weakens the immune system, rendering the person more prone to infection and increasing the duration and severity of illness. One Japanese study found that hypertensive people who slept less than 7.5 hours per night were significantly more likely to experience a cardiovascular event than those who slept longer.[83]

"Sleep debt" is also a major problem when it comes to productivity, presenteeism, and on-the-job safety. A 2003 study by researchers at the University of Pennsylvania School of Medicine concluded that adults sleeping 6 hours per night for 14 consecutive days experienced cognitive impairment equivalent to two nights of total sleep deprivation, even though subjects did not rate themselves as more than mildly sleepy.[84] The study also suggested that these impairments are cumulative over time, and can therefore present even if a person becomes accustomed to operating on reduced sleep.

It's important to note that not all people who are sleep-deprived are insomniacs. In Western medicine, insomnia is considered a sleep disorder, diagnosable by a person's inability to fall asleep or stay asleep, despite having the opportunity to do so. Lack of sleep, on the other hand, is a lifestyle choice, albeit one that many people feel forced to make because of work, family, or other pressures.

Though it may sound like a harsh indictment, I assert that a significant part of the blame for our national sleeplessness can be laid at the feet of the celebrated American work ethic. In its poll, the National Sleep Foundation found that the average American's work day has increased over the past decade from 8 hours to 9 hours, 28 minutes; that 20% of workers spend 10 or more hours per week completing work-related activities at home; and that a full 45% of adults will sacrifice sleep time in order to "accomplish more."

This nationwide workaholic mentality might be evidence of the American drive to succeed, but it's not making us more productive. In fact, it's costing us money: estimates place the total cost of lack of sleep to U.S. employers at $150 billion or more annually. This cost comes from workers' reduced efficiency, inability to concentrate, compromised short-term memory, and poor quality of social interaction.

One study conducted at Cornell University examined the per-patient cost of untreated insomnia over a six-month period. In the under-65 patient group, direct and indirect costs (including medical treatment costs, prescription and non-prescription drugs, absenteeism, and short-term disability) were $1,253 greater over 6 months for insomniacs than for non-insomniacs. In the over-65 patient group, costs were $1,143 greater.[85] While this study—like most of its kind—studied insomniacs, not people with poor sleep habits, it serves to demonstrate the volume of cost associated with poor sleep habits.

Lack of sleep also leads to significant increases in on-the-job injuries, and therefore to workers' compensation and disability claims. Coordination and motor functions are measurably impaired when the body is not adequately rested; this presents a major safety issue for anyone working with machinery or driving an automobile. An analysis by the Associated Professional Sleep Societies concluded that average accrued workers' compensation costs were nearly twice as high for insomniacs as for good sleepers ($483 vs. $280).[86]

While statistics show that the majority of American adults are at least mildly sleep-deprived, approximately 40 million people (about 22% of the adult population) suffer from diagnosable sleep disorders like insomnia and sleep apnea.

Chronic insomnia affects at least 10% of adults. Long term, it can play a major role in the development of conditions like depression, obesity, diabetes, and inflammatory diseases. In fact, the University of Pittsburgh study referenced on the previous page concluded that 40% of people with insomnia also suffer from a psychiatric condition such as depression, and that insomnia precedes depression in a large number of cases. Conversely, in people with anxiety disorders, insomnia is usually a secondary condition. More than 40% of insomniacs suffer from chronic pain (such as low back pain) which prevents them from getting to sleep, and which makes it harder for them to stay asleep.[87] Insomniacs are also more likely to experience migraines or other headache disorders.

Severe insomniacs are twice as likely to miss work, are hospitalized twice as often, and have more medical problems than good sleepers. People who can't sleep are also more likely to self-medicate with alcohol and drugs (both legal and illegal), and are more likely to suffer from addictions to these substances. Direct and indirect costs associated with insomnia were estimated by one study to total $30 to $35 billion in 1994.[89] Based on inflation rates, that number could be as high as $63 billion or more today.

When people have trouble sleeping, they often turn to prescription and non-prescription sleep aids. The problem is that these drugs don't induce the kind of restful, rejuvenative sleep that the body needs. Also, the side effects of sleep aids, especially prescription sleep aids, can be as disruptive as the insomnia itself.

There are two major types of prescription sleep aids: those which help patients fall asleep, and those which help patients stay asleep. The side effects are similar in both, and may include dizziness, prolonged drowsiness, headache, facial swelling, and abnormal sleep behaviors such as walking, eating, or driving while asleep. I don't think I need to expound on the possible dangers of these last, apparently common, side effects.

Some drugs, like the popular Lunesta (Eszopiclone), can adversely affect the liver, and may cause, according to the manufacturer's web site, "abnormal thoughts and behaviors… confusion, agitation, hallucinations, worsening of depression, and suicidal thoughts or actions."[90] If these symptoms do occur, the drug must be stopped gradually, or withdrawal symptoms could include anxiety, sweating, shakiness, unusual dreams, and muscle cramps.

Another commonly prescribed drug, Ambien CR (Zolpidem), presents many of the same side effects as Lunesta, with the added possibility that withdrawal may rarely cause seizures.

Most over-the-counter sleep aids are, in fact, antihistamines, just like allergy medications. These products include diphenhydramine (Nytol, Sominex) and doxylamine (Unisom). Side effects include dizziness, headache, and prolonged drowsiness; also, there are numerous conditions including depression, asthma, and hypertension, in which use of these drugs may be contraindicated.

Rather than correct the problems inherent to a chronic lack of sleep, these medications only saddle patients with a new and different set of potential problems. They do nothing to address the reason why the person is having trouble sleeping in the first place.

Acupuncture has proven highly effective and safe for the treatment of sleeplessness in every study I have read—which, of course, doesn't surprise me at all.

TESTIMONIAL

Insomnia, Depression

PATIENT: Sandra C., RI

AGE (at time of treatment): 54

CONDITION(S): Insomnia, anxiety, depression, arthritis, COPD,
high cholesterol, herniated discs.

CONDITIONS TREATED WITH WESTERN MEDICINE:
All of the above.

PRESCRIPTION AND NON-PRESCRIPTION DRUGS USED BEFORE CCM TREATMENT:
Celebrex, Nexium, Lasix, Advair, Enablex, Benicar, Trazodone, HRT medication,
Zirtec, Xanax, Celexa, Pravachol.

SIDE EFFECTS FROM MEDICATIONS:
I had so many things wrong with me that it was hard to tell which drugs were
causing which side effects. It was a merry-go-round. At one point my doctor
was trying to adjust my depression medication, and I started having anxiety
attacks. My antidepressants caused extreme sweating. I also experienced frequent
diarrhea, headaches, and swelling in my hands and feet.

TOTAL ESTIMATED COST OF WESTERN MEDICAL TREATMENT
OVER THE COURSE OF THE CONDITION(S):
$1,500,000+

HOW DID YOUR CONDITIONS AFFECT YOUR WORK LIFE?
I was out of work a lot, and took a lot of sick days. I'd even take vacation days
because I didn't feel good enough to come in to the office. My performance
reviews were definitely affected, and I wasn't even considered for promotions.

PLEASE DESCRIBE YOUR EXPERIENCE WITH DR. SZTYKOWSKI AND CCM:
When I started seeing Dr. Tad, I was unable to sleep for more than 1½ hours
at a time. I'd had numerous operations, including gall bladder and appendix
surgeries, hernia surgery, bunyonectomies, hammertoe surgeries on both feet,
carpal tunnel release surgery (both wrists), and hip replacement.

I suffered from arthritis in my feet, and my painful gait coupled with the
herniated discs made my back hurt as well. I was depressed, and worry about my
conditions brought on anxiety attacks.

(continued next page)

My personal medical bills totaled more than $250,000 from surgeries alone, and my prescriptions cost more than $1,000 per month.

Worse than the cost, though, was the fact that none of the drugs or surgeries healed me.

When I arrived for my first appointment with Dr. Tad, I could hardly walk. After only three treatments, I was able to dance at my nephew's wedding! Today, I'm feeling better than I have in many years. I'm sleeping 5 or more hours at a time without waking up, eating regular meals instead of snacking, and I have a lot more energy. I've stopped drinking coffee, and I'm eating less meat. So far, I've lost about 32 pounds. I still have things going on in my life, and there are days when I still feel mildly depressed, but overall I have a much easier time dealing with everything.

I found the answer to my health problems with Classical Chinese Medicine and Dr. Tad. I tell everyone to choose acupuncture before surgery: it's far more effective, for just a fraction of the cost!

WHICH PRESCRIPTION MEDICATION(S) ARE YOU STILL TAKING TODAY?
None.

TOTAL COST OF TREATMENT WITH CCM:
$14,362

One study conducted at the Centre for Addiction and Mental Health in Toronto, Canada found that five weeks of acupuncture treatment significantly improved nocturnal melatonin secretion, as well as total sleep time, sleep onset latency, and sleep efficiency (as measured by polysomnogram).[91] A German study used traditional Chinese diagnoses and treatments performed by individual therapists for patients who had trouble falling asleep or staying asleep. The acupuncture group showed "statistically significant" improvement compared to the control group.[92] A third study conducted in São Paulo, Brazil examined the outcome of acupuncture treatment for mild to moderate obstructive sleep apnea. Over the treatment period of 10 weeks, both the apnea index and number of respiratory events decreased significantly in the acupuncture group, and results of two standardized questionnaires showed significant improvement in functional quality of life.[93]

The vast majority of patients I see in my practice are carrying a "sleep debt" for one reason or another. In many cases, the lack of sleep has either caused the conditions from which they are suffering, or has made them worse. They are caught in a vicious cycle of stress and sleeplessness, from which their bodies cannot recover without help.

I don't consider complaints like restless leg syndrome, sleep apnea, or even chronic insomnia to be independent disorders. Rather, CCM has taught me to see them as symptoms of an underlying disharmony which may relate to organ imbalance, poor blood or tissue oxygenation, hormonal imbalance, adrenal fatigue, an over-stimulated nervous system, or any of numerous other imbalances. I have treated patients for insomnia related to conditions as diverse as hyperactivity, gastrointestinal disturbances, major depressive disorder (MDD), back pain, caffeine and/or nicotine addictions, obesity, and upper respiratory conditions. Once the root imbalance is corrected, the sleep disorder will generally disappear by itself.

Unlike with other complaints, where I work to correct the root imbalance in order to alleviate the symptoms, insomnia is nearly always the first thing I address, regardless of where it originates. I may begin to work with it as early as the preliminary 6- to 8-visit treatment course. It's true that treating the insomnia directly may not affect the underlying imbalance, but sleep is one of the most important elements in the healing process, and I know I will have only limited success in treating the underlying condition until the sleep issues are resolved.

The duration of my insomnia treatment ranges from 12 to 40 visits, depending on cause, severity, and related conditions. Costs range from $900 to $3,800—but again may be more or less depending on the cause of sleeplessness. When treating insomnia as a primary complaint, my success rate is 95% or better.

When people come to my office complaining of insomnia, usually what they're looking for is stress relief. Acupuncture can be used to induce a state of calm by encouraging the body to release endorphins and serotonin, and can help overworked adrenals recover more quickly—but over the long term, good self-care is the best remedy for stress-related insomnia. Establishing a bedtime routine, avoiding food and alcohol for at least three hours before bed, listening to soft music or meditation CDs, and utilizing aromatherapy products can all help prepare the body for optimal rest.

Employers can also help sleep-deprived employees with tools like nap rooms, flexible scheduling, and other wellness initiatives. We'll learn more about this in Section 4.

I teach my patients these techniques not only so that they can assist themselves in overcoming their sleeplessness, but so that they can also get the most out of their acupuncture treatment. For certain patients, I might prescribe herbal preparations to ease the body into sleep (where such preparations are not contraindicated by existing conditions or current prescription medications).

Section 3:11—Repetitive Motion Injuries

Ergonomic injuries, also known as repetitive motion injuries, are the most common workplace incidents in America. These may include, but are not limited to: back strain or injury, neck strain, carpal tunnel syndrome, and tendinitis.

In this section, I'll discuss carpal tunnel syndrome—which, while less common than some other types of injury, is the second most costly workplace injury in America behind back strain. (Statistics and treatments for back strain are described in Section 3:2.)

Carpal Tunnel Syndrome

Carpal tunnel syndrome, or CTS, occurs when pressure is exerted on the median nerve which runs through a "tunnel" in the wrist. Pressure can be caused by swelling of the tendons, or by swelling of the synovial membrane within the carpal tunnel (tenosynovitis). CTS occurs at a disproportionately high rate in the manufacturing, assembly, and meat processing industries, and among workers who spend a lot of time at a keyboard.

For years, CTS, especially cases triggered by typing, was more prevalent among older workers, whose wrists had been subjected to years of abuse, but we're now starting to see a rise in CTS cases in younger professionals and even children who spend large amounts of time using a computer, playing video games, or text messaging on cell phones. Women are more prone to the condition than men, presumably because of the smaller internal diameter of their wrists.

According to a 2005 report, carpal tunnel syndrome ranks as the second leading lost-time diagnosis for workers' compensation claims. It's also second for total costs of claims, behind only back injury. The average total lost-time claim per incidence of CTS is more than $20,400 (at 18 months).[94] The average time out of work for CTS patients is 30 days per incident.

The accepted Western treatments for CTS are drugs and surgery—neither of which, in my opinion, are effective. In fact, like so many reactive Western treatments, they can cause more problems than they solve, and cost a lot of money.

Anti-inflammatory drugs like ibuprofen are often recommended to ease tendon or synovial swelling. In some cases, corticosteroids (like Prednisone) are prescribed, or lidocaine may be injected into the wrist. While these medications may provide temporary relief, they do not address the underlying cause of the inflammation, and are therefore ineffective in the long term. Also, corticosteroids have a number of well-documented side effects (see Section 3:2). Of particular concern is that corticosteroids may increase the incidence of osteoporosis in older women, who make up the majority of CTS patients. For CTS or anything else, I do not believe that the risks of corticosteroids are worth the rewards.

Another option for treatment is a procedure known as Carpal Tunnel Release surgery. This method involves cutting the transverse carpal ligament to relieve pressure on the median nerve, and can be performed both as an open surgery and endoscopically. While in theory this procedure should work, in practice it often does not: over 60% of patients experience significant relapse within months of the surgery. The procedure also has the effect of reducing wrist strength, because the transverse carpal ligament is no longer functional. After surgery and rehabilitation, many surgeons advise their patients to limit or cease the activity which triggered the carpal tunnel syndrome in the first place—a course of action which would have provided relief even without a surgery that inflicts permanent damage on the internal structure of the wrist. In total, this ineffective procedure costs an average of $7,000 to $9,000 (or more, if complications occur).

Other surgical methods, including "percutaneous balloon carpal tunnel-plasty" in which a saline balloon is inflated inside the carpal tunnel to lift the transverse carpal ligament off of the median nerve, are being developed, but are too new for their long-term efficacy to be determined. Despite being less invasive than open release surgeries, the potential for complication with these new surgeries still exists. Also, the fact remains that, no matter what their condition, people should avoid inserting foreign objects into their bodies whenever possible.

In recent years, there has been a movement toward active prevention of carpal tunnel syndrome and other repetitive motion injuries in workplaces around the world. Japanese companies, in particular, have implemented mandatory stretching periods to help relieve symptoms among manufacturing workers. The advent of "ergonomic" hand tools and computer equipment here in America is another step in the right direction. Yoga has also been touted as a potential solution to carpal tunnel pain, because it not only stretches the wrist but can increase blood flow to the affected area.

Every business, from manufacturers to corporate officers, should encourage some form of preventive exercise for its employees if carpal tunnel syndrome is a concern. In my consultations with corporations whose workers are at a high risk for repetitive motion injuries, daily stretching exercises are one of the first things I suggest. I tailor this routine to the specific nature of the work performed, and in some cases to the needs of individual workers.

Daily stretches, properly performed, will not only prevent most instances of carpal tunnel syndrome, but will also alleviate pain from existing carpal tunnel syndrome.

For employees who are already suffering from CTS, Classical Chinese Medicine can provide relief, and eventually full recovery, far more quickly and safely than Western methods.

The National Institutes of Health (NIH) asserted in its 1997 Consensus Statement that acupuncture may be helpful as an alternative or complementary treatment for CTS, but there are few reliable published studies which specifically document the success of traditional (non-laser) acupuncture in CTS treatment.

I have had remarkable results with CTS patients in my own practice: my success rate is 92 to 95%.

One of the most important aspects of successful CTS treatment is proper diagnosis. Pain in the wrist and numbness and tingling in the fingers are symptoms of CTS, true—but they can also be caused by other injuries. It is my personal estimate that up to half of all patients diagnosed with CTS do not, in fact, have carpal tunnel syndrome, but rather are suffering from other conditions like degenerative arthritis of the cervical spine, herniated cervical discs, or an injury to the shoulder, all of which can adversely affect the nerves of the arm.

The tendency of neck and shoulder injuries to manifest as wrist pain may also partially explain why Western treatments are statistically so ineffective: if pain is not caused by an injury to the wrist, treating the wrist is not going to solve anything! This seems simple and quite logical, but I've seen dozens of patients who have undergone extensive treatment for CTS, including surgery, when there is, in fact, nothing wrong with their wrists. Once the real problem is corrected, the wrist pain disappears.

I will not treat a patient for CTS unless I am certain that inflammation of the transverse carpal ligament is the actual cause of his pain. If this is indeed the case, treatment is relatively short and simple. Usually, inflammation is eliminated and full wrist function restored within 10 to 12 visits, at a cost of $800 to $1,100. The patient can often return to work within one to three weeks.

CTS patients who have already undergone surgery for their condition may prove harder to treat. Even laparoscopic surgeries will create some scar tissue, the presence of which makes it more difficult for the body to heal. In some cases, scar tissue can actually increase pressure on nerves in the wrist, thereby compounding the problem the surgery was conducted to treat. In these instances, it takes longer to correct the problem—about 20 to 30 visits—but my success rate remains similarly high.

In cases where the cause of wrist pain is an injury to the neck or shoulder, treatment will be longer and more complex, as befits the nature of the injury—about 30 to 40 visits, at a cost of $2,400 to $3,200. Compare this cost to that of surgery, which can reach $9,000 or more (and remember that surgery is guaranteed to be ineffective for patients whose CTS symptoms are caused by neck or shoulder injury) and it's easy to appreciate the benefit of CCM treatment. Also, patients with neck or shoulder conditions can usually return to work within three to four weeks of the start of CCM treatment, whereas surgery may keep them out of work for six weeks or more.

Carpal Tunnel Syndrome

PATIENT: Angela A., Wakefield, RI

AGE (at time of treatment): 56

CONDITION(S): Carpal tunnel syndrome (left side),
rotator cuff impingement (right side),
migraine headaches.

CONDITIONS TREATED WITH WESTERN MEDICINE:
All of the above.

PRESCRIPTION AND NON-PRESCRIPTION DRUGS USED BEFORE CCM TREATMENT:
Naprosyn, Frova as needed at onset of migraine.

SIDE EFFECTS FROM MEDICATIONS:
None.

**TOTAL ESTIMATED COST OF WESTERN MEDICAL TREATMENT
OVER THE COURSE OF THE CONDITION(S):**
$8,000

HOW DID YOUR CONDITIONS AFFECT YOUR WORK LIFE?
The cramping in my left hand, and the pain in my right shoulder, kept me up at
night. The migraines made work and activities very difficult, and at times I was
unable to do anything but lie down and wait for the pain to pass.

PLEASE DESCRIBE YOUR EXPERIENCE WITH DR. SZTYKOWSKI AND CCM:
For two years I experienced cramping in my left hand. I was seeing a neurologist
for my migraines and mentioned this problem to him. He tested my wrists and
found carpal tunnel syndrome. He prescribed Naprosyn and physical therapy. I
did see some small improvement with these, but I didn't like taking the anti-
inflammatory drugs.

I work as a gardener, and the season before I started seeing Dr. Tad was extremely
difficult. In addition to the carpal tunnel syndrome and migraines, I was also
having issues with arthritis pain in my back and knees, and it was getting more
and more difficult to do my job. Western treatment wasn't working for me, and I
was afraid my pain was only going to get worse. I came to Dr. Tad a month after
my visit to the neurologist, and began acupuncture treatment for all of my pain
issues, including the migraines. The results were amazing.

(continued next page)

Three months after I started treatment, I had little, if any, cramping in my hands, and I have not taken an anti-inflammatory for any reason since I started my CCM treatment. My migraines, which used to occur very frequently, now only occur about once a month, and I hope that I will soon overcome them completely. I have no more pain in my back or knees.

Unlike the last work season, this year has been a breeze. I've been able to enjoy my job again! In addition to reducing my pain, I believe that working with Dr. Tad has increased my overall well-being. I would recommend his treatment to anyone who suffers from chronic pain.

WHICH PRESCRIPTION MEDICATION(S) ARE YOU STILL TAKING TODAY?
Occasionally Frova for migraine.

TOTAL COST OF TREATMENT WITH CCM:
$5,053

Like every other course of treatment, Western modalities included, Classical Chinese Medicine requires patient participation to be most effective. Once a person has been successfully treated for CTS, she should practice the prescribed daily stretching routine to prevent any relapse in the future, and schedule maintenance visits for the first year after treatment, in the same way as she would schedule follow-up visits after surgery.

Section 3:12—Gastrointestinal Diseases

Digestive complaints are extraordinarily common among Americans, in part because our national diet is so poor. Over-processed, chemically "enhanced," nutrient-deficient foods, combined with our tendency to eat on the run, make gastrointestinal dysfunction almost inevitable.

Gastrointestinal complaints can take many different forms. Irritable bowel syndrome (IBS), acid reflux disease (GERD), and other disturbances are common among people of all ages and ethnicities. Celiac disease (gluten intolerance) is one of the most under-diagnosed conditions affecting the American population. More severe conditions like peptic ulcers and inflammatory bowel diseases (including Crohn's disease and ulcerative colitis) can render sufferers unable to function normally on a day-to-day basis.

Irritable Bowel Syndrome

IBS is a condition in which abdominal pain, constipation or diarrhea, bloating, and discomfort occur regularly enough to be disruptive. Symptoms can be triggered by stress, certain foods, or infection, but often, no abnormalities are revealed by clinical tests. Some Western doctors theorize that the condition is the result of altered interaction between the brain and the intestinal tract; whether that alteration is neurological or otherwise has not yet been determined.

Some sources estimate that 1 of every 5 people has some form of IBS. It's second only to the common cold as a cause of absenteeism in the United States, and costs an estimated $30 billion per year. According to a recent study of Medicaid patients at the University of Georgia, annual medical costs for people with IBS were estimated to be about $1,600 more than for non-sufferers.[95] Interestingly, it is the cost of treatment for secondary conditions which drives up the per capita cost burden of IBS, and not the IBS itself. Only about 25% of those with IBS actually seek medical treatment for their symptoms, but all of those with the condition are more likely than people without IBS to seek treatment for something else—like fibromyalgia, chronic fatigue syndrome (CFS), or chronic pain.

Unlike Crohn's disease or ulcerative colitis, IBS does not usually lead to serious complications or degeneration of the bowel tissue, and rarely necessitates surgery. However, it can cause chronic pain, fatigue, and stress. Although Western medicine denies the connection, I believe that IBS, particularly IBS with constipation can and does contribute to colorectal cancers, because the body is not able to properly eliminate toxic waste. Patients with chronic diarrhea may become severely dehydrated, and the function of their kidneys, adrenals, small bowel, and other organs may be compromised.

Western treatment of IBS usually includes long-term management of symptoms. Usually, some method of stress reduction or stress management is recommended (including psychotherapy), and certain food triggers may be identified. The importance of exercise is also recognized. Medications like antidepressants, anti-anxiety drugs, and anticholinergics may be prescribed—the former two to help alleviate stress, the latter to reduce cramping.

We have already reviewed the complications inherent to antidepressants; I will discuss anticholinergics in the next section. Whether or not they provide symptom relief, none of these medications are capable of doing anything to address the actual cause of the patient's IBS. In fact, gastrointestinal disturbances are among the most commonly reported side effects of all varieties of antidepressants, and anticholinergics often cause both urinary retention and diminished bowel movements, which makes them doubly dangerous for patients experiencing constipation.

IBS with constipation is often treated with laxatives. The much-touted MiraLax, a powdered form of the polymer polyethylene glycol (PEG-3350), was recently given the thumbs-up for long-term daily use, and is now available over the counter. Variants of PEGs are widely used in cosmetics for their emulsifying abilities. Although evidence suggests that PEG toxicity is low, I would never recommend that a patient swallow plastic on a daily basis! Even natural, herbal laxatives are really nothing but a quick-fix, and prolonged use can lead to dependence and a reduction in the efficacy of natural bowel function and peristalsis (the muscle contractions which move fecal matter through the large bowel and colon).

For those in whom IBS presents with diarrhea, loperamide (Imodium) may be recommended. Loperamide is an opioid receptor agonist, and while studies claim it does not affect the central nervous system, and that it does not cross the blood-brain barrier in "significant" amounts, it does work on neuroreceptors in the large intestine. Long-term use will often cause physical dependence; withdrawal symptoms are consistent with opiate withdrawal and can be quite severe. Side effects can include constipation, dry mouth, dizziness, fatigue, and abdominal pain.

The high incidence of IBS among Americans indicates to me a great need to reduce stress in our everyday lives, both in the workplace and at home, and to improve our national diet. While 20% (or more) of your work force may suffer from IBS in some form, it's not a condition that's often discussed at the conference room table.

Encouraging employees with IBS to receive treatment in an effective modality like CCM can help prevent chronic pain, fatigue, and other secondary conditions associated with IBS from taking a serious toll on their health.

Celiac Disease

People with Celiac disease suffer from two distinct digestive abnormalities. First, they experience an abnormal immune reaction to gluten (a protein found in wheat, barley, and other foods) which causes their immune system to attack and damage villi in the small intestine. Secondly, people with celiac disease are prone to malabsorption as a result of damage to the villi, and are unable to properly digest food.

The immune response to the presence of gluten creates what can become a chronic inflammatory condition, causing extreme pain and discomfort as well as a host of other conditions resulting from ongoing malnutrition. Additionally, people with celiac disease are more likely to have other autoimmune or chronic inflammatory diseases like Type I diabetes, rheumatoid arthritis, and thyroid conditions.

Celiac disease is often mistaken for other digestive complaints like IBS or Crohn's disease, or for other disorders like chronic fatigue syndrome. Common symptoms include nausea, excessive gas, abdominal pain and bloating, and chronic diarrhea; some people also experience cognitive symptoms like mental fuzziness and inability to concentrate, as well as irritability, paleness, and dermatological symptoms. Digestive symptoms experienced upon ingestion of gluten can be sudden and severe, and can last for hours or days.

The number of celiac diagnoses increases every year. At the moment, the total number of people with the disease is reported to be about 3 million, or 1 out of every 133 people. However, some sources estimate that only about 1 out of every 30 people with the disease even know they have it. In support of that theory, I can say that I've met many people who have suffered for years, undergoing all manner of tests for other, less common ailments, before they were finally diagnosed with celiac.

From a clinical standpoint, there are both good and bad points to a diagnosis of celiac disease. The good thing is that once wheat and other sources of gluten are eliminated from the diet, symptoms will almost always improve. The bad news is that damage to the villi caused by the abnormal autoimmune reaction, combined with chronic inflammation and prolonged malnutrition, can wreak havoc on every system in the body, including the circulatory system, nervous system, liver, pancreas, and adrenals. This damage will not always repair itself once gluten is removed from the diet. Also, for some people sensitivity can actually *increase* after cessation of daily gluten intake; although daily symptoms are no longer present, these individuals actually become more likely to suffer a severe attack after ingesting trace amounts of gluten in prepared foods like sauces or dips, or after using personal care products like shampoos or skin creams which contain gluten. This hypersensitivity can be a "fear factor" for many people, and can limit both their professional and social lives.

In terms of cost, celiac disease places an enormous burden on the patient. Besides the battery of tests which is often necessary to diagnose this condition, special "gluten-free" foods are pricey and in some areas can be hard to find. For example, a loaf of gluten-free bread may be up to five times more expensive than a loaf of wheat bread from the local supermarket.

When I treat patients with celiac disease, my job is not just to tell them that they need to stop eating wheat and gluten, but to treat the imbalances in their bodies which cause their immune systems to attack the intestines in the first place. While a complete cure may not always be possible, I can help my celiac patients by reducing systemic inflammation and lessening the severity of their reaction to gluten through stimulation of acupoints related to the digestive organs and nervous system, so that if they do ingest a trace amount of gluten it will not trigger a crippling attack. I can also help to repair damage in the body caused in the period before the person was diagnosed, and ensure that the patient is no longer suffering from malnutrition or its effects.

In patients who are diagnosed soon after symptoms begin, and who have only minimal damage to the digestive tract, I can usually restore balance and relieve residual symptoms in 10 to 12 visits, at a cost of $900 to $1,100. For patients who were ill for a prolonged period before their diagnosis, and who have more extensive intestinal damage as a result, treatment may take up to 50 visits and cost up to $4,000.

This may seem like a lot of money, considering that the best and only treatment for celiac disease offered by Western medicine is a tightly controlled diet—but when you consider the implications of letting imbalances continue untreated, the benefit of preventive measures becomes clear. Patients with celiac disease, even controlled, are five times more likely to be diagnosed with non-Hodgkin's lymphoma than their healthy peers.[96] Intestinal cancers and liver disease are also more common among celiac patients than among non-sufferers.

Other conditions, often occurring as a result of prolonged malabsorption, may include osteoarthritis, bone loss, menstrual problems in women, infertility, and anemia. All of these conditions need to be addressed as soon as possible after the patient is diagnosed with celiac disease, in order to prevent later complications and preserve the patient's quality of life.

Inflammatory Bowel Diseases

Although it's estimated that only about 1 million Americans suffer from inflammatory bowel diseases (with that number split equally between ulcerative colitis and Crohn's disease), these gastrointestinal conditions have the potential to wreak havoc on the personal and professional lives of their sufferers.

Crohn's disease—also called ileitis or enteritis—is an autoimmune condition which causes the body to respond abnormally to food, bacteria, or other materials, causing inflammation and swelling in the digestive tract, particularly in the small intestine. Over time, this damages the delicate structure of the intestines, and can cause tremendous pain and violent digestive upset, as well as a host of secondary conditions including anemia, weight loss, fatigue, arthritis, and skin problems. Inflammation can affect the entire intestinal structure, from the inner mucosal lining to the outer wall; in some patients, certain areas may be profoundly affected while adjacent areas remain normal.

Ulcerative colitis, or UC, is similar to Crohn's disease in that it is an autoimmune condition. But where Crohn's disease usually affects the small intestine, UC causes ulcers to form in the mucosal lining of the large intestine, usually beginning in the rectum and extending upward through the colon. Unlike Crohn's disease, UC only affects this mucosal lining, and not the entire bowel wall. Symptoms at onset may include frequent diarrhea and mild intestinal cramping; as the disease progresses, symptoms may become more pronounced.

Patients with UC are at increased risk of colorectal cancer; this risk becomes greater the longer the person suffers from the condition.

Unlike patients with IBS, who may continue to work even when symptoms present, patient with Crohn's disease and UC are often forced to miss work due to flares. When these flares become severe, hospitalization may be necessary. This results in a major loss of quality of life for the patient, and of money and productivity for the patient's employer. In fact, one study published in the *Journal of Occupational and Environmental Medicine* (JOEM) found that as of 2005, annual medical expenses for commercially insured workers were more than three times as high for patients with Crohn's disease than for those in the comparison group ($18,963 vs. $5,300 annually). Patients with UC generated costs almost as high ($15,020 vs. $4,982).[97]

To date, Western treatments for Crohn's disease and UC have proven profoundly ineffective. Medical treatment (as opposed to surgical treatment) may include anti-inflammatory drugs, corticosteroids, immune system suppressants, antibiotics, and anti-diarrheal medications like loperamide and codeine. The side effects of these drugs have been discussed in detail in previous sections—but all may be made more severe by the fact that the patient's already imbalanced body is ill-equipped to handle the greater imbalance caused by pharmaceuticals. Corticosteroids and immunosuppressants, in particular, present a grave concern, because they increase a patient's risk for infection, and may increase healing time. In one study, patients undergoing medical treatment for UC experienced a 65% steroid-related complication rate.[98] Also, antibiotics, while they may be necessary in some cases, are often used to counter infections contracted because of immunosuppressant use. Over the long term, this back-and-forth can damage the immune system and create greater imbalance in the body, further complicating treatment.

When the body fails to respond to medical treatment, IBD patients may undergo surgery to remove the affected areas of the intestines. Many researchers assert that nearly all Crohn's and UC patients will eventually require surgery—which, to me, merely proves the ineffectiveness of medical treatment. Some proponents use these statistics to tout surgery as a "preventive" measure, claiming that delaying surgical intervention can make surgery riskier and more complicated.

Crohn's Disease

PATIENT: Shannon W., MA

AGE (at time of treatment): 19

CONDITION(S): Crohn's disease

CONDITIONS TREATED WITH WESTERN MEDICINE:
Crohn's disease.

PRESCRIPTION AND NON-PRESCRIPTION DRUGS USED BEFORE CCM TREATMENT:
Entocort EC (budesonide), prednisone, Imuran (azathioprine),
Flagyl (metronidazole), ciprofloxacin, Prilosec (omeprazole), Dipentum
(olsalazine sodium), Asacol (mesalamine), Levaquin and Cephalexin
(antibiotics), Vicodin, Percocet, Motrin, Advil, Tylenol, Tylenol PM.

SIDE EFFECTS FROM MEDICATIONS:
Neuropathy of the hands and feet, depression, nausea, loss of appetite, chronic
fatigue, dizziness, heartburn, thrush, acne, thinning hair.

TOTAL ESTIMATED COST OF WESTERN MEDICAL TREATMENT

OVER THE COURSE OF THE CONDITION(S):
$188,000

HOW DID YOUR CONDITIONS AFFECT YOUR WORK LIFE?
I needed to take painkillers just to make it through a work shift. I was unreliable,
forced to call in sick at the last minute, and unable to cover for coworkers. Also,
I could not complete any job tasks that required lifting or excessive movement,
since those put stress on my abdomen.

PLEASE DESCRIBE YOUR EXPERIENCE WITH DR. SZTYKOWSKI AND CCM:
I was diagnosed with Crohn's disease at age eleven and spent nine years trying to
control the symptoms and consequences. After a long struggle with prescription
medications, I discovered CCM. This medicine has restored my faith in the
healing power of the human body and has challenged me to question the efficacy
of Western medicine for chronic disease.

When I first came to Dr. Tad, I was in the middle of a year-long flare. I was
taking twenty-three pills a day, and my gastroenterologist was pushing for
potent intravenous drugs. A freshman in college, I was forced to spend most of
my time in my dorm room, since even walking to class triggered excruciating
stomachaches. Unable to eat much solid food, I lost an alarming amount of
weight in a short amount of time.

(continued next page)

Hesitant to take any more prescription drugs because I was already experiencing the painful side effects of the ones I was taking, I searched for more natural methods of healing. I used a heating pad on my stomach, experimented with dietary adjustment, and tried meditation and self-hypnosis. All these techniques produced only short-lived effects. The only option I had was to take painkillers and try to sleep through the pain.

Acupuncture was a last resort for me. Unfortunately, by the time I came to Dr. Tad, my condition had already progressed to the point where surgery was inevitable. Since Crohn's disease is an autoimmune condition, surgeries do not cure the illness, only remove damaged tissue. My intestines had fused to form a channel to my bladder, and I was in danger of developing more fistulas. I saw numerous specialists before the surgery, but nobody was able to diagnose my exact problem.

After the surgery, I continued my treatment program with Dr. Tad. Usually, students who undergo this type of procedure have to take a full semester off, but I was able to return to college full-time only two weeks after the operation. And that's not all! I have not had a stomach ache since shortly after my treatment began. For the first time in nine years, I am not on a single prescription medication, and I feel like my body has been purified of lingering toxins. I can eat whatever food I want without experiencing pain. I gained back all the weight I lost, and have been able to exercise for the first time in my life. My SED rate (which measures the amount of infection the body is fighting) dropped from 83 to 5 in a matter of months. Best of all, a recent colonoscopy showed no recurrence of the disease.

Acupuncture has given me a chance to live life in a way I never dreamed possible. When I look back at the condition I was in a year ago, the transformation is unbelievable. Where I was struggling to make it through a single work shift, now I'm picking up extra shifts, and I'm a lot more reliable. I even have the energy to undertake more volunteer work and extracurricular activities! For the first time since I was diagnosed with Crohn's, I feel completely in control of my health. I am endlessly grateful for the peace that CCM has helped to bring into my life.

WHICH PRESCRIPTION MEDICATION(S) ARE YOU STILL TAKING TODAY?
None. I take only a multivitamin, Omega 3 fish oil, and atractylodes (a Chinese herbal preparation).

TOTAL COST OF TREATMENT WITH CCM:
$5,500

I agree that effective treatment should be administered to IBD patients as soon as possible. However, removing a length of one's intestines should never be considered a "preventive" measure. In fact, it should be avoided if at all possible—not only because of the cost, which can equal $30,000 or more per surgery, but because of the impact such a drastic alteration of the digestive structure may have on a patient's life and health. Also, while surgery may eliminate symptoms for some people with ulcerative colitis, for many with Crohn's disease surgery provides at best a temporary respite, and symptoms recur after only a short time.

Unlike Western medicine, CCM addresses inflammatory bowel disease from a whole-body perspective. When treating IBD patients, I work to correct the immune dysfunction causing the symptoms, rather than focusing on the inflamed, ulcerated sections of the digestive tract. Particular attention is paid in CCM therapy to eating habits and food combining, since ease of digestion is imperative in patients with damaged intestines. Also, unlike immunosuppressants and steroids, herbal therapies can help to reduce inflammation while simultaneously permitting the body to heal at its natural rate.

My success rate in working with IBD patients is approximately 80%. Generally, the term of treatment for IBD is between 40–100 visits over the course of four months to one year, at a cost of $3,000 to $8,000—far less than the cost of a single surgery. After my treatment is complete, patients will no longer experience daily symptoms of IBD. Progressive damage to their intestines will be halted, and potentially even reversed.

In some cases, internal damages have progressed too far to be completely corrected; although I can help these people regain a more normal quality of life and halt the progression of damage, they will need to maintain a strict regimen of diet and herbal therapies to prevent future flares. People who have used oral corticosteroids for long periods of time, or who have taken multiple rounds of antibiotics as part of their medical treatment, may experience longer healing times. Also patients who have undergone surgeries to remove part of their bowel may never regain normal digestive function, although CCM can help ensure that they don't become malnourished or nutrient-deficient as a result.

It is interesting to note that many IBD patients are former IBS patients. Whether this is because their symptoms were misdiagnosed at the onset of disease, or because they developed IBD as a result of the imbalances in their bodies, the possibility of such a progression is worth considering. This is not to say that everyone with IBS will be diagnosed with IBD, but it's possible that some of them will. Therefore, it is imperative that people with IBS—especially IBS presenting with diarrhea—receive effective treatment (like CCM) to correct their digestive imbalances as soon as possible after symptoms appear. Such preemptive measures can prevent many long-term complications, protect employees' quality of life, and save employers tens of thousands of dollars annually.

Section 3:13—Chronic Inflammatory Diseases

The descriptive title of "inflammatory disease" covers a broad range of conditions, from gastrointestinal diseases and osteoarthritis to autoimmune disorders, lupus, multiple sclerosis, and even certain cancers. Fibromyalgia is also considered by some Western sources to be a chronic inflammatory disease, despite the fact that it does not cause tissue destruction and anti-nuclear antibodies are generally not found at high levels in testing.

Whether or not the condition is actually deemed "inflammatory," inflammation is a connecting factor for most chronic ailments. When inflammation is present in the body, conditions are ripe for disease to manifest.

There are five distinct signs which indicate inflammation in the body. These are increased body temperature (particularly indicated by high rectal temperature), pain, swelling, impaired blood flow, and decreased function of organs and systems. In most chronic disease, some level of inflammation is present: the above factors connect nearly every condition we've discussed thus far, from obesity to back pain, gastrointestinal disorders to asthma. Whether they knew it or not, most people who suffer from one or more chronic conditions were experiencing low-level inflammation in their bodies for months or years before their symptoms began.

For the sake of brevity, I will constrain this discussion to those chronic inflammatory conditions which are statistically most likely to affect your work force: osteoarthritis, rheumatoid arthritis, and chronic obstructive pulmonary disease (COPD).

Osteoarthritis

Osteoarthritis (OA) is the most common type of arthritis, affecting nearly 27 million Americans. OA is a degenerative condition which results in the breakdown of cartilage in the joints; this can happen anywhere in the body, but is most common in the fingers and thumbs, knees, hips, elbows, shoulders, and spine (particularly the vulnerable lumbar spine).

As OA progresses, the cartilage inside the joint which cushions movement and reduces pressure begins to lose its elasticity, becoming stiff and brittle. The normal movements of the joint, once cushioned by the healthy cartilage, now can begin to cause the stiffened cartilage to wear away, forcing the tendons and ligaments to stretch and causing joint pain and stiffness.

While it's believed that most people over the age of 60 have some degree of OA, the condition is becoming more prevalent among younger people. Obesity places excessive strain on the load-bearing joints of the knees, hips, and spine, and can cause OA to develop in people far younger than retirement age. According to the Johns Hopkins Arthritis Center, just 10 pounds of excess body weight exerts an additional 30 to 60 pounds of pressure on the knee with each step.

The first National Health and Nutrition Examination Survey found that overweight women were 4 times more likely to develop osteoarthritis of the knee than women of healthy weight, while overweight men were 5 times more likely to develop the condition.[99] Another study, conducted in Switzerland, looked at patients scheduled for hip replacement due to osteoarthritis; data suggested that pain increased, and functionality decreased, proportionately to BMI.[100] Poor circulation, a common side effect of obesity and sedentary lifestyle, is believed to contribute to the early onset of arthritis in the smaller joints of the hands and feet. Previous injuries (such as ACL tears) may also contribute to development of OA in younger people.

The cost of osteoarthritis is more than $80 billion per year in the United States. In a 2008 report, the Arthritis Foundation estimated that the average direct cost of OA is about $2,600 per person per year, and that the overall cost of the condition (including absenteeism and lost productivity) is about $5,700. [101] The report also asserted that "loss of joint function as a result of OA is a major cause of work disability and reduced quality of life." The cost burden of OA for employers is nearly as high as for hypertension or depression; one analysis placed the cost of osteoarthritis per eligible employee at $327 per year.[102]

Part of the cost problem is that many insurers do not cover pain medications—but they *will* cover trips to the emergency room, doctor's visits, and knee and hip replacements (at a cost of $32,000 or more for each surgery). This selective coverage—and the recent Vioxx scandal, which scared many patients away from COX-2 inhibitors—may make people with OA more likely to consider expensive and potentially dangerous surgical options.

About 650,000 people a year undergo a type of surgery called knee arthroscopy, in which fragments of bone and/or cartilage are cleaned out of the knee. Only about 50% of patients report actual pain relief from this procedure, which costs $5,000 or more. Arthroplasty, or total joint replacement, has a higher rate of "patient satisfaction," but this rate varies widely depending on whether the surgeon or the patient is the one doing the reporting.[103] Even subjective reports state that 15–30% of arthroplasty patients experience little or no improvement after surgery. In my experience, most people who undergo knee or hip replacement surgeries do not realize the quality of life benefits they were led to expect.

Most of the studies conducted with regard to acupuncture have to do with its pain-relieving and inflammation-reducing capabilities. However, because the majority of these studies have been conducted by Western doctors, acupuncture has been treated as a secondary therapy, used in addition to, rather than in place of, traditional Western methods. For example, a randomized, blinded trial conducted in Israel compared patient knee pain after 8 and 12 weeks of acupuncture plus conventional therapy (including NSAIDs, steroids, intra-articular hyaluronic acid, and COX-2 inhibitors). The patients in the acupuncture group showed statistically significant improvement over their peers in the control group at 12 weeks.[104]

Another study, conducted in Germany, studied the effects of acupuncture in addition to routine care for OA of the knee and hip: after 3 months, OA index scores improved an average of 17.6% in the acupuncture group, while the control group only improved by 0.9%.[105] While these sorts of results bode well for the future use of acupuncture as a complementary therapy, they do not speak to its potential as a primary mode of care for OA. Also, I have found in the 20 years of my own practice that medications, particularly steroids, detract from the positive effects of acupuncture treatment rather than enhance them. My goal is not simply to act as another painkiller for the patient to "manage" his disease: my goal is to improve his OA as completely as possible, so that he no longer needs to take daily pain medications.

As with any chronic disease, OA progresses in stages. And, as with any therapy, the success rate of CCM increases the earlier treatment begins. Unfortunately, in my personal practice, I don't see a lot of those early-stage patients. Most of my patients are in the degenerative stages of this disease and have already sustained major joint damage. They come to me because Western medicine can offer them no further solutions.

Usually, these patients take three or more medications daily (including steroids); some have undergone one or more surgeries before coming to me for treatment. Many can barely walk, and all have a very poor quality of life. Now, not only do I need to stop the progression of the osteoarthritis, I must reverse its effects, improve blood flow, help the body rebuild damaged cartilage, and help prevent the recurrence of inflammation. All of these things can be done, at least to some extent.

In one particular case, a woman came to my office with advanced OA. At the time of her first visit, she was barely able to walk, and was already scheduled for surgery. I had only 4 weeks to improve her condition enough to make surgery unnecessary. At the end of those 4 weeks, she was walking on her own, her pain level was drastically reduced, and she was able to call off the surgery. In another few months, we were able to restore her body to an optimal condition. Her story is a great example of how effective acupuncture can be for OA patients.

If a patient comes to me when OA is in its early stages—where pain is present, but there has been no marked degeneration of the cartilage—I can often reverse the condition and alleviate pain in 20 to 30 visits, at a cost of $1,700 to $2,600. If the patient

maintains good habits after the treatment is completed, and
follows up with maintenance visits at least a few times per
year, he or she will be free of their OA for many years to come.
Considering that drugs used to treat the pain of OA—such as
prescription COX-2 inhibitors—can cost upward of $100 per
month in perpetuity, CCM presents a tremendous cost savings
opportunity. My correction rate for mild to moderate OA is
upward of 90%, whereas the cure rate of prescriptions and over-
the-counter drugs is virtually zero.

Patients in the later stages of OA may require a longer period
of treatment, usually ranging from 6 to 12 months (although
significant pain relief will be achieved after only a few visits), at
a cost of $3,400 to $6,000. This is still a tremendous savings over
the cost of surgery, which as we learned earlier can be upward
of $32,000. Also, the patient will not be subject to any iatrogenic
effects from hospital stays, anesthesia, and surgical complications.

I do not consider my treatment successful unless the patient
regains at least 75% of normal function, and can stop taking his
or her medications completely.

Rheumatoid Arthritis

Rheumatoid arthritis (RA) is an autoimmune condition in
which the body's own immune cells attack the bone and cartilage
of the joints. The tissues around the joints become swollen
and inflamed, resulting in terrible pain. Once the disease
progresses beyond its initial stages, bone erosion and perpetual
inflammation in the synovial membranes can cause permanent
damage to the joints. In people with severe RA, the joints can
become so deformed as to be practically non-functional.

Although RA comes and goes in most patients, with
symptomatic periods known as "flares," some sufferers live with
a constant low level of pain. However, whether accompanied
by pain or not, inflammation is constantly present in the body,
and can cause damage to not only to joints but to organs and
tissues. Chronic inflammation also makes the body more
prone to infection (a situation compounded by many common
pharmaceutical treatments). Patients with RA will be sick more

often, and take longer to recover from illness, than their healthy peers.

An estimated 1.3 million people in the United States suffer from RA. The condition is two to three times more prevalent in women than men. RA can occur at any age, but most people begin to experience symptoms after the age of 40. There is evidence that flares may be triggered by hormonal fluctuations, and some studies suggest that hormone replacement therapy (HRT) may contribute to the onset of RA in women. Despite the fact that RA sufferers account for only about 5% of the total number of arthritis patients in this country, the disease is responsible for 22% of arthritis-related deaths. Also, the standardized mortality ratio for RA patients is 2.26—meaning, they are 2.26 times more likely to die than non-sufferers of the same age.

One study cited by the CDC estimated indirect costs associated with RA at $2,784.90 per year (in 2000 dollars), and direct medical costs at $5,768.32. The same study revealed that the typical work experience of an RA sufferer was far different than that of a non-sufferer, in that RA patients were more likely to reduce work hours, lose a job, be unable to find a job, or retire early.[106] Median lifetime costs for RA patients could be up to $122,000 in 1995—more than $177,000 in 2010 dollars. If the same calculation is made using standard medical cost inflation rates, that number could be more than $217,000!

Western medicine offers no cure for rheumatoid arthritis, and small relief for sufferers. Having worked with RA patients both as an MD and a Doctor of Acupuncture, I can honestly say that I know of no pharmaceutical preparation which has as profound or as lasting an effect on symptoms as Classical Chinese Medicine.

Three types of drugs are commonly used to treat the symptoms of rheumatoid arthritis: non-steroidal anti-inflammatory drugs (NSAIDs), corticosteroids, and disease-modifying anti-rheumatic drugs (DMARDs). Of these, DMARDs present the greatest possibility for damaging side effects.

DMARDs include drugs like Rheumatrex® (Methotrexate), Arava® (Leflunomide), Enbrel® (Etanercept), Humira®

TESTIMONIAL

Arthritis (Knee)

PATIENT: Dorothea T., Pawtucket, RI

AGE (at time of treatment): 57

CONDITION(S): Arthritis (knee), high blood pressure, high cholesterol,
Type II diabetes.

CONDITIONS TREATED WITH WESTERN MEDICINE:
All of the above.

PRESCRIPTION AND NON-PRESCRIPTION DRUGS USED BEFORE CCM TREATMENT:
Metapropol (beta-blocker), Fosamax (for osteoporosis), Lipitor, Advil.

SIDE EFFECTS FROM MEDICATIONS:
Upset stomach, heartburn. The medications did not give me a feeling
of well-being.

TOTAL ESTIMATED COST OF WESTERN MEDICAL TREATMENT
OVER THE COURSE OF THE CONDITION(S):
$50,600

HOW DID YOUR CONDITIONS AFFECT YOUR WORK LIFE?
By the time I began to see Dr. Tad I was retired, but taking care of my
grandchildren on a regular basis. I had problems carrying them up and down the
stairs, and some days it was hard just to pick them up.

PLEASE DESCRIBE YOUR EXPERIENCE WITH DR. SZTYKOWSKI AND CCM:
When I started seeing Dr. Tad, I was at a point with my knee where I couldn't
walk for very long. I had trouble picking up my grandchildren, and carrying
them up and down the stairs. By favoring my knee, I in turn threw my hip out
of line, causing myself more pain. My doctors said that losing weight was the
only way to improve my knee, but I was in too much pain to exercise. Also, I
wasn't comfortable taking all of the drugs I was prescribed due to their potential
side effects, so I often skipped doses for fear that the cure would be worse than
the disease.

In part because of my "white coat syndrome," I was doubtful about what
acupuncture could do for me, but after my first visit I felt good about coming to
Dr. Tad's office. He and his staff were very sensitive to my situation, and I knew
that my health was important to them. I saw an improvement in my pain level
almost immediately.

(continued next page)

I was able to walk further without resting, and be more active. Soon, I was able to get back on the treadmill for 30 minutes a day, and the weight started coming off. As the treatment went on, my blood pressure normalized and so did my glucose levels. I was able to come off all of my medications.

Today, I have more energy for my grandchildren, and I feel that my quality of life has improved. I notice an overall feeling of lightness in my body, and I can move freely without pain.

WHICH PRESCRIPTION MEDICATION(S) ARE YOU STILL TAKING TODAY?
None.

TOTAL COST OF TREATMENT WITH CCM:
$7,915

(Adalimumab), and others. While the mode of action is different for each drug in this class, all are immunosuppressive. Since there is no way for a drug to target the specific action of the immune system against joints and cartilage, these medications inhibit the entire immune mechanism. Side effects can be particularly nasty. Methotrexate can cause cirrhosis of the liver and severe myelosuppression (decrease in bone marrow activity and red blood cell production) in rare cases, and even lymphoma; more common side effects include elevated liver enzymes, nausea and GI upset, ulcers in the mouth, fatigue, alopecia (hair loss), and cognitive impairment (jokingly nicknamed "methotrexate fog"). Leflunomide can have similar side effects, but is used as an alternative in patients who cannot tolerate methotrexate. Etanercept and adalimumab are part of a class of drugs known as tumor necrosis factor inhibitors (TNF inhibitors), and are similar in many ways to chemotherapeutics. These drugs lower white blood cell count and increase the risk of infection, particularly upper respiratory infections (like pneumonia and bronchitis) and urinary tract infections (which may migrate to the kidneys). Early clinical trials of several TNF inhibitors demonstrated an increased risk of tuberculosis, as well as a risk for reactivation of latent disease.

It is apparent that DMARD treatment only addresses the mechanism of rheumatoid arthritis—the aggression of the immune system against the joints—and not its cause. Not only do these drugs fail to cure the condition they were prescribed to treat, they create a whole new class of problems which must be addressed to preserve the health—and in some cases, the life—of the patient, at great cost both in dollars and in personal suffering. To me, it seems as though the cure is almost worse than the disease. RA sufferers using these types of therapies may only be trading one type of pain and discomfort for another as they embark on a lifelong cycle of "disease management."

Since we have already discussed the potential dangers of both NSAIDs and corticosteroids, I will not go into detail here. However, it is important to note that the ill effects of corticosteroids increase with the duration of use, and that long-term use can weaken the immune system: this can become doubly dangerous when steroids are combined with DMARDs.

Weight gain is also a side effect of prolonged steroid use —but as we learned in the section on osteoarthritis, extra pounds increase the load on the joints, which can also increase pain levels. Finally, long-term steroid use can cause loss of bone mass and make bones more brittle, thereby hastening the progression of the very condition the steroids were prescribed to treat. Of course, side effects of this magnitude won't occur in everyone— but the potential is there, and in my experience most people taking corticosteroids will experience at least one serious adverse effect over the course of their use.

Although the results I've seen in my own practice have been overwhelmingly positive, the study of acupuncture for rheumatoid arthritis has thus far been limited, and has not been effective in proving (or disproving) the efficacy of acupuncture for this condition. One analysis, entitled "Exploring Acupuncturists' Perceptions in Treating Patients with Rheumatoid Arthritis," concluded that "Clinical trials of acupuncture in RA may have failed to administer a treatment which reflects that administered in clinical practice."[(107)] Basically, the trials conducted to date have failed to take into account the individualized nature of CCM diagnostic techniques, and have not allowed practitioners to treat study participants as they normally would in private practice. This attempt to homogenize acupuncture treatment has, in my opinion, damaged the outcomes of what might have been relevant —and revealing—studies.

The key to treating rheumatoid arthritis lies not in the suppression of the immune system, which is what Western medicine attempts to do with drugs, but in finding the trigger which prompts the immune system to attack the body in the first place. If that underlying imbalance can be uncovered, the body can be returned to homeostasis and the condition may be corrected.

My success rate in treating RA is 75–80%. Usually, I will need at least 60 to 75 visits to help a patient to be totally flare-free, with no more chronic inflammation and no need for pharmaceuticals. The treatment process is much easier in younger patients, or in those in whom the disease has not progressed to the point where it has crippled the joints. In people with severe bone degeneration, or who have had one or more surgeries, recovery may not be complete; however, I have still had great success in combating chronic inflammation, restoring joint function, and preventing the progression of the disease in these patients. Total cost of treatment can range from $5,000 to $6,500, possibly up to $8,000 for severe cases. This cost, you will notice, is only slightly more than the average RA sufferer incurs in medical bills in a single year. The patients I have treated for RA can stay symptom-free for many years, so long as they follow good lifestyle habits and a preventive maintenance schedule.

COPD

Chronic Obstructive Pulmonary Disease (COPD) is an inflammatory condition of the lungs that affects an estimated 16 to 24 million Americans. It's the fourth leading cause of death in the United States (not counting iatrogenic deaths), taking 120,000 lives each year. Both emphysema (progressive destruction of the walls of the alveoli, or air sacs, and the capillaries which supply them), and chronic obstructive bronchitis are considered factors of COPD, and some patients exhibit symptoms of both conditions. For many, COPD is the end result of a lifetime of struggle with chronic respiratory ailments. According to the American Lung Association, asthma sufferers are 12.5 times more likely than non-asthmatics to develop COPD in their lifetimes.[108]

More than 85% of COPD sufferers are smokers or former smokers. The major distinguishing component of COPD is inflammation. Other contributing factors include environmental pollution, occupational hazards, and second-hand smoke. There is no cure for COPD, and Western medicine has found no way to reverse damage to the lungs in COPD patients, or to prevent its eventual occurrence.

The National Heart, Lung and Blood Institutes estimated that total costs of COPD in America exceeded $42.6 billion in 2007, with direct (medical) costs accounting for $26.7 billion of that amount. The condition ranks second only to heart disease as a cause of disability in people over 40. In 2005, more than 6% of the adult population aged 45–64 suffered from COPD.[109]

Employers are not oblivious to the burden of COPD among their work force. Decrease in daily functionality resulting from COPD doubles a worker's chances of job loss. 70% of people with emphysema and 8% of people with chronic bronchitis report that their condition limits the type and duration of work they can perform. A total of 45% of COPD patients report some restriction in their activity level.[110]

Although Western medicine offers no cure for COPD, it does offer several treatment options, most involving pharmaceuticals. Common drugs include bronchodilators, corticosteroids, mucolytics, and antibiotics. Prescription costs for COPD patients may vary—but since the average patient will be taking several drugs at any given time, costs can reach several hundred dollars per month or more.

Bronchodilators used for COPD may be different than those used in asthma treatment. Common types include beta2-agonists like albuterol, theophyllines, and anticholinergics. I have already detailed the possible consequences of albuterol use in the section on asthma, and theophyllines are not widely used (in part because they are known to cause a number of side effects and dangerous drug interactions).

Anticholinergics are antispasmodic, and relax the bronchial muscles to keep airways open even in the presence of irritants like cigarette smoke. This can give the patient the feeling of easier breathing and increased lung capacity. However, the most popular anticholinergic inhaler, Atrovent (ipratropium bromide), has been associated with increased risk of cardiac death in COPD patients,[111] and the action of the drug on the central nervous system can create effects such as dizziness, drowsiness, inability to concentrate, memory problems, wandering thoughts, incoherent speech, and visual disturbances.

Other dangerous side effects can include cessation of perspiration, increased body temperature, ataxia (loss of coordination), diminished bowel movements, urinary retention, and double vision. In short, anticholinergics may open airways, but they don't necessarily improve quality of life for those who use them. Also, anticholinergics are often combined with albuterol, which can compound side effects.

Mucolytics are drugs used to combat mucus production, which is very high in COPD sufferers and contributes to breathing difficulties already present because of alveolar damage. They are used primarily in COPD patients with a chronic sputum-producing cough. Side effects can include gastrointestinal complaints, fever, and drowsiness, and long-term use can lead to hypothyroidism (underactive thyroid).

Antibiotics are used frequently in COPD patients, who are more prone to bacterial infection of the respiratory tract and elsewhere than non-sufferers. The danger with overuse of antibiotics, of course, is the development of "supergerms," which we have all heard about in recent years. The other potential complication is the systematic compromise of the patient's own immune system, and increased tolerance of systemic infection to the antibiotics.

Oxygen therapy and/or breathing machines may be necessary for those with advanced COPD. Even more than pharmaceuticals, these treatments detract from the patient's already diminished quality of life.

As a last resort, patients with COPD can consider lung volume reduction surgery (LVRS) or lung transplant surgery. One analysis concluded that while patient quality of life reports were positive after LVRS, the 90-day mortality rate was significantly higher for those who received surgery than those who did not (regardless of the surgical technique), and that after 2.5-year and 4-year follow-ups, the mortality rate among LVRS patients was no lower than in the control group.[112] The cost of this ineffective surgery can be $26,000 or more.

Transplant surgery is far more dangerous and costly than LVRS, and while five-year survival rates for patients after bilateral lung transplant are reported by some sources to be as high as 50%, recipients over the age of 50—the vast majority of COPD patients—have a much more rapid decline in survival rates than younger patients after the first year. Transplant surgery is only considered an option for patients with advanced COPD, who are not expected to live without the operation.

Acupuncture has been demonstrated to be effective for patients with advanced COPD. One Japanese case study, published in the Journal of the Japanese Respiratory Society (*Nihon Kokyuki Gakkai Zasshi*), found that acupuncture improved walking distance and respiratory function.[113] A later study conducted by the same researchers found that acupuncture, in addition to conventional treatment, "contributed to the reduction of COPD-related dyspnea" in 100% of matched-pair test subjects.[114] A third small study, conducted in Germany, showed that acupuncture created an "improvement of large magnitude in quality of life and a trend of lower demand of the respiration pump." On the other hand, the control group in this study had a higher demand of the respiratory pump, and experienced a deterioration in lung function.[115]

Like every other chronic disease, COPD progresses in stages. The earlier the stage at which treatment begins, the greater the possibility that the patient will experience complete or near-complete remission. I have been very successful at treating patients with mild to moderate COPD, and have been able to improve not only their performance on lung capacity tests, but their daily functionality and quality of life.

My success rate for the treatment of COPD depends greatly on the situation of the individual patient. Generally, recovery is a long process, and treatment duration can range from 8 months to 1 year or more, at a cost of about $5,000 to $7,000. Despite the extended duration of treatment necessary to restore the person to a healthy or nearly healthy state, the cost of my treatment is equal to or less than the cost of one year of conventional Western treatment, and far less than the cost of LVRS or transplant surgery. I do not consider a COPD patient rehabilitated until they can resume a normal or nearly normal quality of life, and no longer need to use inhalers, oral medications, or oxygen therapy on a daily basis. I try to work directly with my patients' primary care doctors to gradually step down medication levels and use of inhalers, so that the process can be monitored at every stage.

In addition to acupuncture and herbal therapies, I nearly always recommend breathing exercises for my patients with COPD. Breath work helps to calm the central nervous system, increase lung capacity, and decrease recovery time after a coughing attack. For patients whose COPD has not progressed to the point where they need oxygen therapy, and whose mobility is not impaired, I recommend Tai Chi classes—both for breath work and the moderate, gentle exercise. Certain styles of yoga may also be helpful; many studios now offer "chair yoga" for people with mobility issues.

Of course, the first step toward wellness for any COPD patient is to stop smoking. It sounds logical—but a great number of people suffer from so severe a nicotine addiction that they cannot give up cigarettes even when it's obvious that their habit is killing them. Continued smoking will hasten the progression of COPD, and of any chronic disease, lung-related or otherwise. But even people who quit smoking years ago are still susceptible to the ravages of COPD and cancer, because the toxic chemicals in cigarettes are not easily eliminated from bodily tissues. Many of the lung cancer patients I've worked with developed their disease five or even ten years after they quit smoking.

In my practice, I offer a stop-smoking program using acupuncture and Chinese herbal formulas. I can generally reduce cravings enough to help the person stop smoking within 2 to 4 weeks. But while most smoking cessation programs stop at this point—when the patient is no longer smoking—my process actually begins here, with an intense detoxification program aimed at correcting the profound imbalances in the body caused by years of smoking.

This detoxification process generally takes between 25 to 30 visits over the course of 2 to 3 months, and costs about $2,100. While this is more expensive than, let's say, the nicotine patch, my program has an 85% rate of success, while over-the-counter patches help less than 10% of people. Also, unlike patches, gum, or hypnotherapy, my program has the power to help prevent the late onset of smoking-related disease. Cleansing the body of residual nicotine and other chemical components significantly lowers the risk that the patient will start smoking again, and also reduces the risk that the patient will develop lung cancer.

Section 3:14—Prevention is the Best Medicine

I hope that the preceding section has helped you to understand the inefficacy of Western medicine for chronic ailments, and the hope which CCM can offer.

When discussing the thirteen chronic diseases most likely to affect your work force, I did not delve deeply into prevention, because I wanted to demonstrate effectively how CCM can help those already suffering with—and paying for—chronic disease. But the best, least expensive, and most successful way to treat any condition is to keep it from happening in the first place. And that's where a conscientiously-designed corporate wellness program truly comes into play.

There is a story about three brothers in ancient China. They were all doctors, practicing what was then called "the Emperor's medicine"—truly, Classical Chinese Medicine.

The youngest brother was very famous as a doctor, known as the best in China. People came from all over to be treated by him. But when asked about his accomplishments, he replied, "I cure people, yes. But my brother is truly the best doctor in China, because he doesn't just make people better once they are already ill. He treats them when they are only a little sick, and makes them well again."

So the people went to the second brother, and named him the best doctor in China. But he shook his head and said, "I am not the best. My eldest brother is the best, and more learned than either my younger brother or I."

"But we have never heard of this third brother," the people said. The middle brother replied, "That is because he is such a great doctor that he does not allow anyone ever to become sick. He cures none of his patients, because none of them are ill."

And so it was that the eldest brother, the one no one had ever heard of, was named the best doctor in China, because his medicine kept people healthy.

This isn't just a story about modesty; it's a story about the Chinese attitude toward prevention in medicine. The belief of Chinese practitioners and patients in the preventive power of CCM is so powerful that it even permeates their legends.

Classical Chinese Medicine has the ability to prevent disease of any sort from ever gaining a foothold in the human body. While death is part of the human condition, and comes to us all in time, we do not have to accept that the last 25 or 30 years (or more) of our lives will be riddled with pain and disease. There is no reason that a person can't be as active and vital at eighty as they were at fifty, provided they engage in the practice of preventive care.

The concept of prevention is simple. It's the same one you apply to your car, your home, and all the other investments in your life. If you don't change the oil in your car, bad things might happen to your engine. If you don't care for your body inside and out, things start to go wrong. But while most people make oil change appointments every few months without even thinking about it, they don't make time for preventive maintenance in their own bodies.

If we—business owners, practitioners, and others alike—can work together to change our attitudes about the nature of health care and the importance of prevention, the burden of chronic disease in this country will be greatly reduced. If we change our oil every few thousand miles, get regular tune-ups, and change our tires, chances are that our cars will run beautifully for hundreds of thousands of miles. If we take care of our bodies with proper nutrition and exercise; avoid excessive intake of chemicals in our food, water, and personal products; and get regular "tune-ups" to correct any imbalances which may be manifesting under the surface, we will live long, healthy, happy lives, with no chronic disease or pain.

In the next section, you will learn how to harness the power of prevention in your workplace, take steps toward positive change, and move yourself and your employees toward a true culture of wellness.

Section 4: Wellness That Works

IN THE LAST section, we discussed thirteen of the most common chronic conditions plaguing the American work force—*your* work force—and learned exactly how much they may be costing you and those who suffer from them. Now, let's talk about ways in which you can improve overall health in your company, and start chipping away at those enormous costs through proactive, targeted wellness initiatives.

We often speak of businesses in terms of health. When a company reaches its targets for sales and growth, shows a profit at the end of its fiscal year, and performs in accordance with the standards set by its administrators and customers, we say that company is "healthy." When a company bleeds money like a sieve, when its workers are being laid off, or when its management is inadequate or corrupt, we say that company is "sick." Given the current state of American health, even a company that's making money hand over fist can't claim to be "healthy," because its greatest assets—its employees—are not healthy. As we know, unhealthy workers cost a lot of money. They don't produce at the same capacity as healthy workers. They need to take sick time, medical leave, and perhaps even long-term disability leave. They lose their quality of life and their *joie de vivre*.

While having a healthy, balanced, clear-headed, and physically fit work force won't necessarily guarantee your success in the business world, it can truly make a difference in your bottom line, which is important in an economic crunch. When you're able to cut your health insurance, workers' compensation, and sick leave costs for every employee by 10, 20, even 30% or more, you'll have a little more room to roll with the punches.

Forward-thinking companies have offered wellness plans for years. In fact, 70% of Fortune 500 companies have some sort of employee assistance program (EAP). But while exercise programs, nutrition education, weight loss initiatives, and smoking cessation programs can all increase productivity, reduce absenteeism, and help save a company money, the truth is that without a targeted wellness plan which addresses not only poor lifestyle habits but major conditions like hypertension and diabetes, the long-term health benefits to the company and its employees cannot not be fully realized.

My plan for corporate wellness includes three areas of concern:

1. Stress reduction
2. Education
3. Treatment

While most wellness programs concentrate on the second area, with some attention to the first, almost no one is attempting to address the primary and immediate health concerns of a chronically ill population. A good diet and regular exercise will go a long way toward preventing the onset of Type II diabetes, but will they cure someone of the disease once they have it? In many cases, the answer will be no: once the body has reached the level of imbalance where major symptoms present (remember the iceberg principle from Section 1?) it will probably need a bit of extra help to regain the perfect balance of true health, homeostasis. Classical Chinese Medicine can offer that help.

To date, insurers, hospitals, and other providers have been reluctant to offer patients with chronic conditions access to CCM and other "alternative" medicines. As a result, many patients shy away from these treatments, even when they believe they might be helped by them; instead, they turn to the far more expensive, less effective Western treatments covered by their insurance. When these treatments fail to correct their conditions, they become part of the revolving door culture of reactive medicine.

Businesses also become caught in this vicious cycle, because they often foot the bill for health insurance. Small businesses take the worst hits: if just one or two employees experience major illness, rates for the whole company can skyrocket. The spike will often be enough to force a small company to eliminate its insurance coverage altogether.

On the other hand, if companies of every size are willing to invest in the health of their workers by implementing company-wide "treatment plans" in addition to prevention-centric wellness initiatives, the potential returns could be staggering.

In this section, I'll discuss the three facets of my plan for corporate wellness, and outline the steps every company should take to create a culture of wellness.

Section 4:1—Wellness Area #1: Stress Reduction

Stress is so common in our lives, and can take so many different forms, that it often seems intangible. So let's start this discussion with a question: *What is stress?*

Stress is the body's response to an exterior factor (called a stressor) which requires the person to make an adjustment either physically, mentally, or emotionally.

Stress is not always bad. In fact, some stress is good for us. When demands are placed upon us, we perform better, and try harder. But those periods of acute stress, which can push us to our maximum capacity as people, should be precisely that: acute. After they're over, we should be able to take a step back, relax, and revel in our achievement. In today's complex, fast-paced, technology-saturated world, stressors come at us from all directions, and make the "fight or flight" response a normal part of daily life. When this happens, healthy acute stress becomes unhealthy *chronic* stress.

Living in a constant state of alertness and arousal—of "readiness," if you will—places an enormous strain on a person, both physically and emotionally. No one can function properly when they're in a constant state of physiological panic. Eventually, the dam will burst, and some sort of breakdown, mental or physical, will occur.

It is at this point—the point of breakdown—that most people call their primary care doctor. While this doctor will of course do his or her best to stem the tide of disease, reactive Western medicine offers no way for doctors to return their patients to homeostasis. The stress-induced imbalance in the patient's body will continue to grow, even if symptoms are less visible after pharmaceutical treatment.

Stress is tremendously expensive, both financially and in terms of human cost. It's hard to place a firm dollar amount on the problem, as stress is a contributor to nearly every major chronic condition, and in many cases it's difficult to prove whether stress caused a condition or simply worsened it. However, we can make an educated guess: averaging recent national estimates gives us a probable annual cost of *$300-$400 billion per year*. This number includes absenteeism, loss of productivity, workers' compensation, legal costs, medical costs, and on-site accidents.

Here are more facts and figures about stress in the workplace:

- According to surveys compiled by NIOSH, 40% of workers say their job is "extremely stressful," and 26% say they often feel "burned out" by their job.[113]
- One in four people report that they've missed work because of stress.
- A survey of 800,000 employees in 300 companies found that stress-related absenteeism increased tripled between 1996 and 2000.
- An estimated 1 million workers are absent each day in the U.S. because they are too stressed to come to work.
- The American Institute of Stress estimates that the annual cost of "unanticipated absenteeism" was more than $750 per employee per year in 1994. That's $1,398 in 2010 dollars!
- 40% of job turnover is attributable to stress. Replacing an employee can cost $3,000-$13,000 or more.
- The average workers' compensation claim for stress-related disability in California exceeds $15,000. [114]

In 2004, the American Psychological Association estimated that *75% of all patient visits to general and family practitioners were for stress-related ailments*. This is a staggering number, but easy enough to understand when you consider the nature of our biological stress reaction. The power of the stress response over the body is so strong that even people with no predisposing risk factors can become seriously ill when subjected to prolonged periods of intense stress.

Our stress reaction is left over from our days as hunter-gatherers, when we roamed the trackless forests and plains of Africa, Europe, and Asia. When we perceived ourselves to be in a dangerous situation—if, for example, we happened to notice a mountain lion staring at us from ten yards away, or a wooly rhinoceros readying itself to charge—our bodies would react in a very particular way. Our pupils would dilate to enhance our vision. Our heart rate would increase. Glucose would flood into our brain and central nervous system, putting our neuroreceptors on high alert, and sharpening our senses. Blood would draw away from the digestive tract and concentrate itself in our arms and legs, preparing us for a mad dash to safety. Blood clotting factors would increase in anticipation of injury, and our immune system would be temporarily suppressed. All of these responses would be triggered by our adrenal glands—or rather, by the particular hormones the adrenals produce—in order that we might survive a potentially life-threatening situation.

In our modern world, all of these physiological events still occur when a stressor triggers our "fight or flight" response. But today, those stressors don't go away after a few minutes or hours. They are perpetual and ongoing—and therefore our bodies' responses to them are also perpetual and ongoing. The adrenals can only continue to function at maximum capacity for so long. After a while, they start to function at a reduced capacity, and eventually sputter out altogether: this is state known as "adrenal exhaustion." At this point, a person will likely become physically incapacitated, either by an external infection (like a cold of flu), or will suffer emotional burnout as his hormone levels plummet and his neuromediators go berserk.

Even if a person never experiences full adrenal exhaustion, it's easy to see how chronic stress causes so many problems. Increased heart rate, combined with enhanced clotting factors, eventually leads to the formation of blood clots and arterial plaque, and can cause or worsen hypertension. A 2004 analysis by Blue Cross Blue Shield of Massachusetts found that people living with chronic stress are twice as likely to have a heart attack. Reduced immunity makes the body a magnet for infections ranging from pneumonia to toenail fungus.

Limited blood flow to the digestive tract equals poor digestion and elimination, low nutrient absorption, elevated blood sugar (resulting in increased hunger and eventually weight gain), and overgrowth of harmful bacteria in the intestines. Imbalances of liver enzymes and "mediator" hormones like serotonin, norepinephrine, and dopamine—all essential to mental and emotional balance—are also common in people who suffer from chronic stress.

Most of the conditions we discussed in Section 3 can be worsened or even created by chronic stress. To demonstrate how this happens, let's formulate a hypothetical scenario, using a 44-year-old male with no pre-existing conditions as our subject.

Let's say that he has recently been under a lot of pressure at home (maybe his wife lost her job, or his teenage son is acting out), and his job responsibilities have just been increased. He's working longer hours, and not sleeping well. His brain, burning 30% more glucose than normal because of his lack of recuperation time, signals his body to crave larger amounts of fatty, sugary foods—the kinds of foods that can be converted into quick energy. His constant tiredness prompts him to increase his caffeine intake.

After a few weeks (or perhaps even less), the excess calories he's consuming, combined with dehydration from too much caffeine, begin to impair his digestion. He may suffer from intestinal cramps, bloating, or constipation. Because his back hurts (a common symptom of dehydration), he finds he can't sleep even when he has time, so he drinks more coffee, which dehydrates him even further. He's eating on the run now, because stress has impaired his cognitive function and he's falling behind at work no matter how many hours he puts in. He's stopped exercising. He's gained a few pounds.

After six months of living in this state of chronic stress, our hypothetical employee has gained ten to fifteen pounds. His blood pressure and serum cholesterol are elevated, his insomnia is getting worse, and his back pain is intensifying. Now, he's beginning to worry about his health, which adds another level to his emotional stress. He may be experiencing tension headaches a few times a month. He may be feeling achy in his joints, both from the extra weight and his lack of regular exercise.

I don't think I have to carry this story any further, because you can surely see where it's going. If nothing changes for this hypothetical person, by the end of a year, he will likely be 20 pounds or more overweight, hypertensive, and hyperlipidemic. He will suffer from regular tension headaches and IBS with constipation due to his lack of "decompression time" and poor diet of convenience foods. He may be depressed; perhaps his doctor has already prescribed an SSRI to help him cope, and sleeping pills to combat his insomnia. He may be scheduled for a $50,000 back surgery to alleviate his lower back pain, which keeps intensifying even though x-rays show no spinal defect. He may take over-the-counter analgesics, or a prescription pain medication, every day.

And it all started with stress.

While this progression of illness may seem rather rapid, it's not unusual for symptoms to cascade in just such a manner; I've seen it happen over and over again. Other people linger for years in a state of stress-induced imbalance. They become accustomed to living with their symptoms, and even start to think of them as normal. By the time they acknowledge the harmful impact of chronic stress on their bodies, they're already sick.

The goal of a wellness program should be not only to help employees alleviate their stress, but to help them recognize it in the first place. When people are aware of their bodies' natural responses to stressors, they can learn to sense the minute changes stress creates before those changes manifest as disease. In the same way that we must teach beginners to ski on the "bunny trail" before they can tackle the K-12, we must teach employees to tame their stress before it becomes a monster too big for them to wrestle. That way, when challenges come along, no one is left stranded on the mountain without their poles.

Before we examine the ways in which you can help your work force cope with stress, let's look at the three major types of stress. They are:

- Physical Stress
- Emotional Stress
- Chemical Stress

Most people deal with a combination of these three stressor groups every day. Each affects the body differently, but all can lead to imbalance and disease if not controlled. You cannot effectively reduce a person's—or a company's—stress level without addressing all three of these areas because, by the very definition of health, they are intertwined.

Physical Stress
This is the easiest type of stress to understand. Physical motions and external environments affect different areas of the body in different ways. If those stressors can be eliminated, or at least counteracted, the body experiences less chronic strain, and is therefore less prone to injury and disease.

Repetitive motions—such as typing, twisting, or bending—create strain in one or more areas of the body, and cause the body to respond in predictable ways. For example, people who lift heavy objects all day usually end up with overdeveloped back muscles on their dominant side, herniated lumbar discs, and perhaps nerve damage if the spine becomes misaligned or compressed. Despite their imbalances, these people don't usually need back surgery: what they need are ways to restore proper alignment, and to become more conscious of the way they move their bodies.

Another type of physical stress is related to insomnia. While this condition is often a result of the action of mental stressors, it does have a marked effect on the body. As we learned in Section 3:10, chronic pain, sugar imbalances (including diabetes), headache, digestive dysfunction, and numerous other conditions can be worsened or even created by lack of healing sleep. Adrenal function in insomniacs is typically low; this renders the body less capable of dealing with any type of stressor. Also, insomniacs are more likely to abuse caffeine, nicotine, and other stimulants, the use of which compounds their sleeping problems.

Emotional Stress

Emotional stress is what we typically think of when we hear the word "stress." It includes feelings of being overwhelmed, of having too little time, or of being underequipped to deal with problematic situations.

Grief, anger, and other strong emotions can also be considered emotional stressors, as they change a person's reaction to everyday occurrences and occupy large amounts of his attention, but it is fear which is usually the underpinning of emotional stress—fear of failure, fear of losing money or material goods, fear of public embarrassment, fear of disappointing a loved one or authority figure. When fear comes into play, the "fight or flight" response is activated. When fear becomes a part of daily life, the result is chronic emotional stress.

Emotional stress is different for every person. Some people can become very agitated by small events—like getting cut off on the freeway, or when someone takes too long at the drive-thru—while to others these things are only small annoyances, like bumps in the road. People who are truly healthy (by the World Health Organization's definition) tend to be less reactive than those who are chronically ill or imbalanced.

Chemical Stress

The third type of stress has not been widely discussed until recently, and is only now starting to enter the vocabulary of mainstream America. But regardless of their comparative invisibility, chemical stressors have a profound effect on your employees every day, both at home and in the workplace.

There are more than 80,000 chemicals in production in America today. Only about 20% have ever been tested for acute toxicity, and less than 10% have been tested for long-term toxicity. These chemicals—many of which were originally invented for use in chemical warfare—are present in our food, our cleaning products, our furniture, our personal care products, even our clothing. Hundreds of minute doses of these compounds every day can add up to a hefty dose of toxins over the course of a year. When the body is exposed to more chemicals than it can eliminate via the normal means of detoxification (urine, feces, and sweat), the result is chemical stress.

Chemical stress affects every area of the body, and can cause a vast array of symptoms. Asthma, indoor allergies, hypertension, cardiac arrhythmia, cognitive impairment, memory loss, lupus, chronic joint pain, IBS, and cancers are all possible results of long-term exposure to common household and workplace chemicals.

Many of the chemicals we ingest are stored in our liver and in our fat cells. Taking into account the physiological connection of the liver to a person's emotional state (i.e., the liver's production of mood-regulating hormones), it becomes clear how an excessive toxic load in the body can negatively impact mood, and even lead to symptoms resembling dysthymia and depression. Because they have more fat cells, overweight people also tend to carry a bigger toxic load. This can make losing weight difficult, because as fat cells are burned as energy their toxic contents are released into the bloodstream, triggering the "swings" we discussed in Section 3:1.

Chemicals stressors are everywhere, and can be hard to avoid unless definitive steps are taken to eliminate them. Common chemical stressors you'll find in nearly every workplace include: chlorine beach (sodium hypochlorite), benzene, formaldehyde, phthalates, dioxins, and VOCs (volatile organic compounds). These are present not just in chemical plants and manufacturing facilities, but in offices, spas, art studios, banks, and restaurants.

That's not all: before most people leave for work, they apply two hundred or more potentially dangerous chemicals to their body in the form of personal care products like body wash, shampoo, toothpaste, deodorant, perfumes, and moisturizing creams. That's a lot of toxins to ingest before breakfast! And that number doesn't include all the other chemicals present in the average home—like those in cleaning products, carpeting, insulation, furnishings, textiles, and the public water supply.

Most people are shocked to discover the extent of their daily chemical exposure. But once they begin to take steps to eliminate the most "stressful" chemicals from their homes and office surroundings, they start to feel better almost immediately. I'll give some specific examples of how this can be done later in this section.

Every one of us is exposed to all three types of stress in some way every day, or nearly every day. A comprehensive stress reduction program should address not only the way people cope with stress outside the workplace (as most wellness programs strive to do), but aim to reduce stressors *within* the workplace. After all, most people spend one third of their life at work: that's a significant amount of time in which to create positive change.

Problem: Repetitive motion injuries
Solution: Stretching routines

Repetitive motion injuries such as back strain, neck strain, carpal tunnel syndrome, and tendonitis are among the most common causes of temporary disability among American workers. As we learned in Section 3, back strain accounts for 333,000 lost time workers' compensation claims every year. Carpal tunnel claims are far fewer in number than those for back or neck strain, but result more often in lost time claims. For people working in physical trades like construction or manufacturing, proper alignment and flexibility is extremely important to prevent injury. This is also true to a lesser degree for office workers, although white-collar workers tend to experience more generalized discomfort, such as non-specific lower back pain, rather than acute sprains and strains.

As we discussed briefly in Section 3:11, stretching programs have been popular in countries like Japan for years. Workers take breaks at specified intervals to stand up, bend forward, stretch their shoulders and wrists, and perform other exercises to address the particular physical vulnerabilities created by their repetitive motions. One study examined workers in a Japanese nursing home who were given a 5-minute daily standing stretch routine to perform over a period of three months. At the end of the study, 21% reported that their low back pain had disappeared completely. Fully half of participants were still performing the stretches (and reporting relief) a year later.[115]

Companies of any size and type can implement stretching routines for their employees. While some manufacturing plants can shut down assembly lines every few hours so workers can perform a 5-minute stretching routine, most do not find this to be practical; therefore, group stretching at the commencement and conclusion of each shift may be a better approach. In an office situation, the possibilities are endless: stretching breaks mid-morning and mid-afternoon (either formal or informal), mats laid out in the break room, even in-house yoga or tai-chi classes. The goal of any stretching routine is to get your employees to move their bodies in way that counteracts their most common repetitive motions. If breathing exercises are incorporated into the stretching routine, the benefits will be even greater.

Educational brochures on the benefits of stretching at work may be a valuable tool. These can be created in-house, or by a wellness consultant such as myself. The objective is to address the particular physical stressors inherent to your business, and provide solutions in a clear and easy-to-understand way. If, for example, people in your workplace are prone to lower back pain, a routine incorporating standing forward bends and hamstring stretches will be extremely helpful. Assembly line workers may benefit from wrist, neck, and shoulder stretches. Once your workers understand the best stretches to perform, they can do them on their own.

Problem: Insomnia
Solution: Nap rooms

While it still seems like a novel idea in this country, the idea of the mid-day nap is not a new one. Take, for example, the Spanish *siesta*, which allows workers a mid-day break to refresh and recuperate. Some pioneering American companies are beginning to take this idea to the next level by constructing "nap rooms" or "quiet rooms" on their premises for employee use.

Recent studies have suggested that human beings have a biological need for rest in the middle of the day—particularly between 2:00 and 4:00 p.m., when most people experience the dreaded "afternoon crash."

For businesses that require long hours and undivided attention from their employees, nap rooms present an enormous benefit. When people are mentally and physically exhausted, they have no extra energy for innovation; it's all they can do to perform the tasks in front of them. A 20–30 minute "recharge" can spark their creative thinking and give them the energy they need to get through the rest of the day (or the rest of the 24-hour cram session). Even a 10-minute afternoon nap has been proven to enhance memory and cognitive function, and can restore vision and reaction time to mid-morning levels. Naps can also make up for sleep lost during nighttime hours because of stress or a hectic schedule.

There are many effective ways to implement a napping policy. Some companies prefer to make naps an accepted part of the daily routine; some have gone so far as to schedule napping "shifts" for their workers. Others prefer to make napping an off-the-clock activity, much like an exercise break. This approach also serves to alleviate some of the guilt workers feel about sleeping on the job.

I would suggest that nap breaks be no longer than 30 minutes, and no shorter than 10 minutes. Of course, measures should be implemented to make sure this privilege is not abused—but giving employees the option to nap when they feel they need will not only increase productivity but build loyalty and improve morale.

A nap room doesn't need to be a fancy lounge with reclining chairs and mood lighting. You don't even need cots or blankets. Mats (like those used for yoga) and pillows with washable covers strewn on the floor will work just fine. Alarm clocks are a nice addition, but only if your workers are napping in shifts. Also, be sure the lights can be lowered or switched off inside the room. Post a sign on the door reminding people to be quiet and courteous, and you have a nap room.

An alternative to nap rooms is flexible scheduling. Work-from-home options allow workers greater flexibility with their sleep schedule. A "siesta" of two hours in the middle of the day will allow employees enough time to eat lunch, take a power nap, and freshen up before returning to complete their day. In order to accommodate this longer mid-day break, workers can come in an hour earlier, or leave an hour later.

Problem: Mid-afternoon crash
Solution: Curb the caffeine

Stimulants like caffeine can be useful in some situations—like on an all-night drive—but daily abuse of caffeine, especially when taken in the form of coffee, can actually reduce productivity and impair mental processes. Employees who drink more than 2 cups of coffee in the morning are more likely to experience a mid-afternoon crash. To boost their energy, they drink more coffee, and then wonder why they have trouble sleeping at night.

You don't need to remove all coffee from the premises in order to curb coffee drinking. In fact, I wouldn't recommend trying it; you might cause a riot. But supplementing your stock of exotic flavored coffees with a selection of herbal teas can help your employees make healthier choices. Black and green teas are also good choices: while they do contain moderate doses of caffeine (about 30 to 50 mg per 6-ounce cup, versus 120 to 150 mg per cup of brewed coffee), teas don't usually cause people to become jittery or irritable in the way that coffee can.

Also, encourage your employees to abstain from caffeine after 2:00 p.m., to prevent nighttime wakefulness.

Problem: Overwhelmed, irritable, emotionally exhausted workers
Solution: Meditation and breathing exercises

In the days before a big deadline or the launch of a new product, notice how the mood in your workplace shifts. While there may be some excitement, chances are that nearly everyone will be snappish, irritable, and on edge. These are common signs of emotional stress.

In some workplaces, nearly every day is like this. Of course, deadlines are unavoidable—but when workers feel overwhelmed, they tend to shut down, and the results can look and feel a lot like panic. Just as with lack of sleep, employees' cognitive functions and memories are impaired, they lose focus and motivation, and their reaction times slow. All of this happens because their adrenals are working overtime, causing their bodies to shift into "fight or flight" mode.

Interestingly, when this boiling point is reached, most people don't become stressed out over the task at hand; rather, they worry about the days ahead, about events and situations which have not yet occurred. Meditation can help them focus their thoughts and redirect their energy to the here and now.

One of the oldest and most successful stress relief techniques, meditation is often associated with spirituality—particularly with Buddhist, Hindu, and Zen practices—but it does not need to be religious in nature. For some, meditation is simply the practice of sitting quietly and breathing deeply. Tai Chi Chuan and Qigong are forms of moving meditation practiced extensively in China and the world over; in these practices, the motions of the body coincide with the inhale and exhale of breath, creating a state of relaxation and inner calm, and allowing chi to circulate freely within the body. Yoga also links breath with movement to create a synchronicity between body and mind.

Meditation is not only good for the mind: it's good for the body. Study after study has shown that meditation is an effective way to lower blood pressure and heart rate, restore adrenal function, reduce chronic pain, and improve sleep quality. One study looked at trials involving transcendental meditation (a technique involving a mantra, or thought-sound, created by

Maharishi Mahesh Yogi in 1958) in reduction of hypertension. After review, it was discovered that regular use of this technique reduced systolic and diastolic blood pressure by 4.7 and 3.2 mm Hg respectively. [116]

Another study examined the effect of "contemplative meditation" over 8 weeks in subjects not concurrently treated with pharmaceuticals. Researchers determined that meditation reduced heart rate and systolic and diastolic blood pressure, and improved results of ambulatory monitoring and mental stress tests.[117]. A third study tested the effect of mindfulness meditation combined with cognitive-behavior therapy for insomnia. Results indicated significant improvements in several areas, including sleep effort and dysfunctional sleep-related cognitions.[118].

My personal experience with meditation has been overwhelmingly positive. I've been practicing Zen meditation since 1978, and have found that it clears my mind, puts me at ease, reduces tension throughout my body, and calms racing thoughts. Twenty minutes of meditation per day is enough to keep me focused and productive, no matter how busy my schedule becomes.

The highest goal of any meditation is to ease the mind out of a frantic, disordered state and into a state of balance. The exact technique is truly a matter of personal preference. Once learned, meditation techniques can be used anytime, anywhere, to quiet a racing mind and restore emotional equilibrium. In my consulting practice, I often recommend that companies install a meditation room on the premises; this space can double as the nap room, if space does not permit both.

A meditation room can be used during the course of the day by individuals seeking a few minutes of peace and quiet, or for guided group meditations. Incense, candles, and Laughing Buddha figurines are nice touches, but they're completely unnecessary, and may even detract from the experience for some. More important are pillows and/or mats for sitting, gentle lighting—and above all, quiet.

When meditation isn't practical, breathing exercises can help to bring the body and mind back into equilibrium. Simply deepening the inhale and exhale can increase oxygenation of the brain and vital organs, reducing the "panicky" feeling that often accompanies stress. Breath retention exercises, which involve holding the breath at the top of the inhale for as long as comfortable, then exhaling slowly, can normalize an elevated heartbeat in only 1–2 minutes. Breath accompanied by motion, as with Tai Chi Chuan or yoga, is particularly effective for calming a racing mind. A simple exercise in which the arms are raised overhead on inhale and lowered on exhale is usually enough to help a person release mental tension and regain focus.

Employers who wish to incorporate meditation and/or breathing exercises into the daily workplace routine would do well to bring in an independent consultant or practitioner to teach specific techniques to employees, and answer any questions employees might have. While there's really no wrong way to meditate, most people feel more comfortable beginning with a basic knowledge of the practice. Also, when people learn the potential benefits of meditation, they are more likely to practice it.

Problem: Chronically stressed, discontented workers
Solution: Empowerment techniques

I'll be the first to admit that not all people are good workers. I myself set high standards for my employees, and have been disappointed more than once by someone's failure to live up to my expectations. But while workers at all levels of a company need both supervision and direction to be productive and stay on track, there is such a thing as too much control. Unless you own a fast-food franchise staffed by sixteen-year-olds, your workers are more than likely adults, and therefore capable of making their own decisions. They do not need to be micromanaged or babysat; in fact, if your management is doing so, they may be contributing to the level of emotional stress in your workplace.

One of the leading causes of workplace stress is the feeling of having no control. When people feel that they have no autonomy in their job, that they're being watched by "big brother" all the time, that they have no say in their workload or their responsibilities, or that they don't have the freedom to request changes or improvements in their work situation, they become stressed. It's the professional equivalent of claustrophobia: when there's no room for growth or change, people panic. While this phenomenon has historically been more common among blue-collar workers, it is steadily becoming more prevalent in the corporate world, and the effects are felt by everyone.

Now, I'm a doctor: I'm not interested in telling you how to run your business, nor am I qualified to do so, save by the fact that I'm a business owner myself. But if you're interested in reducing stress in your workplace while simultaneously increasing employee loyalty and health, consider reviewing your management policies to ensure that they allow everyone in your company, from the CEO to the janitor, to feel involved, important—and most of all, appreciated.

Classical Chinese Medicine, and the education I provide as an accompaniment to my services, gives my patients a sense of control over their health and well-being. They feel that they can take charge of their health—that they can *own* it. In the same way, a management style which welcomes constructive criticism, respects personal needs, and works to create a positive job experience has the effect of empowering workers, and in doing so instills a sense of responsibility for, and loyalty to, the company. In other words, it's "healthy" management.

Problem: Indoor allergies
Solution: Improve air quality

Many people who suffer from indoor allergies, tension headaches, and even depression are in fact reacting to chemical stressors. As we discussed earlier in this section, we're surrounded by dangerous chemicals every day. Making small changes within your workplace can significantly improve air quality and reduce symptoms associated with chemical exposure. Here are some suggestions:

Restrooms
- Replace chemical deodorizing sprays and/or disinfectants with non-toxic air fresheners which use essential oils instead of chemical fragrances.
- Replace antibacterial or harshly-scented soaps with natural and/or certified organic products made from soy, coconut oil, or other non-toxic lathering agents.
- Use toilet paper and paper towels whitened without chlorine bleach. Buy recycled where possible.

Air Filtration
- Be sure that your air filtration system is clean and working properly.
- In areas where allergens are particularly concentrated, or where new furniture has been placed or paint applied, use portable ionic or HEPA air filters to remove chemicals and particulates from the air.

Furniture and Décor
- Inexpensive office furniture made with particleboard uses formaldehyde-based glues and finishes. Purchase hardwood furnishings finished with non-toxic stains or varnishes when possible to reduce chemical off-gassing.
- Cubicles are made from several components which emit chemicals like formaldehyde, naphthalene, benzene, perchloroethylene, and 1,4 dioxane, all of which are hazardous to human health. Look for cubicle modules classified as "low-emitting."
- Avoid stain-resistant carpet and upholstery; the treatments used to make it stain-resistant contain formaldehyde and other noxious chemicals.

Cleaning
- Cleaning products are a major source of chemical allergens. Ask your cleaning staff or cleaning company to replace their current products with industrial "green"

cleaners which use plant-based surfactants, and which contain no bleach, ammonia, or phosphates.
- Carpet cleaners are among the most toxic chemicals used in an office environment. When it's time to have your carpets shampooed, ask about non-toxic options that do not contain perchloroethylene (a.k.a. PERC, used in dry cleaning and recently banned in California), naphthalene, or 1,4 dioxin (a known carcinogen).
- If you're renovating, choose wood or tile floors over carpeting where possible. Carpeting traps dust, chemicals, and other allergens, and redistributes them into the air every time someone walks across the floor.

Personal Hygiene
- Ask employees not to wear heavy perfumes or colognes in the office. Many people are allergic to these chemical fragrances.
- If you allow the burning of candles in your workplace, ask that they be of the soy-based variety: paraffin wax, used in most commercial candles, emits 11 known toxins when burned.

I'm not encouraging these changes in your workplace to get you to "go green"—although it may be beneficial, from a public relations standpoint, for you to do so, and it's certainly an ecologically and socially responsible choice. Rather, it's because non-toxic, eco-friendly products are *healthy* products. Only now are researchers starting to address the damage which common chemical stressors can cause in the long term. Making positive changes in your work environment today will be of enormous benefit in years to come.

These are just a few examples of how a company can address physical, mental, and chemical stressors in the workplace. Beyond the common stressors I've discussed, every company will have unique stressors determined by the nature of its business, its physical location, and its employee demographic. When I work with a company to create a wellness program, I look at all possible stressors which may be acting upon employees, and come up with individualized solutions to reduce their effects.

If you decide to implement any of the stress reduction techniques discussed above without the assistance of a wellness coach or medical professional, remember that the most important factor in any stress reduction plan is balance. The action of physical, emotional, and chemical stressors on employees must all be reduced in order for any stress reduction program to be truly effective.

Section 4:2—Wellness Area #2: Education

Education is an important factor in any wellness program. It's also where the majority of wellness programs concentrate their efforts. When people know what to do to become healthier, many wellness coaches theorize, they'll do it, because no one *wants* to be sick.

I agree with this approach, to a point: education equals empowerment, and my goal is to empower people both as patients and as human beings. However, in my experience, the impact of health education in the workplace to date has been mediocre at best, because it has been based on accepted Western ideals of health and health care.

Also, educational programs sponsored by drug companies or health care providers are inherently biased in favor of those providers, just as good marketing materials should be. This is not to say that information about common conditions or traditional Western care should be eliminated in the workplace; rather, such materials should be balanced by information about other care options, so that employees can make an informed choice.

Areas of education which should be a part of every corporate wellness program include:

- Nutrition
- Physical fitness
- Weight loss
- Smoking cessation
- Caffeine addiction and its dangers
- The preventive power of daily stretching
- Healthy sleep habits and schedules
- Easy stress relief and relaxation techniques
- Easy breathing techniques and the benefits of breath work
- Common prescription drugs and their side effects
- Over-the-counter medications and their side effects
- Effects of overweight and obesity on the body and mind

How you choose to present this material to your employees is entirely up to you. What's important is that the information you provide is accurate, accessible, and unbiased. Remember that when people don't know any better, they'll believe what they're told: just look at how effective television ads have been for the pharmaceutical companies! Beware of information marketing the latest fad diets, detox products, "miracle pills," or surgeries. Such information will not, in the end, benefit you or your employees.

Education can take many forms. Some of the most effective for use in the workplace are seminars, printed materials, and group activities.

Seminars
It's a great idea to have an outside expert comes to your place of business to conduct a class or lecture for your employees. This format allows employees to passively participate in the wellness program, without feeling that they've been put on the spot. If it's feasible, make attendance to presentations mandatory, or at least "strongly suggested." Small companies might also choose to send

their employees to outside seminars or classes, in order to keep costs low.

Presentations should be short enough to hold employees' interest, but long enough to convey a significant amount of information. I myself prefer the "Lunch and Learn" format. Employees can eat while I talk, and ask questions at the end of the session. This also allows employees to feel that they're multitasking, so they don't need to feel stress over lost productive hours. Another good option for smaller companies is to hold presentations at a restaurant or banquet hall, so employees can enjoy a lunch hour away from the office and feel that they're being spoiled.

Printed Materials

These can take the form of pamphlets, posters, postcards, or any other visual materials designed to make people think and talk about wellness. A poster touting the benefits of whole grain, or a chart listing the Environmental Working Group's "Dirty Dozen" (fruits and vegetables most likely to contain dangerous pesticides) could be a great addition to the lunch room. Passed pamphlets might include information on company-sponsored smoking cessation or addiction recovery programs, information about an upcoming seminar, brochures for exercise studios (some studios offer substantial discounts to companies who sponsor employees memberships), or illustrations of stretches to ease low back pain.

Group Activities

Many companies encourage employees to participate in outside activities like charity walks, weight-loss programs, or nutrition programs. When these activities are performed by groups of employees, they not only have a positive effect on employee wellness, but also encourage team-building, and can be especially helpful if your workers are isolated for most of the day in cubicles or offices. Examples of wellness-promoting group activities include a company hiking excursion, a rafting trip, a walking race, or even a beach day. What activity you choose is not important, so long as it fosters a sense of well-being among participants, and helps employees feel valued.

When introducing education as part of a wellness program, the trick is to make it appealing to your employees. Many people view extracurricular activities as impedances to their social and family lives. To put it simply, you need to make it worth their while. Therefore, all educational seminars and group activities should take place during normal working hours. If activities do fall outside normal hours, offer employees incentives for participation. Cash prizes have been demonstrated to be most effective, but goods and services (perhaps donated by local businesses in a cooperative marketing arrangement) also work well.

For all its merits, education alone is not always enough to cause major changes; after all, every smoker knows that cigarettes are bad for them, yet they continue to smoke. Therefore, education needs to be implemented as an adjunct to action. Concrete changes in the workplace which complement educational services demonstrate that your company is serious about helping employees achieve balanced, total health. When they see steps being taken on behalf of their personal well-being, employees are more likely to take action themselves.

Section 4:3—Wellness Area #3: Treatment

Stress reduction programs and educational initiatives will benefit everyone in your company. They should be the first parts of your wellness program to be implemented, precisely because they benefit everyone equally. They will also get people used to the idea that wellness is about to become part of their workday.

But sometimes, universal programs are not enough to make the kind of difference you're hoping for. In order to create a true culture of wellness in your company, you need targeted initiatives which address the needs of specific groups of employees.

The reason most wellness and preventive programs fail to show the spectacular results they should is because many employees are already sick. Adding whole grains isn't going to cure the emotional eating disorders which affect so many obese people. Switching to honey or stevia-based sweeteners isn't going to cure the sugar addictions from which so many Type

II diabetics suffer. When people's bodies are too imbalanced for them to correct the problem on their own, they need what I call *corrective treatment*—meaning, treatment performed by a qualified practitioner—before they can benefit from *active treatment* in which they themselves participate.

It's a fact that 20% of the American population generates 80% of our health care costs. This isn't just due to the gross operating expenses innate to our health care industry. It's because chronically ill people spend a lot of time in doctor's offices, hospitals, and operating rooms, and take a lot of prescription drugs. Since reactive Western treatment don't work to cure their conditions, these costs are not one-time expenses, but rather compound over time.

Your company may not be in a position to send every single employee to a qualified CCM practitioner for treatment, but when it comes to cutting health care costs and increasing productivity, company-sponsored treatment for certain workers—your "twenty-percenters"—can actually be cost-effective.

The best way to find out who these unhealthy workers are, and how they can be helped, is to invite a health professional (such as myself) to visit your workplace and perform a comprehensive company-wide health screening.

When I perform a health screening, I request that all workers—not just volunteers—participate in this process. In order to present a clear picture of the health of the company, and outline a wellness program that will generate real results, I need to have information about everyone who works there.

The screening process I use is simple. First, I ask that participants fill out a "stress survey," which questions how stress affects their performance, mood, sleep habits, and other factors. The answers help me to determine the overall level of daily stress within the company culture, and can even reveal stress-related disorders within individual employees. Next, each person receives basic adrenal testing, which includes blood pressure tests at different points on the body, and an examination of the pupils.

If a person is experiencing symptoms of adrenal dysfunction due to stress, these tests will usually uncover them.

Finally, each participant receives Chinese Pulse Diagnosis. I have uncovered many latent or undiagnosed conditions, including liver disease, kidney disease, hypertension, and IBS, using nothing more than this test.

Once I have finished the initial company-wide screening, I assign each employee to one of three groups. These are:

- Minimal-Risk group
- Low-Risk group
- High-Risk group

The percentage of employees who fall into each category generally accords with the bell curve. Of your workers, 20–25% will be minimal-risk, 20–25% will be high-risk, and 50–60% will fall somewhere in the middle. Every company will, of course, be different, and your group "membership" may vary due to factors including the mean age of your work force, type(s) of labor performed, mean economic status of employees, and even the region of the country where your company is located.

Since the health status of your employees will likely encompass a wide spectrum, a truly effective wellness program must be a tiered system which addresses the problems common to each of these three groups.

First, let's meet the *Minimal-Risk* group.

This group comprises about 20% of your employees, and includes your health nuts, weekend warriors, and others to whom good health is a major priority. These people probably take the least number of sick days (unless it's really beautiful outside, and the trails are calling), and they're comparatively productive. They take few, if any, prescription drugs, have the fewest addictions, and spend the least amount of time in doctors' offices. Their health care costs are generally limited to yearly checkups, OTC medications like aspirin, and the occasional prescription antibiotic.

This is not to say that everyone in the Minimal-Risk group is healthy. In fact, most are probably *not* healthy by the standards I set for my patients; they are simply non-symptomatic. Upon examination, most will exhibit minor imbalances caused by poor sleep, emotional stressors, or minor addictions (such as to caffeine), all of which can be corrected through simple lifestyle changes.

The Minimal-Risk group is the group most likely to benefit from a traditionally structured wellness program which focuses on education, healthy eating, and physical fitness, because they are not struggling to overcome preexisting conditions. They are still at the stage where they can self-correct, and are more likely than other groups to implement healthful changes because they already make personal well-being a priority. This is not to say that Minimal-Riskers should not receive corrective treatment, but such treatment will not necessarily be required for them to benefit from other company wellness initiatives.

Next, the *Low-Risk* group.

This group includes the bulk of your work force, about 50–60% of your employees. This is a symptomatic group, but by Western standards they are still considered fairly healthy. This is where you'll find a lot of people with the conditions we explored in Section 3. They're still functional, but they're probably less productive than their coworkers in the Minimal-Risk group, due to chronic health issues, sleep problems, poor diet, or other factors.

Like the Minimal-Risk group, these people will also benefit from education, fitness, and nutrition programs, as well as detox programs aimed at issues like smoking cessation and weight loss, but may have trouble obtaining results without corrective treatment to address their chronic conditions.

Finally, the *High-Risk* Group.

This group is made up of the final 20–25% of your employees. These are the people whose chronic health issues prevent them from living a "normal" life. They are part of that 20% of the American population who generate 80% of health care costs—the twenty-percenters, as I've nicknamed them.

The High-Risk group will likely take more sick days than other employees, and require more time away from work for doctor's visits. They will also be less productive than their healthy coworkers, either because their conditions or medications impact their cognitive and mechanical function, or because they're distracted by pain and worries about their own poor health.

For these people, achieving balanced health is a process that can take months or years of commitment. Traditional wellness programs simply don't work for them. Telling your twenty-percenters to walk for thirty minutes four times a week and eat more salad isn't going to solve their health problems. In fact, refusing to acknowledge the difficulties they will face in the course of becoming healthy may be counterproductive. As these people are already under tremendous emotional stress due to their illnesses, the pressure to "get healthy" may actually make them sicker.

The High-Risk group will need more corrective treatment than the Low-Risk group, and will take longer to show results. However, this is also the group which will show the greatest improvement in productivity and presenteeism if their problems are addressed correctly, and they are *allowed to become healthy*.

I use the words "allowed to become healthy" purposefully. Members of the High-Risk group are our health care lifers. They're the biggest burden on America's health care system—but they're also its best customers, the "managees" in "disease management," Big Pharma's golden geese.

Regardless of whether their diseases are lifestyle-related, work-related, or genetic, America's twenty-percenters are not entirely to blame for their current state. They have become dependent on medical care that doesn't work. They have been disempowered, because they no longer believe that they have the ability to heal their own bodies. They have been conditioned to rely on drugs, surgeries, and tests which will either mask their problems, make the problems worse, or create entirely new problems. Through my recommendations at the initial screening, and corrective treatments performed thereafter, I try to restore to these people a sense of responsibility for their own health and well-being, and encourage them to take charge of their lives.

I urge companies to invest in CCM treatment for their twenty-percenters whenever possible. Healthy, dedicated workers are a company's best asset, but sick, disempowered workers are a liability. Therefore, in the same way you invest in repairs for your expensive equipment, you should invest in "repairs" for your sickest employees. While this will involve a moderate expenditure at first, in the long run, it will reduce your health care costs dramatically, and bolster your bottom line.

Once the overall health of your work force has been analyzed, you can begin to formulate a plan of action. Your ultimate goal is to reduce the amount of money you spend on health care, and redirect those extra profits to fuel the growth of your company. In order to do this, you will have to invest some money. How *much* money is up to you—but whatever your budget, be sure to convey importance of your wellness program to your employees, and encourage them to participate. Every little bit helps, but half-hearted investments will always produce half-hearted returns. Make sure your employees realize how important their good health is to you, and how important it should be to them.

By offering incentives to employees who choose holistic treatment modalities like CCM, companies have the opportunity not only to help their twenty-percenters regain their good health, but also to help all employees regain a sense of control over their health and health care.

There are several different ways in which you can make it possible for your employees to explore CCM treatment options, while simultaneously reducing your health care expenses. Your plan will need to be tailored to your business and your budget, of course, but here are some methods to consider.

- Health Savings Accounts (HSAs) and Medical Savings Accounts (MSAs)
- Flexible Spending Accounts (FSAs)
- Health Reimbursement Accounts (HRAs)

Health Savings Accounts and Medical Savings Accounts

HSAs are offered as additions to high-deductible insurance policies. For example, a policy with a $2,500 deductible will often be accompanied by an HSA, which allows the employee to pay for medical services tax-free until their insurance kicks in. Accounts can be managed through banks, or even through your existing health insurance company. Blue Cross Blue Shield, for example, offers the "MySmartSaver HSA" through Bankcorp Bank. Enrollees receive a debit card which they can use for all of their medical purchases—including acupuncture, yoga, meditation classes, and more. Recent increases now permit pre-tax contributions by the individual, the employer, or both of up to $3,000 for individuals and $5,950 for families. People over 55 can contribute an additional $700 per year.

MSAs differ from HSAs in that the employer cannot contribute to an MSA. These accounts can only be opened by those who are self-employed, or who work for a company with fewer than 50 employees. MSA contribution limits are based on a person's gross income and insurance deductible, rather than a set dollar amount. Also, MSAs may be harder to qualify for than HSAs, and often have a higher tax penalty for non-medical early withdrawal (15% vs. 10%).

Once money goes into an HSA or MSA, the account acts like an IRA. The monies in it can be invested. At the end of the year, the money in the account does not revert to income and is not subject to taxes. Rather, it stays in the account, and continues to earn compound interest (also tax-free) until it is withdrawn. This not only makes an HSA or MSA a good way to reduce health insurance costs, but a good way for healthy individuals to save additional monies for retirement. Most CCM modalities are considered "qualified medical costs" by Section 213(d) of the Internal Revenue Code.[119] Herbal supplements and daily vitamins are not considered qualified unless prescribed by a licensed practitioner, and purchases of these items from vendors other than a practitioner may be subject to taxation. Also, while money can be left in the HSA account after the close of the calendar year, it cannot be rolled into an IRA or 401(k).

Unlike HSAs or MSAs, *Flexible spending accounts (FSAs)* are not tied to any insurance policy. Employees can opt to create flex accounts even if they have no insurance coverage. Like HSAs and MSAs, FSAs allow pre-tax deposits. The amount of these deposits is not limited, however if deposited amounts are not used by the end of the year, the leftover funds are lost: the company offering the FSA gets to keep them, or (depending on the arrangement) the funds are added to the employee's gross income and taxed accordingly.

Some employees choose to deposit money into their FSA only when they expect to incur a qualified medical expense. This allows them to pay for services tax-free without losing any money at the end of the year. Also, unlike HSAs, FSAs can be used to pay for child care, which makes them very valuable to families.

Only under certain circumstances are employees allowed to have both an HSA and an FSA. Generally, it's better for employer and employees to choose one or the other.

Health Reimbursement Accounts

Large employers are often self-insured; meaning, they create and fund their own insurance plans, which may be managed by an outside insurance company. All medical and administration costs are paid by the company; this saves money by eliminating an outside insurer's profit margin. Companies who self-insure are in a unique position to design coverage for their employees, and can choose to include coverage for CCM and other alternative modalities in their insurance plans.

Most companies in America, however, are not financially able to self-insure. Just one or two big claims could wipe them out. Therefore, they are forced to work with insurance companies to provide coverage for employees. Some insurers offer acupuncture riders and coverage for alternative modalities, but most of the time this coverage is minimal. Rather than attempt to persuade your insurance company to change its ways (although I wish you luck if you decide to try!) consider creating Health Reimbursement Accounts, or HRAs, to cover CCM and other treatments that might benefit your employees.

HRAs, unlike HSAs and flex accounts, are funded entirely by the employer, and managed by a third-party administrator. They can be offered in conjunction with FSAs and other health plans, can be rolled over from year to year, and are separate from employee income. There is no limit to how much an employer can contribute to an HRA, although in order to be eligible for full tax benefits the employer must comply with certain requirements.[120]

What types of services will be reimbursed by an HRA plan is entirely up to the employer, as long as those services fit the IRS guidelines for qualified medical costs. Traditionally, these types of accounts have been used to help employees with co-payments and deductibles, but since specific parameters are set by the employer, they can easily be modified to include CCM and other alternative modalities. Employees simply provide documentation of qualified expenses, and the employer reimburses them. The employer is not required to pre-pay into a fund, but rather can reimburse expenses as they occur, greatly reducing costs.

Companies that utilize HRAs can streamline the reimbursement process and save money by negotiating agreements with practitioners. Personally, I work with companies on contract rates: if a certain number of employees come to my office for treatment, the company receives a discount on services for all employees. If an employee signs up for an entire course of treatment—for example, a smoking cessation program—the discount will be even greater. Purchasing treatment programs, as opposed to individual sessions, is not only beneficial to employers, but also makes it more likely that employees will finish their entire course of treatment and receive the full benefits.

Implementing HRAs, HSAs, or flex accounts doesn't have to be a monumental expense. When you create these types of accounts, you can often modify insurance coverage at the same time. HSAs, for example, are usually tied into a high-deductible plan. Even if your company doesn't currently offer health insurance to employees, you can still create HRAs and qualify for the accompanying tax benefits.

Getting started with these new methods of coverage may require an investment. But if your investment is combined with a comprehensive wellness initiative which includes exercise and nutrition programs, effective stress reduction policies, and proper education, I firmly believe that you will see a tremendous return in 12 to 24 months. You will find that your employees have better health, higher morale, and increased productivity, and that you no longer have to approach the purchase of health insurance like a "sick" company. Working with the types of tax-favored accounts described above, you may be able to save as much as 40% on insurance premiums for each participating employee!

A Note on Sickly Ducklings

You might encounter resistance from employees who are not interested in participating in a wellness initiative. This is where the old "carrot and stick" approach comes into play. Of course, there should always be a lot more carrot: people respond far better to incentives than to punishment. But some need a little more of a nudge, and that's when you might have to bring out the stick.

Compulsory participation in wellness programs has not been widely implemented, but it should be. Of course, the *degree* of participation required must be reasonable, but it should be required all the same. If employees choose not to become part of your company's new culture of wellness, that is their decision—but like all decisions, it should have a consequence.

Ironically, the people most resistant to participating in a wellness program will likely be your twenty-percenters, members of the high-risk group who generate 80% of total medical costs. There are many psychological reasons why someone would resist action which could help them get well—far too many to discuss here—but suffice to say that some will resist, and will continue to cost your company vast amounts of money until something happens to make them take action.

Since the overall goal of any wellness program is to reduce health care and absenteeism costs, employee behavior which increases costs is counterproductive, and should be addressed.

For example, if your (fictional) employee, Jane Smith, consistently refuses to take part in company wellness initiatives, consider offering her a clear choice.

Explain that preventable health conditions among her coworkers are major factors in the rising cost of health benefits. Tell her that you understand her reluctance to participate in some of the wellness programs being offered, but that her health is important to you, and you'd like to see her become more involved. Therefore, if she participates in the upcoming company-wide weight loss challenge, attends 3 out of 6 lectures in the next 6 months, and attends a stop-smoking seminar, you will give her a bonus of $XXX at the end of this fiscal year. If she does not do these things, however, you will be forced to increase her monthly health care contribution from 25% to 50% of the total premium as of the next renewal date. If, after 2 years, she continues to choose not to participate in the programs recommended for her by her doctor and the medical professional supervising the company wellness program, her contribution will increase to 75%.

Now, Jane has a clear choice, must assume responsibility for both her health and her finances. There is no discrimination taking place against her, since all employees will be offered the same choice. Her bonus and continued health care coverage are not dependent on her success in the recommended programs, only on her participation.

While this strategy may cause some employees to curse and moan, most will be excited by the prospect of a bonus, and by the chance to engage in health-related activities on company time. After two to three years, you may notice that the "sickly ducklings," as I call them, have either become more healthy or fallen away, and that your company has started to attract motivated, healthy people who are drawn to the culture of wellness you've created.

Make your opinions on wellness known to prospective employees during interviews. Include it in your internet job postings. Make wellness a vital part of your corporate culture, and watch what happens.

Section 5: Creating Results

DESPITE WHAT INSURANCE companies and Big Pharma might say about Classical Chinese Medicine, I'm no witch doctor. I don't make promises I can't keep—not to my patients, and not to the companies I work with. I won't tell you that you'll see miraculous returns on your wellness plan tomorrow, or next week. But gradually, over the course of a year or two, you will start to notice that a culture of wellness is evolving within your company. And the benefits will just keep growing.

The benefits of your wellness plan will be realized in three steps:

1. Increased productivity and employee morale
2. Reduced health care costs
3. Health benefit restructuring

Even mainstream wellness programs, which use Western medical testing and Western "preventive" treatments (like cholesterol drugs to "prevent" heart attacks), report incredible results. For example, a report published in 2008 cited ROIs for major companies' lifestyle programs. Returns are per dollar invested: Coors ($6.15); Citibank ($4.56); General Mills ($3.90); Motorola ($3.15), PepsiCo ($3.00).[121] You can't argue with the numbers.

Now, imagine how much greater those returns could be for a program which offered not only lifestyle education and health screenings, but true preventive care. By giving people a chance to go beyond managing their diseases and imbalances to actually reversing them, companies can realize the true potential of a wellness program.

Benefit #1: Increased Productivity and Employee Morale

The first effects of a well-designed wellness plan will be increased worker productivity and loyalty, higher employee morale, and reduced absenteeism.

As we learned in Section 3, chronic diseases generate exorbitant absenteeism costs and hours of lost productive time. As your employees become healthier, these costs will be greatly reduced, but the *psychological* boost will often precede physical healing. By taking steps to regain their own good health, employees regain a sense of control and empowerment in their lives. As an employer, you want to be the one who gives that to them.

Benefit #2: Reduced Health Care Costs

Bottom line: Fewer claims mean lower premiums. If you can lower the number and cost of claims your employees generate, your costs will go down. How large your particular decrease is will depend on your insurance plan and your employees' "before" and "after" health statistics, but you can probably expect to see at least some savings as of your first renewal.

"Healthy" companies can qualify for commitment plans, like the *BlueChip for Healthy Options* plan currently offered by Blue Cross Blue Shield here in Rhode Island. These plans offer discounted rates to businesses, groups, and individuals who sign a pledge to perform certain wellness-related activities over the course of the year, and who agree to submit to periodic evaluation by their primary care physician. These activities are relatively basic, and might include smoking cessation programs and weight management classes. Non-smokers and people of healthy weight must pledge to maintain their current status. As long as participants do their part, their rates stay low. Some plans also offer different discount levels to match various levels of participation.

Larger companies can work with insurers to reduce rates based on the improving health of their employees. If a reduction in claims can be proven, it should be possible to negotiate a discount. Unfortunately, small businesses are often assigned to "pools" which include other companies in the same geographical area or business field. Unless other companies in the pool are also working to get their employees healthy, rates will continue to be determined by the average cost to insure everyone in the pool, therefore other options (like HSAs or HRAs, which also cover CCM and other preventive treatments) might be considered.

When you're ready to implement your wellness program, talk to your agent about possible discounts offered by your insurer. Your company may be eligible for savings from the moment your wellness initiatives begin. Ask for your policies to be reviewed with each renewal.

Benefit #3: Health Benefit Restructuring

A year or two after commencing your wellness program, you will probably be able to make some major changes to your benefits programs.

Healthy employees don't need anywhere near as much insurance coverage as sick employees do. In fact, most people in your Minimal-Risk group will only see their primary care physician once a year, and will take few if any prescriptions. So why should you (and your employees, if they contribute to their benefit plans) pay thousands of extra dollars a year for coverage that's not being used?

If you currently provide 100% benefits coverage for your employees, consider the difference in cost between your current coverage (which could exceed $14,000 per year per family), and the cost to fill your employees' HSAs or flex accounts with the maximum contribution and switch to high-deductible insurance coverage. Even if every dime in these accounts comes from you, this approach may still save you 10% or more on your insurance costs. It will also be perceived as a great benefit to employees, who will gain the freedom to explore alternative health care options like CCM, and experience no practical reduction in benefits. If you choose Health Reimbursement Accounts (HRAs), you may save even more, since statistically a percentage of your employees will not utilize the full amount allotted to them for medical expenses in a given year, and the extra funds can be rolled over or revert back to the company at the end of the year.

Small businesses can save money by opting to increase deductibles without investing in HSAs or FSAs. One business owner I know recently increased deductibles for covered employees to $2,500. While they now have to pay for prescriptions and doctor's visits out of pocket, his 18 employees are happy with the arrangement because their monthly contributions have been slashed so dramatically. Also, in lieu of implementing flex accounts for his staff to bridge the coverage gap, this business owner has agreed to pay 100% of the $2,500 deductible for his employees in the event that they require emergency treatment, hospitalization, or expensive testing. Employees can create flex accounts on their own if they wish, or simply track their qualified medical expenses (including CCM and other "alternative" treatments) and claim the tax deduction at the end of the year.

Because most of his employees are under 55, relatively healthy, and rarely utilize services other than checkups and the occasional specialist visit, this business owner is saving thousands of dollars per employee per year. In fact, he's saving so much that even if *every employee* used their entire $2,500 deductible in the same year, he'd still pay less than he did for his old health coverage!

There are downsides to this type of reduction in coverage, of course. When looking for a new job, employees consider health benefits one of the top selling points, and may pass over a company whose benefits do not look appealing at first glance. Also, employees used to receiving full prescription drug coverage may balk at paying out of pocket; in these cases, prescription discount plans might be considered.

Before you make any changes to your existing health care program, be sure to explore all of the options available to you. Most important is to consider how you can help your employees utilize preventive treatments and services, which offer true benefits and greater health, instead of sticking with the status quo.

Growing Your Culture of Wellness

Let's leap into the future, and take a look at what's going on in your company two years from now.

Twenty-four months ago, you implemented your total wellness program, which includes CCM treatment initiatives for some or all of your employees. Many of your twenty-percenters aren't twenty-percenters anymore. You've been able to shift almost all of your employees to high-deductible insurance plans with HSAs, HRAs, or FSAs. You've noticed a vast improvement in employee morale, and productivity is through the roof. You've recouped all of the money you initially invested in your wellness plan, and according to the projections you'll see a sizeable return this year.

But now that the newness and excitement of your wellness initiative has worn off, how do you keep your healthy company going strong?

Here are some ways to keep interest in your wellness programs high, and keep your employees on the track to total health.

- Management participation
- Family participation
- Ongoing prevention
- Ongoing incentives
- Redistribution of savings

Using the techniques outlined in this section, you should be able to keep your wellness program on track for years to come.

Management Participation

There's an old Russian saying that translates something like this: "The fish stinks from the head." In essence, this means that if you try to implement change at the bottom and work up, you're in for a bumpy ride. Every change in your company's policy needs to start at the top and trickle down. In order to lead effectively, you've got to "walk the talk." Why would your employees place value on something you don't value yourself?

Make sure that you, your officers, and your managers are visible at every wellness-related activity. If you choose to create a nap room or meditation room, make sure management uses it regularly. If you provide educational seminars, yoga classes, group meditation, company fitness programs, or any other group activities as part of your wellness plan, make sure your top dogs are there, and that they're participating with enthusiasm.

Above all, hold your executives to the same high standards that you hold your employees. If they choose not to participate in your wellness program, they should be subject to the same consequences as any other employee. If necessary, build wellness participation into your contracts. Make getting healthy part of the job.

Family Participation

Your employees don't spend all of their time at the office. And often, the same bad habits from which your workers suffer are also present in their spouses, children, and other household members. Your employees may bring some of their new healthy habits home with them, but it may not be enough. Remember, worry about a loved one's illness can cause nearly an employee nearly as much emotional stress as worry about his or her own illness, and can have a similar impact on productivity and absenteeism. The same incentives for education and treatment offered to employees should also be offered to everyone covered under their insurance policy.

Bring your employees' families into the fold through company-sponsored activities and family-centric education. Working with kids is especially important: our kids are America's future work force, and if they learn healthy habits at a young age, they'll be less likely to suffer the same health problems as their parents.

Ongoing Prevention

Helping sick employees get well is the first task of your wellness program. Ongoing prevention to keep them well is the second.

It is a good idea to work with a health care professional such as myself on an ongoing basis to keep your company on the path to prevention. A yearly or bi-yearly company-wide health screening will help identify imbalances before they manifest as disease. While these maintenance visits will constitute an additional expense above and beyond your usual medical costs, health screenings can save you thousands of dollars per employee (or more) in the long run.

Remember that story about the three doctors from Section 3? Ultimately, you want your workers to be like the patients of that third doctor—the patients who were so well-cared-for that they never become ill enough to "cure."

Ongoing Incentives

Tangible prizes are big motivators. Cash or goods prizes for those who reach predetermined targets will never fail to achieve results. For example, you might consider a cash reward of $250 for those who complete a smoking cessation program. Or, you might give employees $50 for every 50,000 steps they walk (using pedometers to track their progress). Set yearly, quarterly, and monthly wellness goals within the company, and create deadlines by which certain company-wide objectives are to be achieved. As your company gets progressively healthier, increase the difficulty of the challenges. A drive toward constant improvement will keep your employees interested, motivated, and above all, healthy.

Redistribution of Savings

We're still two years into your hypothetical wellness program, and your restructured insurance coverage is working beautifully. Your calculations show a savings of $2,800 per covered employee per year. Where should all that savings go?

It's entirely up to you. I'm not going to tell you where to put your money; I'm a doctor, not an accountant. But if employees perceive that their new healthy lifestyles are benefitting their financial health as well, they will work harder to stay on track.

One place to put your savings is into ongoing wellness incentives, the benefits of which are discussed above. You might consider directing another portion toward tangible improvements like "greening" up your building, upgrading your equipment, or purchasing specialty items your employees can use every day. Above all, be vocal about where the capital for these improvements is coming from. If your employees know that their fancy flat-screen monitors were paid for by the health benefit savings they helped create, they'll feel a sense of pride and ownership in the company, and therefore take more responsibility for its future growth and success.

Conclusion

Over the course of this book, we've learned how reactive, compartmentalized Western medicine is ineffective for the treatment of chronic disease. Western health care costs a lot of money, and for people with common chronic conditions, it doesn't provide results of a caliber consistent with its cost. On the other hand, Classical Chinese Medicine, in the hands of a skilled practitioner, can be a cost-effective and safe way to treat, and cure, the chronic conditions affecting your work force.

At the time of this writing, Congress is debating a bill for national health care reform. While it's clear that something needs to be done, a government insurance option isn't going to answer the larger question of the efficacy of Western medical care for chronic disease. It's simply not enough to change the way we pay for health care: we have to change health care itself.

It's obvious that Big Pharma and insurance companies have their hooks in our Washington representatives, since issues involving prevention and restrictions on unnecessary treatments have been repeatedly shot down, both on the House floor and in the media. Therefore, as I stated at the beginning of this book, the impetus for reform must rest with businesses and individuals, not policymakers.

We can no longer entrust responsibility for our well-being to the medical industry. By cultivating a mindset of prevention, we can begin to regain control of our health and our financial future.

Good health should not be a commodity: it should be something everyone can enjoy. By helping your employees discover balanced health—true health—through preventive treatments and an effective wellness plan, you and your company will propel American health care into the future.

Endnotes

Section 1

(1) Centers for Disease Control and Prevention, National Center for Health Statistics. *Health, United States, 2007, with Chartbook on Trends in the Health of Americans, 2007*; Table 70, p. 297. Library of Congress Catalog Number 76–641496. http://www.cdc.gov/nchs/data/hus/hus07.pdf#070 (Accessed May 2010)

(2) Figures according to the American Heart Association, "Cholesterol Statistics," http://www.americanheart.org/presenter.jhtml?identifier=536 (Accessed May 2010)

(3) U.S. Department of Health and Human Services. *Fact Sheet: HHS Targets Efforts on Diabetes*, Jan 2006. http://hhs.gov/news/factsheet/diabetes.html (Accessed May 2010)

(4) The National Institutes of Health/National Cancer Association. "Activity 1: The Faces of Cancer," *Cell Biology and Cancer Teacher's Guide*, 1999. http://science.education.nih.gov/supplements/nih1/cancer/guide/activity1–1.htm (Accessed May 2010)

(5) From the *Preamble to the Constitution of the World Health Organization as adopted by the International Health Conference*, New York, 19–22 June, 1946; signed on 22 July 1946 by the representatives of 61 States (Official Records of the World Health Organization, no. 2, p. 100) and entered into force on 7 April 1948.

(6) From the *Huang Di Nei Jing* (the Yellow Emperor's Inner Cannon), Su Wen 8, p.1°-b.

Section 2

(7) The National Coalition on Health Care. *Health Insurance Costs: Facts on the Cost of Health Insurance and Health Care.* View the report at http://whitehouse2.org/documents/305-facts-from-the-national-coalition-on-health-care (Accessed May 2010)

(8) Centers for Disease Control and Prevention; National Center for Health Statistics. *Health, United States, 2007, with Chartbook on Trends in the Health of Americans,* 2007; Table 132, p. 394. Library of Congress Catalog Number 76–641496. http://www.cdc.gov/nchs/data/hus/hus07.pdf#070 (Accessed May 2010)

(9) The Kaiser Family Foundation and Health Research and Educational Trust. "Employer Health Benefits: 2008 Annual Survey." Publication #7790, released September 2008. http://ehbs.kff.org/pdf/7790.pdf (Accessed May 2010)

(10) Boden, William E., M.D., Robert A. O'Rourke, M.D., Koon K. Teo, M.B., B.Ch., Ph.D., Pamela M. Hartigan, Ph.D., David J. Maron, M.D., William J. Kostuk, M.D., Merril Knudtson, M.D., et al. for the COURAGE Trial Research Group. "Optimal Medical Therapy With or Without PCI for Coronary Artery Disease," *The New England Journal of Medicine,* Vol. 356, no. 15, pp.1503–1516, April 17, 2007. http://content.nejm.org/cgi/content/abstract/356/15/1503 (Accessed May 2010)

(11) Preston, Samuel H., PhD. "Deadweight?—The Influence of Obesity on Longevity." *The New England Journal of Medicine*; Vol. 352, pp. 1135–1137, March 2005. PMID# 15784667. Read an extract at http://content.nejm.org/cgi/content/extract/352/11/1135 (Accessed May 2010)

(12) *Shape Up RI* is a state-wide weight loss challenge founded by Brown University medical student Rajiv Kumar. For more information please visit www.shapeupri.org or call 401–421–0608.

(13) Abramson, John MD, *Overdo$ed America*. New York: HarperCollins Publishers, 2004. "The Commercial Takeover of Medical Knowledge," Ch. 7, p.95.

(14) Null, Gary, PhD; Carolyn Dean MD, ND; Martin Feldman, MD; Debora Rasio, MD; Dorothy Smith, PhD. "Death by Medicine." Published online by *Life Extension Magazine*, March 2004. http://www.lef.org/magazine/mag2004/mar2004_awsi_death_01.htm (Accessed May 2010)

(15) "In-Hospital Deaths from Medical Errors at 195,000 per Year, HealthGrades' Study Finds," *HealthGrades* (press release), July 27, 2004. http://www.healthgrades.com/aboutus/index.cfm?fuseaction=mod&modtype=content&modact=Media_PressRelease_Detail&&press_id=135 (Accessed May 2010)

(16) U.S. Department of Health and Human Services, Health Resources and Services Administration. *National Practitioner Data Bank 2003 Annual Report*. http://www.npdb-hipdb.hrsa.gov/pubs/stats/2003_NPDB_Annual_Report.pdf (Accessed May 2010)

Section 3

Obesity

(17) Davis, Karen, Ph.D.; Sara R. Collins, Ph.D.;
Michelle M. Doty, Ph.D.; Alice Ho; Alyssa L. Holmgren.
"Health and Productivity Among U.S. Workers,"
The Commonwealth Fund, August 2005. http://www.
commonwealthfund.org/publications/publications_show.
htm?doc_id=294176 (Accessed May 2010)

(18) Centers for Disease Control and Prevention, National
Center for Health Statistics. *Health, United States, 2007, with
Chartbook on Trends in the Health of Americans*, 2007; Table
58, p. 255. Library of Congress Catalog Number 76–641496.
http://www.cdc.gov/nchs/data/hus/hus07.pdf#070
(Accessed May 2010)

(19) United States Department of Health and Human Services.
"Prevention Makes Common Cents." September 2003.
http://aspe.hhs.gov/health/prevention/
(Accessed May 2010)

(20) Cawley, John PhD; John A. Rizzo, PhD; Kara Haas, MD,
MPH, FACS. "Occupation-Specific Absenteeism Costs
Associated With Obesity and Morbid Obesity." *Journal
of Occupational & Environmental Medicine*; Vol. 49(12),
pp.1317–1324, December 2007. http://www.joem.org/pt/re/
joem/abstract.00043764–200712000–00007.htm;jsessionid=J
mGcJZk7VH2CgZ2F92wxsZ0X85mK4DhByTZWkJryJMFjL
GZLJBdy!1329102805!181195628!8091!-1
(Accessed May 2010)

(21) Chenowith, David, Ph.D., FAWHP. "Topline Report—
The Economic Costs of Physical Inactivity, Obesity, and
Overweight in California Adults: Health Care, Workers'
Compensation, and Lost Productivity." California
Department of Health Services, Public Health Institute, April
2005. http://www.cdph.ca.gov/HealthInfo/healthyliving/
nutrition/Documents/CostofObesityToplineReport.pdf
(Accessed May 2010)

(22) Christou, Nicolas V. MD, PhD; Didier Look, MD; Lloyd D. MacLean, MD, PhD. "Weight Gain After Short- and Long-Limb Gastric Bypass in Patients Followed for Longer Than 10 Years." *Annals of Surgery*; Vol. 244(5), pp.734–740, November 2006. http://www.annalsofsurgery.com/pt/re/annos/abstract.00000658–200611000–00018.htm;jsessionid= JmvDHyCC1p5lDJNHThsHw22hYxyvC5j94xCbqcKttD6mfj dv9QLv!-482373940!181195629!8091!-1 (Accessed May 2010)

(23) Cho, S.H.; J.S. Lee; L. Thabane; J. Lee. "Acupuncture for obesity: a systematic review and meta-analysis." *International Journal of Obesity* (London); Vol. 33(2), pp.183–96, January 2009. PMID# 19139756. http://www.ncbi.nlm.nih.gov/ pubmed/19139756 (Accessed May 2010)

(24) Tan, Steven; Kirsten Tillisch; Emeran Maye. "Functional Somatic Syndromes: Emerging Biomedical Models and Traditional Chinese Medicine." *Evidence-Based Complementary and Alternative Medicine*; Vol. 1(1), pp. 35–40, June 2004. http://www.pubmedcentral.nih.gov/ articlerender.fcgi?artid=442118 (Accessed May 2010)

Back Pain

(25) "Cost Effectiveness of Spinal Surgery Analyzed", *Science Daily*, December 29, 2008. http://www.sciencedaily.com/ releases/2008/12/081229200744.htm (Accessed May 2010)

(26) Mortimer, Marina. "Fortune 500s Waste Over $500 Million a Year on Unnecessary Back Surgeries for Workers." Published online by *Business Wire*, June 24, 2008. Read the article on Reuters.com at http://www.reuters.com/article/pressRelease/ idUS111418+24-Jun-2008+BW20080624 (Accessed May 2010)

(27) Chobok J.; I. Vrba; I. Stetkárová. "Selection of surgical procedures for treatment of failed back surgery syndrome (FBSS)." *Chirurgia Narzadów Ruchu i Ortopedia Polska*; Vol. 70(2), pp 147–153, 2005. PMID# 16158875. http://www.ncbi. nlm.nih.gov/pubmed/16158875 (Accessed May 2010)

(28) Wehrwein, Peter; J. Tobias Nagurney, M.D. "Too Much Radiation." *Newsweek* (web exclusive), December 6, 2008. http://www.newsweek.com/id/172697/page/1 (Accessed May 2010)

(29) "FDA ALERT [06/2006]: Development of Serious and Sometimes Fatal Nephrogenic Systemic Fibrosts/Nephrogenic Fibrosing Dermopathy." June 2006. http://www.fda.gov/Drugs/DrugSafety/ PostmarketDrugSafetyInformationforPatientsandProviders/ ucm142911.htm (Accessed May 2010)

(30) Cai, Chunbo, MD. "Treatment Of Chronic Back Pain And Neck Pain Using Scalp Acupuncture: A Case Study." *Medical Acupunture*; Vol. 18(1), pp. 24–5, September 2006. http:// www.medicalacupuncture.org/aama_marf/journal/v18n1.pdf (Accessed May 2010)

(31) Thomas, K.J.; H. MacPherson; J. Ratcliffe; L. Thorpe; J. Brazier; M. Campbell; M. Fitter; M. Roman; S. Walters; J.P. Nicholl. "Longer term clinical and economic benefits of offering acupuncture care to patients with chronic low back pain." *Health Technology Assessment*; Vol. 9(32); pp. iii-iv, ix-x, 1–109; August 2005. PMID# 16095547. http://www.ncbi.nlm. nih.gov/pubmed/16095547 (Accessed May 2010)

(32) Weidenhammer, W.; K. Linde; A. Streng; A. Hoppe; D. Melchart. "Acupuncture for chronic low back pain in routine care: a multicenter observational study." *The Clinical Journal of Pain*; Vol. 23(2), pp. 128–35, February 2007. PMID# 17237661. http://www.ncbi.nlm.nih.gov/ pubmed/17237661 (Accessed May 2010)

(33) Sawazaki, K.; Y. Mukaino; F. Kinoshita; T. Honda; O. Mohara; T. Togo; K. Yokoyama. "Acupuncture can reduce perceived pain, mood disturbances and medical expenses related to low back pain among factory employees." *Industrial Health*; Vol. 46(4), pp. 336–40, August 2008. PMID# 18716381. http://www.ncbi.nlm.nih.gov/pubmed/18716381 (Accessed May 2010)

Hypertension

(34) The American Heart Association. *Heart Disease and Stroke Statistics—2005 Update*, pp.22–23, 2005. http://www. americanheart.org/downloadable/heart/1105390918119HDS Stats2005Update.pdf (Accessed May 2010)

(35) U.S. Department of Health and Human Services, National Institutes of Health. *Prevent and Control High Blood Pressure: Mission Possible Call to Action Paper*. http://hp2010.nhlbihin. net/mission/refs.htm#sources (Accessed May 2010)

(36) The American Hospital Association. *Trendwatch: Healthy People are the Foundation of a Productive America*; p. 6, 2007. www.aha.org/aha/trendwatch/2007/twoct2007health.pdf (Accessed May 2010)

(37) Goetzel, Ron Z., PhD; Stacey R. Long, MS; Ronald J. Ozminkowski, PhD; Kevin Hawkins, PhD; Shaohung Wang, PhD; Wendy Lynch, PhD. "Health, Absence, Disability, and Presenteeism Cost Estimates of Certain Physical and Mental Health Conditions Affecting U.S. Employers." *Journal of Occupational and Environmental Medicine*; Vol. 46(4), pp. 398–412, April 2004. PMID# 15076658. DOI: 10.1097/01.jom.0000121151.40413.bd. http:// healthproject.stanford.edu/publications/health_absence_ disability_goetzel.pdf (Accessed May 2010)

(38) Salvi, Dr. Rosane Maria. "Hydrochlorothiazide", International Programme on Chemical Safety (INCHEM), 0.9994, PIM 265, January 1992. http://www.inchem.org/ documents/pims/pharm/hydrochl.htm (Accessed May 2010)

High Cholesterol

(39) The National Heart, Lung and Blood Institute. "ALLHAT—
Information for Health Professionals: Quick Reference Guide
for Health Care Providers." http://www.nhlbi.nih.gov/health/
allhat/qckref.htm (Accessed May 2010)

(40) Fidan, D.; B. Unal; J. Critchley; S. Capewell. "Economic
analysis of treatments reducing coronary heart disease
mortality in England and Wales, 2000–2010." Published
in *QJM: An International Journal of Medicine*; Vol. 100(5),
pp. 277–89, May 2007. PMID# 17449875. DOI:10.1093/
qjmed/hcm020 http://qjmed.oxfordjournals.org/cgi/content/
abstract/100/5/277 (Accessed May 2010)

Asthma

(41) National Center for Health Statistics, Centers for Disease
Control and Prevention. "Current Asthma Prevalence
Percents by Age, United States: National Health Interview
Survey, 2006." Table 4–1, March 18, 2008. http://www.cdc.
gov/asthma/nhis/06/table4–1.htm
(Accessed May 2010)

(42) The American Hospital Association. *Trendwatch: Healthy
People are the Foundation of a Productive America*; pp. 1,4;
2007. www.aha.org/aha/trendwatch/2007/twoct2007health.
pdf (Accessed May 2010)

(43) Chen, Hubert, MD, MPH; Paul D. Blanc, MD, MSPH;
Mary L. Hayden, FNP, AE-C; Eugene R. Bleecker, MD;
Anita Chawla, PhD; June H. Lee, MD; for TENOR
Study Group. "Assessing Productivity Loss and Activity
Impairment in Severe or Difficult-to-Treat Asthma,"
Value In Health; Vol. 11(2), pp. 231–239, July 2007. DOI#
10.1111/j.1524–4733.2007.00229.x.
http://www3.interscience.wiley.com/journal/119414387/
abstract (Accessed May 2010)

(44) Stanford, Richard; Trent McLaughlin; Lynn J. Okamoto. "The Cost of Asthma in the Emergency Department and Hospital." *The American Journal of Respiratory and Critical Care Medicine*; Vol. 160(1), pp.211–215, July 1999. http://ajrccm.atsjournals.org/cgi/content/full/160/1/211 (Accessed May 2010)

(45) Jobst, K.A. "A critical analysis of acupuncture in pulmonary disease: efficacy and safety of the acupuncture needle." *Journal of Alternative and Complementary Medicine*; Vol. 1(3), pp. 221–3, 225–7; 1995. PMID# 9395603. http://www.ncbi.nlm.nih.gov/pubmed/9395603. (Accessed May 2010)

(46) Maa, S.H.; M.F. Sun; K.H. Hsu; T.J. Hung; H.C. Chen; C.T. Yu; C.H. Wang; H.C. Lin. "Effect of acupuncture or acupressure on quality of life in patients with chronic obstructive asthma: a pilot study." *Journal of Alternative and Complementary Medicine*; Vol. 9(5), pp. 659–70, October 2003. PMID# 14629844. http://www.ncbi.nlm.nih.gov/pubmed/14629844 (Accessed May 2010)

(47) Chu, K.A.; Y.C. Wu, M.H. Lin, H.C. Wang. "Acupuncture resulting in immediate bronchodilating response in asthma patients." *Journal of the Chinese Medical Association*; Vol. 68(12), pp. 591–4, December 2005. PMID# 16379344. http://www.ncbi.nlm.nih.gov/pubmed/16379344 (Accessed May 2010)

(48) Feng, Jun Tao; Cheng Ping Hu; Xiao Zhao Li. "Dorsal Root Ganglion: The target of acupuncture in the treatment of asthma." Published in *Advances in Therapy*; Vol. 24(3), pp. 598–602, May 2007. DOI 10.1007/BF02848784. http://www.springerlink.com/content/p621250708038g22/ (Accessed May 2010)

Allergies

(49) Smith, Sandy. "Allergies Contribute to Economic
Implications in the Workplace." *EHS Today*, March 10, 2003.
http://ehstoday.com/news/ehs_imp_36219/
(Accessed May 2010)

(50) Schoenwetter, William F.; Leon Dupclay, Jr; Sireesh
Appajosyula; Marc F. Botteman; Chris L. Pashos. "Health
Economics of Allergic Rhinitis and Asthma," *Medscape* Curr
Med Res Opinion; Vol. 20(3), pp.305–17, 2004. http://www.
medscape.com/viewarticle/472667_6
(Accessed May 2010)

(51) Xue, Charlie Changli; Robert English; Jerry Jiansheng Zhang;
Cliff Da Costa; Chun Guang Li. "Effect of Acupuncture in
the Treatment of Seasonal Allergic Rhinitis: A Randomized
Controlled Clinical Trial." *The American Journal of
Chinese Medicine*; Vol. 30(1), pp. 1–11, 2002. http://www.
traditionalacupuncture.com.au/files/acupuncture%20
and%20SAR.pdf (accessed May 2010)

Depression

(52) *Merriam-Webster's Medical Dictionary* (accessed online). ©
2009 Merriam-Webster, Incorporated. http://mw1.merriam-
webster.com/medical/depressions (Accessed May 2010)

(53) Agency for Healthcare Research and Quality, Rockville, MD.
"Antidepressant Prescriptions Climb by 16 Million," *AHRQ
News and Numbers*, July 24, 2008. http://www.ahrq.gov/news/
nn/nn072408.htm (Accessed May 2010)

(54) Dworkin, Ronald W., M.D., Ph.D. *Artificial Happiness: the
Dark Side of the New Happy Class.* Carroll & Graf Publishers,
New York, NY, 2006. ISBN# 978–0786717149.

(55) Cohen, Elizabeth. "CDC: Antidepressants Most Prescribed Drugs in U.S." CNN Medical News, July 9, 2007. http://www.cnn.com/2007/HEALTH/07/09/antidepressants/index.html (Accessed May 2010)

(56) Centers for Disease Control and Prevention, National Center for Health Statistics. "Antidepressant Use: Adults." *Health, United States, 2007, with Chartbook on Trends in the Health of Americans*; p. 88, 2007. (Library of Congress Catalog Number 76–641496). http://www.cdc.gov/nchs/data/hus/hus07.pdf#070 (Accessed May 2010)

(57) Stewart, Walter F., PhD, MPH; Judith A. Ricci, ScD, MS; Elsbeth Chee, ScD; Steven R. Hahn, MD; David Morganstein, MS. "Cost of Lost Productive Work Time Among U.S. Workers with Depression." *Journal of the American Medical Association (JAMA)*; Vol. 289(23), pp.3135–3144, June 18, 2003. http://jama.ama-assn.org/cgi/content/abstract/289/23/3135?eaf (Accessed May 2010)

(58) Pratt, Laura A., Ph.D.; Debra J. Brody, M.P.H. "Depression in the U.S. Household Population, 2005–2006 / Key Findings: Data from the National Health and Nutrition Examination Survey, 2005–2006." *NCHS Data Brief*, No. 7, September 2008. http://www.cdc.gov/nchs/data/databriefs/db07.htm (Accessed May 2010)

(59) Simon, Gregory E., MD, MPH; Michael Von Korff, ScD; Kathleen Saunders, JD; Diana L. Miglioretti, PhD; Paul K. Crane, MD, MPH; Gerald van Belle, PhD; Ronald C. Kessler, PhD. "Association Between Obesity and Psychiatric Disorders in the US Adult Population." *Archives of General Psychiatry*; Vol. 63(7), July 2006. http://archpsyc.ama-assn.org/cgi/content/abstract/63/7/824 (Accessed May 2010)

(60) Greenberg, Paul; Patricia K. Corey-Lisle; Howard Birnbaum; Maryna Marynchenko; Ami Claxton. "Economic Implications of treatment-resistant depression among employees." *PharmacoEconomics*; Vol. 22(6), pp. 363–373, 2004. http://pharmacoeconomics.adisonline.com/pt/re/phe/abstract.00019053-200422060-00003.htm;jsessionid=J7zGmhLQNrXH04JZ0Sv7yDQCp1FNsfGLsvysHYv7w01kMW6GHy2l!97158217!181195629!8091!-1 (Accessed May 2010)

(61) Fu, W.B.; L. Fan; Q. He; L. Wang; L.X. Zhuang; Y.S. Liu; C.Z. Tang; Y.W. Li; C.R. Meng; H.L. Zhang; J. Yan. "Acupuncture for the treatment of depressive neurosis: a multicenter randomized controlled study." *Zhongguo Zhen Jiu* (Chinese Acupuncture and Moxibustion); Vol. 28(1), pp. 3–6; January 2008. PMID# 18257177. http://www.ncbi.nlm.nih.gov/pubmed/18257177 (Accessed May 2010)

Diabetes

(62) Olendorf, Donna; Christine Jeryan, Karen Boyden, editors. *The Gale Encyclopedia of Medicine*. Farmington Hills, MI: Gale Research, 1999. Vol. 2, p. 939. ISBN 0–7876–1870–5.

(63) "Direct and Indirect Costs of Diabetes in the United States," published on the website of the American Diabetes Association. http://www.diabetes.org/diabetes-statistics/cost-of-diabetes-in-us.jsp (Accessed May 2010)

(64) The National Diabetes Information Clearinghouse (a service of the National Institute of Diabetes and Digestive and Kidney Diseases (NIDDK), NIH. "Diabetic Neuropathies: The Nerve Damage of Diabetes." http://diabetes.niddk.nih.gov/dm/pubs/neuropathies/ (Accessed May 2010)

(65) Huber, J.D. "Diabetes, Cognitive Function, and the Blood-Brain Barrier." *Current Pharmaceutical Design*; Vol. 14(16); pp. 1594–600; 2008. PMID# 18673200. http://www.ncbi.nlm.nih.gov/pubmed/18673200 (Accessed May 2010)

(66) Yeung, Sophie E.; Ashley L. Fischer; Roger A. Dixon. "Exploring Effects of Type 2 Diabetes on Cognitive Functioning in Older Adults." *Neuropsychology;* Vol. 23(1), pp. 1–9, 2009. DOI:10.1037/a0013849. http://www.apa.org/journals/releases/neu2311.pdf (Accessed May 2010)

(67) "Highlights of Prescribing Information: Avandia® (rosiglitazone maleate) Tablets," GlaxoSmithKline 2008. http://us.gsk.com/products/assets/us_avandia.pdf (Accessed May 2010)

(68) Chang, S.L.; J.G. Lin; T.C. Chi; I.M. Liu; J.T. Cheng. "An insulin-dependent hypoglycaemia induced by electroacupuncture at the Zhongwan acupoint in diabetic rats." *Diabetologia;* Vol 42(2), pp. 250–55, 1999. http://www.nhicb.gov.tw/nhicbe00/herb/h_ebm13.pdf (Accessed May 2010)

(69) "Effects of acupuncture on mood and glucose metabolism in the patient with type 2 diabetes." *Zhongguo Zhen Jiu* (Journal of Chinese Acupuncture and Moxibustion), Vol. 27(10), pp. 741–3, October 2007. PMID# 18257350. http://www.ncbi.nlm.nih.gov/pubmed/18257350 (Accessed May 2010)

(70) Liu, C.F.; L.F. Yu, C.H. Lin, S.C. Lin. "Effect of auricular pellet acupressure on antioxidative systems in high-risk diabetes mellitus." *Journal of Alternative and Complementary Medicine;* Vol. 14(3), pp. 303–7, April 2008. PMID# 18399759. http://www.ncbi.nlm.nih.gov/pubmed/18399759 (Accessed May 2010)

(71) Abuaisha, BB; JB Costanzi, AJ Boulton. "Acupuncture for the treatment of chronic painful peripheral diabetic neuropathy: a long-term study." Published in *Diabetes Research and Clinical Practice*, Vol. 39(2); pp. 115–21; February 1998. PMID# 9597381 http://www.ncbi.nlm.nih.gov/pubmed/9597381. (Accessed May 2010)

(72) World Health Organization Fact Sheet No. 277 (March 2004): Headache Disorders. http://www.who.int/mediacentre/factsheets/fs277/en/ (Accessed May 2010)

(73) Hu, X.H. "Burden of migraine in the United States: disability and economic costs." *Archives of Internal Medicine*; Vol 159(8), pp. 813–8, April 26, 1999. PMID# 10219926. http://www.ncbi.nlm.nih.gov/pubmed/10219926 (Accessed May 2010)

(74) "Employers Who Recognize The Impact Of Migraine May Help To Improve Workplace Productivity," *Medical News Today*, May 4, 2007. http://www.medicalnewstoday.com/articles/69829.php (Accessed May 2010)

(75) Saunders, K, JD; K. Merikangas, PhD; N.C.P. Low, MD, MSc ; M. Von Korff, ScD; R.C. Kessler, PhD. "Impact of comorbidity on headache-related disability." *Journal of the American Academy of Neurology*; Vol. 70(7), pp. 538–47, February 12, 2008. PMID# 18268246. http://www.ncbi.nlm.nih.gov/pubmed/18268246 (Accessed May 2010)

(76) Mulleners, W.M.; E.P. Chronicle. "Anticonvulsants in migraine prophylaxis: a Cochrane review." *Cephalalgia : An International Journal of Headache*; Vol 28(6); pp. 585–97; June 2008. PMID# 18454787. http://www.ncbi.nlm.nih.gov/pubmed/18454787 (Accessed May 2010)

(77) Silva, Hilton Mariano, Jr.; Roberta P. Garbelini; Simone O. Teixeira; Carlos A. Bordini; José G. Speciali. "Effect of episodic tension-type headache on the health-related quality of life in employees of a Brazilian public hospital." *Arquivos de Neuro-Psiquiatria*; Vol. 62, no. 3; September 2004. doi: 10.1590/S0004–282X2004000500005
Print ISSN: 0004–282X
http://www.scielo.br/scielo.php?pid=S0004–282X2004000500005&script=sci_arttext (Accessed May 2010)

(78) Maizels, Morris MD. "The Patient with Daily Headaches." *American Family Physician*; Vol. 70, pp. 2299–306, 2313–14; September 2004. http://www.aafp.org/afp/20041215/2299. html (Accessed May 2010)

(79) Coeytaux, R.R.; J.S. Kaufman; T.J. Kaptchuk; W. Chen; W.C. Miller; L.F. Callahan; J.D. Mann. "A randomized, controlled trial of acupuncture for chronic daily headache." *Headache*; Vol. 45(9), pp. 1113–23, October 2005. PMID# 16178942. http://www.ncbi.nlm.nih.gov/sites/16178942 (Accessed May 2010)

(80) Allais, G.; C. De Lorenzo; P.E. Quirico; G. Lupi; G. Airola; O. Mana; C. Benedetto. "Non-pharmacological approaches to chronic headaches: transcutaneous electrical nerve stimulation, lasertherapy and acupuncture in transformed migraine treatment." *Neurological Sciences: Official Journal of the Italian Neurological Society and of the Italian Society of Clinical Neurophysiology*; Vol. 24 (Supp. 2), pp. S138–42, May 2004. PMID# 12811613. http://www.ncbi.nlm.nih.gov/ sites/12811613 (Accessed May 2010)

(81) Streng, A.; K. Linde; A. Hoppe; V. Pfaffenrath; M. Hammes; S. Wagenpfeil; W. Weidenhammer; D. Melchart. "Effectiveness and tolerability of acupuncture compared with metoprolol in migraine prophylaxis." *Headache*; Vol. 46(10), pp. 1492–502, Nov-Dec 2006. PMID# 17115982 http://www. ncbi.nlm.nih.gov/sites/17115982 (Accessed May 2010)

Insomnia

(82) Cappuccio, F.P.; F.M. Taggart, N.B. Kandala, A. Currie, E. Peile, S. Stranges, M.A. Miller. "Meta-analysis of short sleep duration and obesity in children and adults." Published in *Sleep*; Vol. 31(5), pp. 619–26, May 2008. PMID# 18517032. http://www.ncbi.nlm.nih.gov/pubmed/18517032 (Accessed May 2010)

(83) Eguchi, K.; T.G. Pickering; J.E. Schwartz; S. Hoshide; J. Ishikawa; S. Ishikawa; K. Shimada; K. Kario. "Short sleep duration as an independent predictor of cardiovascular events in Japanese patients with hypertension." *Archives of Internal Medicine*; Vol. 168(20), pp. 2225–31, November 10, 2008. PMID# 19001199 http://www.ncbi.nlm.nih.gov/sites/19001199 (Accessed May 2010)

(84) Van Dongen, H.P.; G. Maislin; J.M. Mullington; D.F. Dinges. "The cumulative cost of additional wakefulness: dose-response effects on neurobehavioral functions and sleep physiology from chronic sleep restriction and total sleep deprivation." *Sleep*; Vol. 1:26(3), pp. 247–9, May 2003. PMID# 12683469. http://www.ncbi.nlm.nih.gov/pubmed/12683469 (Accessed May 2010)

(85) Brook, Richard A., MS, MBA; Nathan L. Kleinman, PhD; Arthur K. Melkonian, MD; Justin F. Doan, MPH; Robert W. Baran, PharmD. "Cost of Illness for Insomnia: Medical, Pharmacy, and Work Absence Costs in Employees With and Without Insomnia," Study conducted by the Associated Professional Sleep Societies; *APSS 2007*, #242. Accessed via the Human Capital Management Services Group at http://www.hcmsgroup.com/hcms/research/papersnew/APSS.Cost%20Insomnia.2007.pdf (Accessed May 2010)

(86) Ozminkowski, R.J.; S. Wang; J.K. Walsh. "The direct and indirect costs of untreated insomnia in adults in the United States," *Sleep*; Vol. 30(3), pp. 263–73, March 1, 2007. PMID# 17425222 http://www.ncbi.nlm.nih.gov/pubmed/17425222 (Accessed May 2010)

(87) Mai, Evelyn, MD; Daniel J. Buysse, MD. "Insomnia: Prevalence, Impact, Pathogenesis, Differential Diagnosis, and Evaluation." *Sleep Medicine Clinics*; Vol. 3(2), June 2008. PubMed Central ID# PMC2504337 DOI: 10.1016/j.jsmc.2008.02.001. http://www.pubmedcentral.nih.gov/articlerender.fcgi?artid=2504337 (Accessed May 2010)

(88) Léger, D.; C. Guilleminault; G. Bader; E. Lévy; M. Paillard. "Medical and socio-professional impact of insomnia." *Sleep*; Vol. 25(6), pp. 625–9, September 15, 2002. PMID# 12224841 DOI: 10.1016/j.jsmc.2008.02.001 http://www.ncbi.nlm.nih.gov/pubmed/12224841 (Accessed May 2010)

(89) Chilcott, L.A.; C.M. Shapiro. "The socioeconomic impact of insomnia. An overview." *Pharmacoeconomics*; Vol. 10 (Supp.1), pp. 1–14, January 1996. PMID# 10163422 http://www.ncbi.nlm.nih.gov/pubmed/10163422 (Accessed May 2010)

(90) From the official Lunesta website, www.lunesta.com; ©2008 Sepracor Inc. http://www.lunesta.com/aboutLunesta/lunesta-side-effects.html?iid=LHC_sideEffects (Accessed May 2010)

(91) Spence, D.W.; L. Kayumov; A .Chen; A. Lowe; U. Jain; M.A. Katzman; J. Shen; B. Perelman; C.M. Shapiro. "Acupuncture increases nocturnal melatonin secretion and reduces insomnia and anxiety: a preliminary report." *Journal of Neuropsychiatry and Clinical Neurosciences;* Vol. 16(1), pp. 19–28, winter 2004. PMID# 14990755. http://www.ncbi.nlm.nih.gov/pubmed/14990755 (Accessed May 2010)

(92) Montakab, H. "Acupuncture and insomnia." *Forschende Komplementarmedizin*; Vol. 6 (Supp. 1), pp. 29–31, February 1, 1999. PMID# 10077713. http://www.ncbi.nlm.nih.gov/pubmed/10077713 (Accessed May 2010)

(93) Freire, A.O.; G.C. Sugai; F.S. Chrispin; S.M. Togeiro; Y. Yamamura; L.E. Mello; S. Tufik. "Treatment of moderate obstructive sleep apnea syndrome with acupuncture: a randomised, placebo-controlled pilot trial," *Sleep Medicine*; Vol. 9(2), p. 211, January 2008. PMID# 1703212. http://www.ncbi.nlm.nih.gov/pubmed/17023212 (Accessed May 2010)

Carpal Tunnel

(94) Shuford, Harry; Tanya Restrepo. "Carpal Tunnel Claims Rank Second Among Major Lost-Time Diagnoses." *The NCCI Research Brief;* Vol. 3, April 2005. http://www.ncci. com/media/pdf/research-may05-carpal-tunnel.pdf (Accessed May 2010)

Gastrointestinal Diseases

(95) Martin, Bradley C., PhD., et al. "New Data Shows Patients With Irritable Bowel Syndrome (IBS) Cost The North Carolina Medicaid System $1,600 More Per Year Than Similar Non-IBS Medicaid Recipients." Contact: Chantal Beaudry/JohnMcInerney; 212–593–6400. Read the press release at http://www2.prnewswire.com/cgi-bin/ stories.pl?ACCT=104&STORY=/www/story/05–22– 2002/0001733097&EDATE (Accessed May 2010)

(96) Gao, Ying; Sigurdur Y. Kristinsson; Lynn R. Goldin; Magnus Björkholm; Neil E. Caporaso; Ola Landgren. "Increased Risk for Non-Hodgkin Lymphoma in Individuals with Celiac Disease and a Potential Familial Association." *Gastroenterology;* Vol. 136(1), pp. 32–34, January 2009. DOI:10.1053/j.gastro.2008.09.031. http://www.gastrojournal. org/article/S0016–5085(08)01700–9/abstract (Accessed May 2010)

(97) Gibson, Teresa B. PhD; Eliza Ng, MD; Ronald J. Ozminkowski, PhD; et al. "The Direct and Indirect Cost Burden of Crohn's Disease and Ulcerative Colitis." Published in the *Journal of Occupational and Environmental Medicine;* Vol. 50(11); pp. 1261–72; November 2008. http://www.joem. org/pt/re/joem/abstract.00043764–200811000–00007.htm (Accessed May 2010).

(98) Sher, M.E.; E.G. Weiss, J.J. Nogueras, S.D. Wexner. "Morbidity of medical therapy for ulcerative colitis: what are we really saving?" *International Journal of Colorectal Disease*; Vol. 11(6), pp. 287–93, 1996. http://www.ncbi.nlm.nih.gov/pubmed/9007625 (Accessed May 2010)

Inflammatory Diseases

(99) Anderson, J.J.; D.T. Felson. "Factors associated with osteoarthritis of the knee in the first national Health and Nutrition Examination Survey (HANES I). Evidence for an association with overweight, race, and physical demands of work." *American Journal of Epidemiology*; Vol. 128(1), pp. 179–89, July 1988. PMID# 3381825. http://www.ncbi.nlm.nih.gov/pubmed/3381825 (Accessed May 2010)

(100) Lübbeke, A.; S. Duc, G. Garavaglia, A. Finckh, P. Hoffmeyer. "BMI and Severity of Clinical and Radiographic Signs of Hip Osteoarthritis," *Obesity* (Silver Spring); February 5, 2009. PMID# 19197252. http://www.ncbi.nlm.nih.gov/pubmed/19197252 (Accessed May 2010)

(101) "Osteoarthritis Fact Sheet," from the Arthritis Foundation, 2008. http://www.arthritis.org/media/newsroom/media-kits/Osteoarthritis_fact_sheet.pdf (Accessed May 2010)

(102) Goetzel, Ron Z. PhD; Stacey R. Long, MS; Ronald J. Ozminkowski, PhD; Kevin Hawkins, PhD; Shaohung Wang, PhD; Wendy Lynch, PhD. "Health, Absence, Disability, and Presenteeism Cost Estimates of Certain Physical and Mental Health Conditions Affecting U.S. Employers." *Journal of Occupational and Environmental Medicine*; Vol. 46(4), pp. 398–412, April 2004. PMID# 15076658. DOI: 10.1097/01.jom.0000121151.40413.bd. http://healthproject.stanford.edu/publications/health_absence_disability_goetzel.pdf (Accessed May 2010)

(103) Jones, C. Allyson, PhD; Lauren A. Beaupre, PhD; D.W.C. Johnston, MD, FRCS(C); Maria E. Suarez-Almazor, MD, PhD. "Total Joint Arthroplasties: Current Concepts of Patient Outcomes after Surgery." *Rheumatic Diseases Clinics of North America*; Vol. 33(1), February 2007. PMID# 17367693. http://www.ncbi.nlm.nih.gov/pubmed/17367693 (Accessed May 2010)

(104) Miller, E.; Y. Maimon, Y. Rosenblatt, A. Mendler; A. Hasner, A. Barad, H. Amir, S. Dekel, S. Lev-Ari. "Delayed Effect of Acupuncture Treatment in OA of the Knee: A Blinded, Randomized, Controlled Trial." *Evidence-Based Complementary and Alternative Medicine*; Jan 5, 2009 (e-published ahead of print). PMID# 19124552. http://www.ncbi.nlm.nih.gov/pubmed/19124552 (Accessed May 2010)

(105) "Acupuncture in patients with osteoarthritis of the knee or hip: a randomized, controlled trial with an additional nonrandomized arm." CM Witt; S Jena; B Brinkhaus; B Liecker; K Wegscheider; SN Willich. Published in *Arthritis and Rheumatism*; Vol. 54(11); pp. 3485–93; November 2006. PMID# 17075849. http://www.ncbi.nlm.nih.gov/pubmed/17075849 (Accessed May 2010)

(106) Hughes, J.G.; J. Goldbart, E. Fairhurst, K. Knowles. "Exploring acupuncturists' perceptions of treating patients with rheumatoid arthritis." *Complementary Therapies in Medicine*; Vol. 15(2), pp. 101–8, June 2007. PMID# 17544860. http://www.ncbi.nlm.nih.gov/pubmed/17544860 (Accessed May 2010)

(107) Centers for Disease Control, National Center for Chronic Disease Prevention and Health Promotion. "Arthritis Types— Overview: Rheumatoid Arthritis." http://www.cdc.gov/arthritis/arthritis/rheumatoid.htm (Accessed May 2010)

(108) The American Lung Association. "COPD Fact Sheet."
http://www.lungusa.org/site/c.dvLUK9O0E/b.252866/k.
A435/COPD_Fact_Sheet.htm (Accessed May 2010)

(109) National Institutes of Health, National Heart, Lung and
Blood Institute. "Morbidity and Mortality: 2007 Chart Book
on Cardiovascular, Lung, and Blood Diseases." http://www.
nhlbi.nih.gov/resources/docs/07a-chtbk.pdf (Accessed May
2010)

(110) Ward, M.M.; H.S. Javitz; W.M. Smith; M.A. Whan.
"Lost income and work limitations in persons with chronic
respiratory disorders." *Journal of Clinical Epidemiology*;
Vol. 55(3), pp. 260–8, March 2002. PMID# 11864797.
http://www.ncbi.nlm.nih.gov/pubmed/11864797
(Accessed May 2010)

(111) Lee, T.A.; A.S. Pickard; D.H. Au; B. Bartle; K.B. Weiss.
"Risk for death associated with medications for recently
diagnosed chronic obstructive pulmonary disease." *Annals
of Internal Medicine*; Vol. 149(6), pp. 380–90, September
16, 2008. PMID# 18794557. http://www.ncbi.nlm.nih.gov/
pubmed/18794557 (Accessed May 2010)

(112) Mora, Jorge I.; Denis Hadjiliadis. "Lung volume reduction
surgery and lung transplantation in chronic obstructive
pulmonary disease." *International Journal of COPD*;
Vol. 3(4); pp. 629–635; 2008. This article can be found at
http://74.125.95.132/search?q=cache:Xx2q5KGDFD0J:www.
dovepress.com/getfile.php%3FfileID%3D4244+cost+of+lu
ng+transplant+for+COPD&hl=en&ct=clnk&cd=8&gl=us
(Accessed May 2010)

(113) Suzuki, M.; Y. Ohno, K. Namura, T. Asai, H. Yuugetu,
M. Sawada, S. Akao, K. Gotou, H. Fujiwara. "A case of
chronic obstructive pulmonary disease (COPD) successfully
treated by acupuncture." *Nihon Kokyuki Gakkai Zasshi*;
Vol. 43(5), pp. 289–95, May 2005. PMID# 15969210.
http://www.ncbi.nlm.nih.gov/pubmed/15969210
(Accessed May 2010)

(114) Suzuki, M.; K. Namura, Y. Ohno, H. Tanaka, M. Egawa,
Y. Yokoyama, S. Akao, H. Fujiwara, T. Yano. "The effect
of acupuncture in the treatment of chronic obstructive
pulmonary disease." *Journal of Alternative and
Complementary Medicine*;
Vol. 14(9), pp. 1097–105, November 14, 2008. PMID#
19055335. http://www.ncbi.nlm.nih.gov/pubmed/19055335
(Accessed May 2010)

(115) Neumeister, W.; H. Kuhlemann, T. Bauer, S. Krause,
G. Schultze-Werninghaus, K. Rasche. "Effect of acupuncture
on quality of life, mouth occlusion pressures and lung
function in COPD." *Medizinische Klinik* (Munich);
Vol. 94 (1 spec no), pp. 106–9, April 1999. PMID# 10373752.
http://www.ncbi.nlm.nih.gov/pubmed/10373752
(Accessed May 2010)

Section 4

(116) Sauter, Steven; Lawrence Murphy, Michael Colligan, Naomi Swanson, Joseph Hurrell, Jr., Frederick Scharf, Jr., Raymond Sinclair, Paula Grubb, Linda Goldenhar, Toni Alterman, Janet Johnston, Anne Hamilton, Julie Tisdale. "Stress… At Work." NIOSH Publication #99–101. http://www.cdc.gov/niosh/docs/99–101 (Accessed May 2010)

(117) American Institute of Stress. "Job Stress." http://www.stress.org/job.htm (Accessed May 2010)

(118) Kimihiko, Tokumori, et al. "A new standing stretch approach to treatment of low back pain for care workers." *Japanese Journal of Occupational Medicine and Traumatology*; Vol. 53(3), pp. 171–5, 2005. http://sciencelinks.jp/j-east/articl e/200522/000020052205A0882507.php (Accessed May 2010)

(119) Anderson J.W.; C. Liu, R.J. Kryscio. "Blood pressure response to transcendental meditation: a meta-analysis." *American Journal of Hypertension*; Vol. 21(3), pp. 310–6, March 2008. PMID# 18311126. http://www.ncbi.nlm.nih. gov/pubmed/18311126 (Accessed May 2010).

(120) Manikonda J.P.; S. Störk, S. Tögel, A. Lobmüller, I. Grünberg, S. Bedel, F. Schardt, C.E. Angermann, R. Jahns, W. Voelker. "Contemplative meditation reduces ambulatory blood pressure and stress-induced hypertension: a randomized pilot trial." *Journal of Human Hypertension* (U.K.); Vol. 22(2), pp. 138–40, February 2008. PMID# 17823597. http://www.ncbi.nlm.nih.gov/pubmed/17823597 (Accessed May 2010)

(121) Ong, J.C.; S.L. Shapiro, R. Manber. "Combining mindfulness meditation with cognitive-behavior therapy for insomnia: a treatment-development study." *Behavior Therapy*; Vol. 39(2), pp. 171–82, June 2008. PMID# 1850225010. http://www.ncbi.nlm.nih.gov/pubmed/18502250 (Accessed May 2010)

(122) "Medical and Dental Expenses (Including the Health Coverage Tax Credit)". IRS Publication 502. http://www.irs.gov/pub/irs-pdf/p502.pdf

(123) "Health Savings Accounts and Other Tax-Favored Health Plans." IRS Publication 969. http://www.irs.gov/pub/irs-pdf/p969.pdf

Section 5

(124) "Company Wellness Programs: What Is the Return on Investment?" Wellness Program Blog; posted March 12, 2009. http://wellnessprogramblog.com/company-wellness-programs-what-is-the-return-on-investment/ (Accessed May 2010)

Recommended Reading List

Abramson, John, M.D. *Overdo$ed America*.
HarperCollins Publishers, New York, NY, 2004
ISBN #978–0–06–134476–3

Dworkin, Ronald W., M.D., Ph.D. *Artificial Happiness:
The Dark Side of the New Happy Class*.
Carroll & Graf Publishers, New York, NY, 2006.
ISBN# 978–0–78671–933–4

Brownlee, Shannon. *Overtreated*.
Bloomsbury USA, New York, NY, 2007.
ISBN# 978–1–58234–579–6

About the Author

DR. TADEUSZ SZTYKOWSKI, MD (LIC. IN EU), D.AC. is the founder and owner of the Center for Preventive Medicine in Providence, RI. Born in Poland, he graduated from medical school in Gdansk in 1982. His postgraduate training included studies in emergency medicine, pediatrics, surgery, and internal diseases, and he obtained board certification as an OB/GYN in 1987. Shortly thereafter, he was nominated by the Polish government to conduct preventative health studies on over 10,000 women residing within 18 miles of the planned construction site of Poland's first nuclear power plant.

Early in his medical career, Dr. Tad developed a keen interest in Traditional Chinese Medicine (TCM), and in September 1987 he moved with his wife Kasia and four children to the United States. Immediately, he commenced studies at the highly respected New England School of Acupuncture (NESA) in Watertown, MA where, in addition to the regular curriculum, he completed training in Japanese acupuncture, pulse diagnosis, palpitation, and disease pattern differentiation.

In September 1990, Dr. Tad founded the Center for Preventive Medicine in Providence, RI. Since then, he and his staff have since helped over 9,000 patients overcome health issues ranging from allergies and headaches to cancer and diabetes. Today, the Center is one of the most respected and successful wellness clinics in the United States.

Dr. Tad is also heavily involved with local and national health and wellness organizations. In 1995, he became the first president of the newly formed Rhode Island Society of Acupuncture and Oriental Medicine. In 1999, he formed the Foundation for the Development of Integrative Medicine. In 2000 he became a board member of the American Association of Oriental Medicine (AAOM); he was made Vice President of the organization two years later. In 1992 he founded the first acupuncture alcohol and drug detox center in Rhode Island, and in 1993 he founded the first program in the United States to educate medical residents about complementary and alternative medicine at Rhode Island Hospital, in cooperation with Brown University Medical School.

Dr. Tad is also active in health care legislature, and strongly supports the right of every patient to choose the health care services and providers which best suit their personal needs. In 1998, he championed the passage of a bill increasing educational standards for acupuncturists; in 1999, he lobbied for and helped to pass a bill requiring coverage for acupuncture treatment via insurance riders, increasing acupuncture coverage from none to about 20% of Rhode Islanders. As a board member of AAOM, he lobbies nationally for the inclusion of acupuncture in Medicare coverage.

In 2007, Dr. Tad renewed his medical license in Poland; he is currently in the process of obtaining licensing in 26 other European countries. His vision is to bring an understanding of Traditional Chinese Medicine to the world, so that people everywhere can enjoy the mental, physical and social harmony which true wellness brings.